HYDROMECHANICAL ASPECTS AND UNSATURATED FLOW
IN JOINTED ROCK

Hydromechanical Aspects and Unsaturated Flow in Jointed Rock

BUDDHIMA INDRARATNA
Professor of Civil Engineering, University of Wollongong, NSW 2522, Australia

PATHEGAMA RANJITH
*Lecturer, Civil Engineering, Nanyang Technological University, Singapore
(formerly Research Fellow, University of Wollongong)*

A.A. BALKEMA PUBLISHERS / LISSE / ABINGDON / EXTON (PA) / TOKYO

Library of Congress Cataloging-in-Publication Data

Indraratna Buddhima.
  Hydromechanical aspects and unsaturated flow in jointed rock / Buddhima Indraratna, Pathegama Ranjith.
    p. cm.
  Includes bibliographical references and index.
  ISBN 9058093093 -- ISBN 9058093107 (pbk.)
    1. Multiphase flow. 2. Rock mechanics. 3. Groundwater flow. 4. Gas flow. I. Ranjith, Pathegama. II. Title.

TA357.5.M84 I48 2001
551.49--dc21

2001046102

Cover design: Studio Jan de Boer, Amsterdam, The Netherlands.
Typesetting: TechBooks, New Delhi, India.
Printed by: Krips, Meppel, The Netherlands.

ISBN 90 5809 309 3 (hardback)
ISBN 90 5809 310 7 (paperback)

Contents

Preface

At present, due to the lack of proper understanding of unsaturated or multi-phase flows, it is often difficult to predict the risk of sudden groundwater inundation and gas outbursts in complex hydro-geological environments prevailing in underground mines and tunnels. For instance, in Australia, large reserves of coal and other minerals lie under tidal water, man-made reservoirs, rivers and the Pacific Ocean. Often, large volumes of groundwater are stored in sandstone aquifers above proposed mines and transport tunnels, while existing joints and fractures created by blasting and excavation provide ideal conduits for water to flow. The occurrence of methane pockets in coal seams and igneous activities that develop large amounts of gas such as carbon dioxide within complex hydro-geological regimes present obvious safety hazards, including human fatalities reported in the past due to methane explosion. The delays caused by water and gas ingress to underground sites carry adverse consequences in terms of stability, productivity and safety. In rock tunnels and mine openings, the cost of associated support systems is also considerable, and their installation can be time consuming.

During the past two decades, a number of numerical, analytical and experimental models have been developed for single-phase flow (i.e. water or gas only) through a single joint in rock. The single-phase flow theories can only provide rough estimates of flow rates and joint pressures. The compressibility of air and its sensitivity to temperature and its solubility in water make the analysis of two-phase flows a more challenging task. A realistic two-phase flow must be described by the characteristic components of each phase present in the mixture, their volume and mass ratio. Depending on the surface roughness and aperture along a rock joint, distinctly different water – gas flow patterns may result.

The fundamental aspects of rock joints have been addressed in a recent book, "Shear Behaviour of Rock Joints" by B. Indraratna and A. Haque, also published by Balkema, Rotterdam, in 2000. In this follow-up volume, the authors have provided a comprehensive background to hydro-mechanics of single- and two-phase flows with associated theories, mathematical and semi-empirical models. The book also includes a vivid description of the type of laboratory equipment utilized to simulate two-phase flows through jointed rock. The high pressure, two-phase triaxial equipment designed and built in-house is essentially the result of a five year project partly supported by industry, and the book describes the salient features of this novel facility, as well as highlighting the key experimental findings. A new mathematical model for two-phase flows developed by the authors is also presented in its initial form, while further modification of the theory is currently in progress. The experimental data obtained through laboratory testing have complemented some of the significant and fundamental concepts addressed in the book. Given the importance of construction and mining activities in jointed rock all over the world, this book will provide a useful reference for practicing geotechnical and mining engineers and researchers alike.

The completion of this book was made possible through the assistance of several national and international colleagues, who have reviewed the subject content presented in this volume on various occasions. The authors are particularly grateful to Dr. Winton Gale, Managing Director, Strata Control Technology (SCT), Wollongong and for the Australian Research Council for providing financial assistance for conducting research on unsaturated flow through rock joints. The authors are also thankful for the comments and criticisms of various colleagues, including Associate Prof. N. I. Aziz and Prof. R. N. Singh (Wollongong), Dr. A. Bhattacharya and Dr. V.S. Vutukuri (University of New South Wales), Dr. Chris Haberfield (Monash University), Dr. D.G. Toll (University of Durham) and Prof. P.H.S.W. Kulatilake (University of Arizona).

Buddhima Indraratna
Pathegama Ranjith
University of Wollongong,
Faculty of Engineering
Wollongong, NSW 2522
Australia

CHAPTER 1

Characterisation of jointed rock mass

1.1   INTRODUCTION

In this introductory Chapter, a brief description of the rock types and properties, and joint characteristics is given. Various attempts have been made in the past to classify rock mass for important engineering applications underground and on the surface, such as design of mines, highway tunnels, slopes, dam foundations and nuclear waste storage plants. Classification of rocks can be made based on their mineral composition and their origin or types for engineering applications (Fig. 1.1). The geological classification provides mineralogical and chemical data, but it does not directly provide stress-strain characteristics of rocks, which are the most relevant in engineering applications. The existing rocks may be subjected to high pressure and temperature changes resulting in the metamorphic alteration of properties, structurally, mineralogically and texturally. Depending on the geological origin, i.e. stages of weathering, transportation, consolidation and alteration (Tucker, 1993), rocks are usually classified into three major groups, (a) igneous (b) sedimentary and (c) metamorphic rocks. Figure 1.2 shows the common rock types, which come under these principal categories. A simplified geological process is shown in Fig. 1.3.

Clayey and silty sediments are deposited and consolidated in layers to form sedimentary rocks, which are often stratified. Such interfaces are usually identified as 'bedding planes', along which movements and fluid flow can occur under certain circumstances. In underground coalmines, planes of weakness can be identified in inter-layered shale and sandstone deposits.

The principal chemical composition of different rocks can be found in the literature (Nockolds et al., 1978; Smith & Erlank, 1982). Depending on chemical composition and their grain size, some igneous rocks such as basalt display columnar fractures with hexagonal cross-sections. During the metamorphism process, fractures may develop within the rock mass and their extent depends on the temperature and pressure change. Based on the mineral grain shape, size and arrangement, metamorphic rocks have different structures such as banded, slaty and foliated (Santosh, 1991). Banded rocks such as gneiss permit relatively easy split through layers. When sedimentary rocks undergo metamorphism, the existing

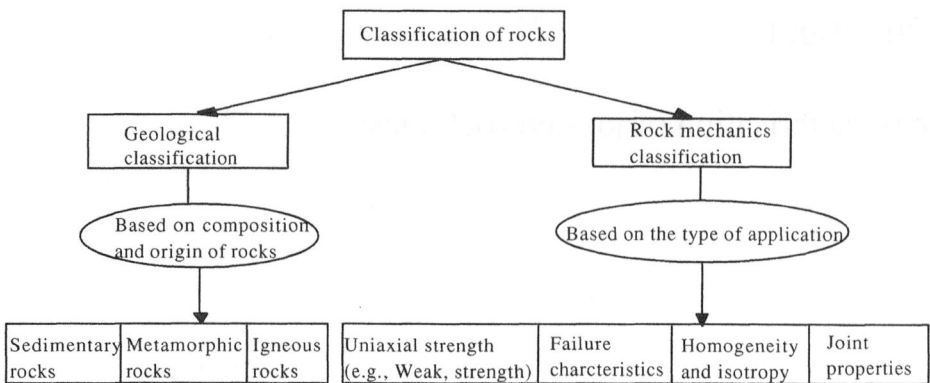

Figure 1.1. Principal classification systems of rocks.

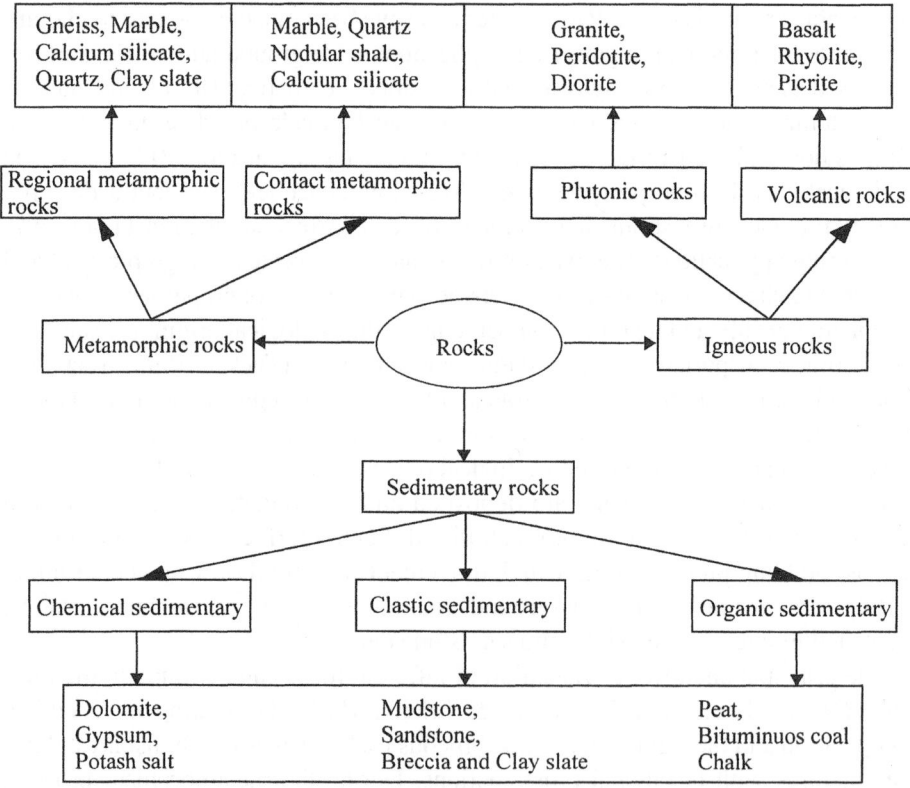

Figure 1.2. Most common rock types on the earth.

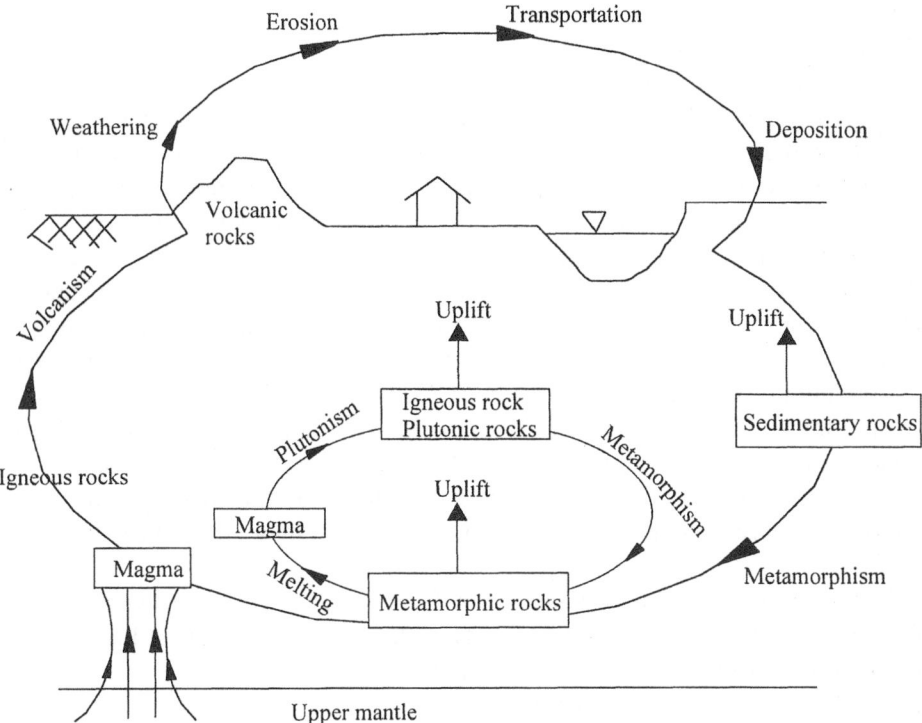

Figure 1.3. Simplified geological processes.

bedding planes in the sedimentary rocks may also be visible in the newly formed metamorphic rock. However, the bond between these layers is stronger than in sedimentary rocks. It is important to note that during excavation or blasting process, fractures may easily develop through bedding planes, which may then provide paths for fluid flow.

1.2 CLASSIFICATION SYSTEMS FOR ENGINEERING APPLICATIONS

Terzaghi (1946), Baron et al. (1960) and Coates (1964) attempted to classify rocks for different applications. Each classification system is based on the strength of rock mass and the geological features/structures such as joints, but little is usually said about rock-mass permeability. In Terzaghi (1946) rock mass classification for tunnels, 7 categories have been identified, as given below:
(1) Stratified rocks;
(2) Intact rocks;
(3) Moderately jointed rocks (e.g., vertical walls require no support);
(4) Blocky rocks (e.g., vertical walls require support);
(5) Crushed rocks;

(6) Squeezing rocks (low swelling capacity); and
(7) Swelling rocks (high swelling capacity).

As described by Coates (1964), one of the main deficiencies of the above classification system is that it does not give any information on the strength or permeability of rock mass. At shallow depth, blocky rocks might not require any support for vertical walls, whereas moderately jointed rocks may require support at greater depths, as well as effective water-proofing to prevent inundation.

Another classification system proposed by US Bureau Mines (1962) for underground openings, and recorded by Coates is described below. Rocks in underground mines have mainly been divided into two groups, namely (a) competent rocks (i.e., no support is required) and (b) incompetent rocks (i.e., support is required to prevent failure of an opening). Competent rocks are further subdivided into three classes, as massive elastic rocks (e.g., homogeneous and isotropic rocks with no significant jointing), bedded elastic rocks (e.g., homogeneous, isotropic beds with the bed thickness less than the span of the underground excavation) and massive plastic rocks (i.e., weak rock that may yield under relatively low stress).

The classification system suggested by Coates (1964) gives a better picture of the strength and failure characteristics of rocks. His method is based on 5 main characteristics of rock mass, as given below:
(1) Uniaxial compressive strength
      weak (<35 MPa)
      strong (between 35 and 175 MPa – homogeneous and isotropic rocks)
      very strong (> 175 MPa – homogeneous and isotropic rocks);
(2) Pre-failure deformation behaviour of rocks
      elastic
      viscous;
(3) Failure characteristics of rocks
      brittle
      plastic;
(4) Gross homogeneity
      massive
      layered (e.g. sedimentary rocks); and
(5) Continuity of rock mass
      solid (joint spacing >1.8 m)
      blocky (joint spacing <1.8 m)
      broken (passes through a 75 mm sieve).

It is clear that, while different classification systems are available, one must employ geological and rock mechanics classification models in conjunction to get a better understanding of the nature of a particular type of rock. The degree of jointing is the primary factor affecting permeability, hence the flow in to an excavation in a rock mass. Saturation of joints increases the internal hydraulic pressures, reducing the effective stress, hence the overall strength of rock mass.

1.3 FORMATION OF DISCONTINUITIES ON EARTH CRUST

In order to appreciate the hydro-mechanical aspects of jointed rock, a thorough understanding of the type of joints and discontinuity characteristics is required. Geological features in rock mass can be broadly divided into two groups, namely, (a) primary and (b) secondary. Primary features include original structures in the rock mass, which after being subjected to deformation and/or metamorphism, they are usually referred to as secondary structures.

Rocks during the process of development and the later formation of rock masses are usually subjected to a number of forces within the earth crust. These forces may be a single force or a combination of forces resulting from ground stresses, tectonic forces, hydrostatic forces, pore pressures, and temperature stresses. As a result of these forces and their magnitudes, rocks continuously undergo varying degree of deformation, resulting in the formation of different kinds of structural features (Fig. 1.4). For example, fractures or joints may initially develop within a rock mass, followed by dislocation of the fractured rock blocks. In some circumstances, these dislocated rock blocks move faster than the adjacent blocks, resulting in larger deformation between each rock block. Such structural features are referred to as faults. Both faults and joints are the result of the brittle behaviour of rock mass. Joints and faults can be easily identified from the component of displacement parallel to the structure. Joints usually have very small normal displacement, referred to as joint aperture. The estimation of joint aperture using various techniques under different stress conditions is discussed in detail in Ch. 3.

1.3.1 Joints and bedding planes

Joints in a rock mass are usually developed as families of cracks with probably regular spacing, and these joint families are referred to as joint sets. While the formation of joints is associated with the effect of differential stress, some joints are more prominent and well developed, extending for a considerable length (several kilometres), while others are minor joints having a length of only a few centimetres. In order to characterise joint sets it is necessary to consider their properties such as spacing, orientation, length, gap length and the apertures. Relative movement of joints is negligible in comparison to fault movements. Joints in a rock mass may be open or close or filled with some other material such as clay and silt. Open joints often provide fluid flow paths to other connected joints in the rock mass, and the quantity of fluid carried by each joint depends on the separation between joint blocks (joint aperture), joint geometry, hydraulic gradient and properties of fluid.

The fundamental geometrical observation of a joint includes the measurement of strike and dip. Based on the dip and the strike, joints are grouped into three classes: (a) dip joints, (b) oblique joints and (c) strike joints. However the classification based on the origin of joints is often more useful in various engineering

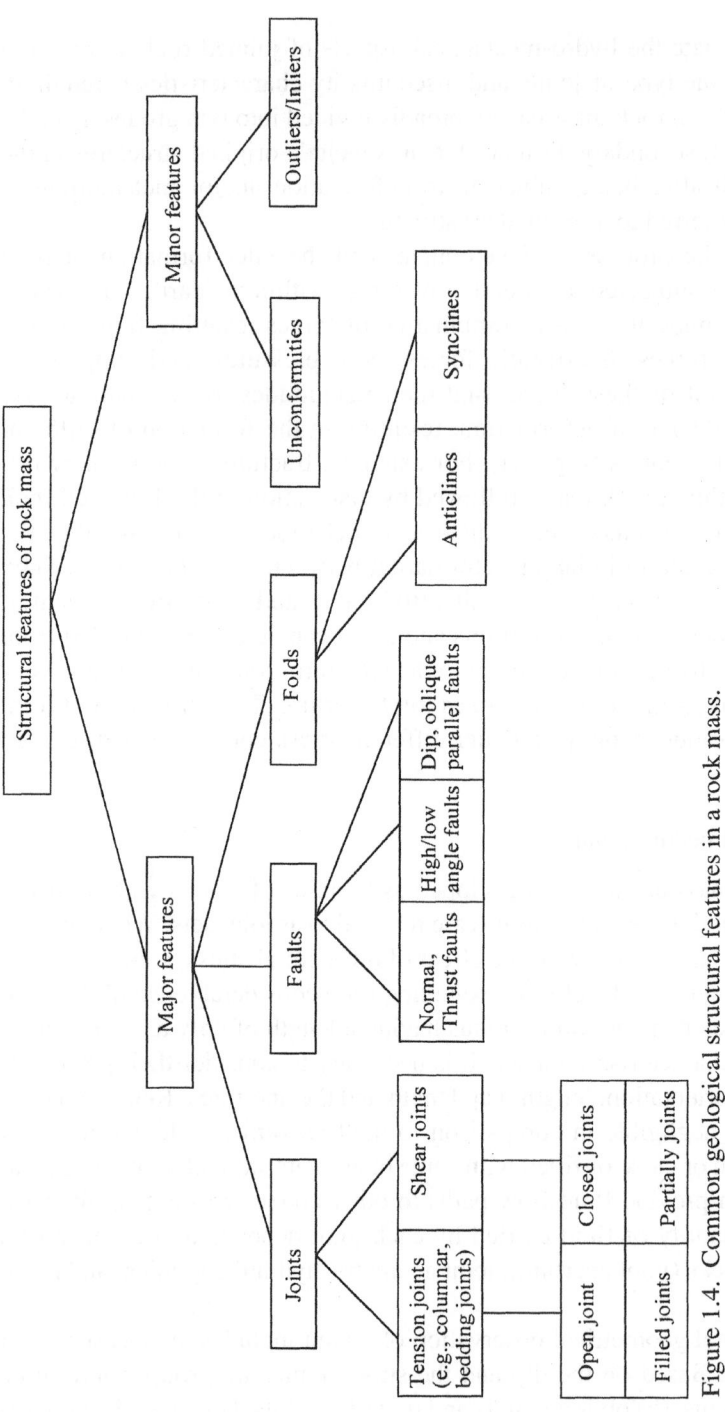

Figure 1.4. Common geological structural features in a rock mass.

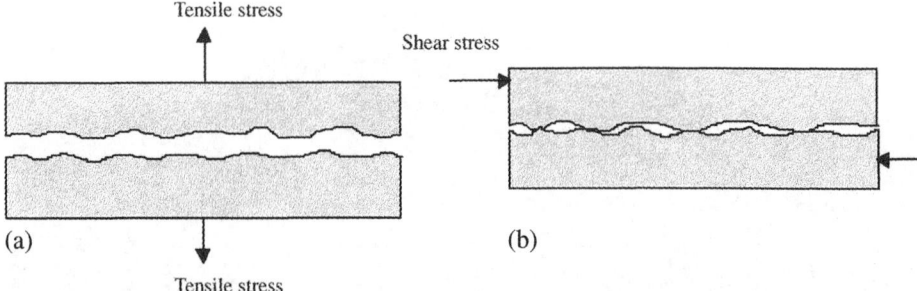

Figure 1.5. Genetic classification of joints. (a) Tension joint (e.g. igneous rocks) and (b) Shear joint (e.g. sedimentary rocks).

applications (Fig. 1.5). Based on the origin of joints, they are mainly grouped as follows:

(1) Tension joints; and

(2) Shear joints.

Tension joints may develop either during the formation of rocks or after their formation, which can be attributed to the tensile forces acting on the rock mass. Columnar, sheeting and mutually perpendicular joints are some common types of tension joints. Columnar joints generally occur in hexagonal shape and are often found in basalt rocks, whereas sheet joints may develop in granite. Figure 1.6a and b show columnar joints in basalt and sheeted joints in granite rocks. The mechanism of formation of columnar joints in basalt is associated with the stress caused by the cooling of magma. Theoretical aspects of the development of joints and faults have been discussed by Price (1966). He proposed that the joints in horizontally bedded rocks were mainly due to uplifting forces developed on the earth's crust. It can sometimes create both shear and tension joints on either side of the limbs of the fold, as shown in Fig. 1.7. Apart from geological processes, fracturing may occur on the surface rocks due to human activities, and the natural weathering accelerated by temperature changes and the action of surface water. According to the definition given by Pettijohn et al. (1972), cross joints are defined as a structure confined to a single sedimentation unit, and characterised by internal discontinuities inclined to the principal bedding plane.

The most common feature of sedimentary rocks is the occurrence of bedding planes, in which each bed is fairly homogeneous and differs in properties from the other beds, in relation to the texture and composition. In some circumstances, significant variations of material properties within a layer may occur because of the varied lithological conditions due to random, uniform or systematic deposition. When the sediments are deposited uniformly, homogenous layers are formed. Between each stratum, a new discontinuity is developed, and the properties of this discontinuity depend upon the properties of the adjacent layers. Typical bedding planes in a sedimentary rock are shown in Fig. 1.8.

(a)

(b)

Figure 1.6. (a) Columnar joints at a basalt quarry in Kiama, NSW, Australia. (b) Closely packed sheet joints, National Park, Western Australia.

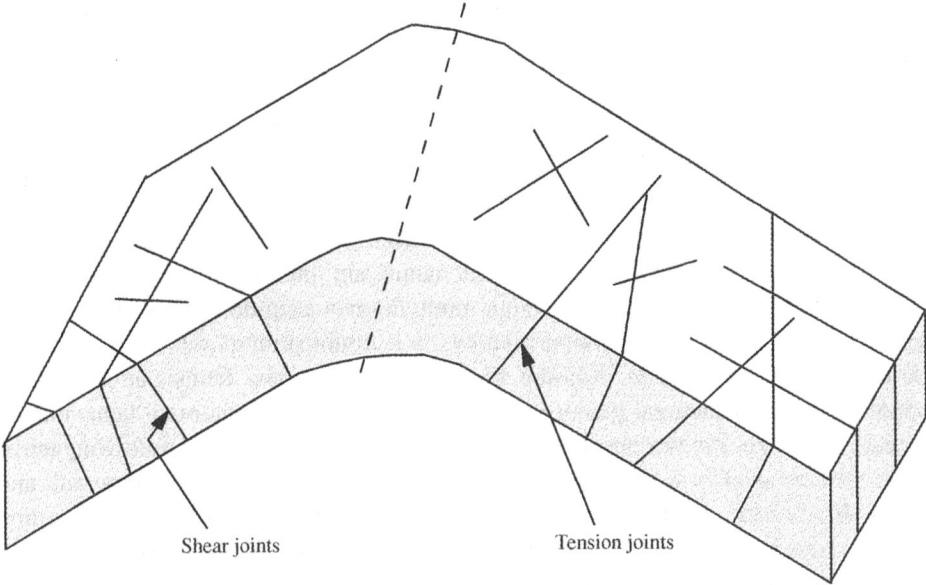

Figure 1.7. Development of shear and tension joints on the limbs of fold (modified after Price, 1966).

Figure 1.8. Typical bedding planes in a sedimentary rock (Southern Freeway, NSW, Australia).

1.3.2   Faults

When the shearing stress exceeds the shearing resistance of rocks, fractured rock blocks undergo considerable displacement along a favourable shear plane resulting in the formation of a new discontinuity, which is referred to as a fault. Depending on the internal stresses, and the rock properties, these relative displacements may vary from few centimetres to several kilometres. Priest (1993) reported the extent of some major faults, such as the San Andreas Fault in California, the Great Glen Fault in Scotland and the Alpine Fault in New Zealand. The nature and the magnitude of the internal forces developed on the earth's crust are rarely known accurately. These forces may give rise to tension, compression or rotation, or a combination of them. For example, thrust faults are believed to form due to compressive stresses. Faults can be broadly identified by the structural geology (e.g., shear zone, gouge, abnormal behaviour of strata), landscape and morphology. To describe a fault geometrically, following terms (Fig. 1.9) are usually used: (a) strike and dip, (b) fault plane, (c) hanging wall and footwall, (d) hade, (e) heave and (f) throw. In the literature, different classification systems have been employed to identify various faults based on the following:

(1) The type of fault (e.g., thrust fault, normal fault, vertical fault), for example see Fig. 1.10;
(2) The dip angle of the fault (e.g., low-angle fault, high-angle fault); and
(3) The direction of slip (e.g., strike fault, oblique fault).

1.3.3   Folds

Planar rock structures on the earth's crust have been subjected to deformation and producing curved or non-planar structures. These new geological structures are referred to as folds, and usually these rocks behave as a ductile material. The

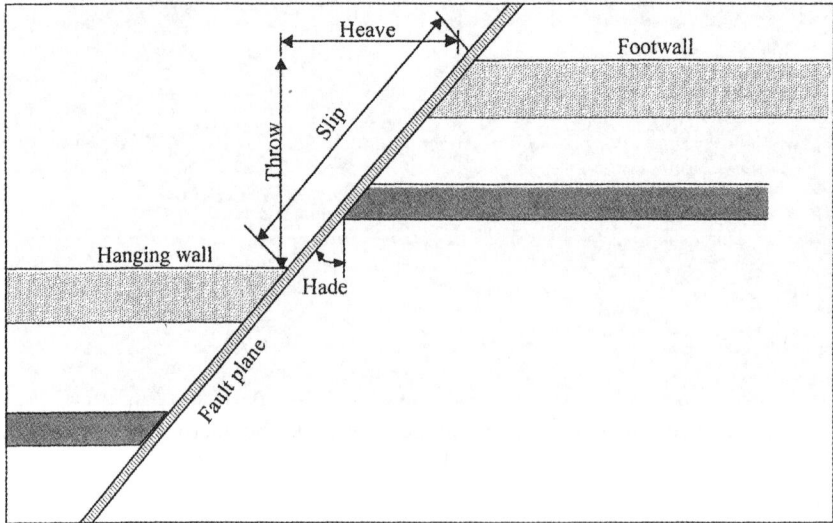

Figure 1.9. Simplified schematic diagram of a typical fault.

Figure 1.10. Normal fault at Wombarra drainage tunnel, NSW, Australia.

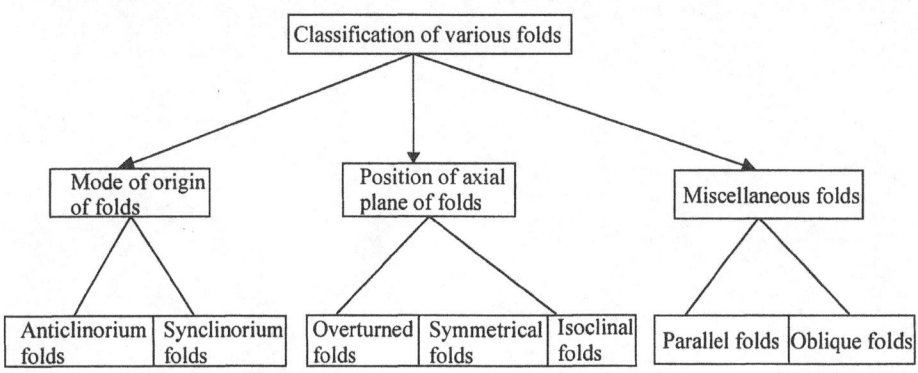

Figure 1.11. Classification of folds.

extent of folding and the ultimate shape of the fold depend on the intensity and the duration of internal forces, as well as the properties of the rock material. In any type of rock, folds may develop, however, in sedimentary and volcanic rocks, these structures are common. The study of the mechanism of folding is a complex subject, and it is not discussed within the scope of this book. Briefly, folding of rock might occur due to the development of tectonic stresses (e.g., lateral compressive forces caused by shrinkage) and non-tectonic stresses caused by landslides, differential compaction and glaciations. Folds are broadly grouped into anticlines and synclines. A detailed classification of folds is given in Fig. 1.11.

Figure 1.12. Symmetrical fold at top of the Southern Freeway, NSW, Australia.

Figure 1.12 shows a symmetric fold on the rock surface, Southern Freeway, NSW, Australia. For more details, the reader is encouraged to consult the work of Hobbs et al. (1976) and Biot (1957, 1959, 1961).

1.4 MEASUREMENT OF DISCONTINUITY CHARACTERISTICS

For successful design and construction of engineering structures on the surface as well as underground, it is necessary to carry out detailed investigation of the properties of soil or rock materials, planes of weakness, the groundwater conditions and existing stresses. Such comprehensive study minimises extensive delays, catastrophic failures, loss of human lives and the overall cost of the project. As described by Franklin and Dusseault (1989), a detailed site investigation can be carried out in two phases (Fig. 1.13).

Measurement of discontinuity characteristics provides identification of the different geological structures, and facilitates input data required for numerical modelling and development of analytical and empirical models. Analyses of groundwater flow, and the roof stability in longwalls or other underground structures are governed by the accuracy of field measurements. During the preliminary stage, published geological reports and topographic maps usually provide valuable information on soil and rock types, their properties and the groundwater conditions. Air photos and remotely sensed data are also useful in mapping geological features such as faults, folds and surface morphology. Remote sensing

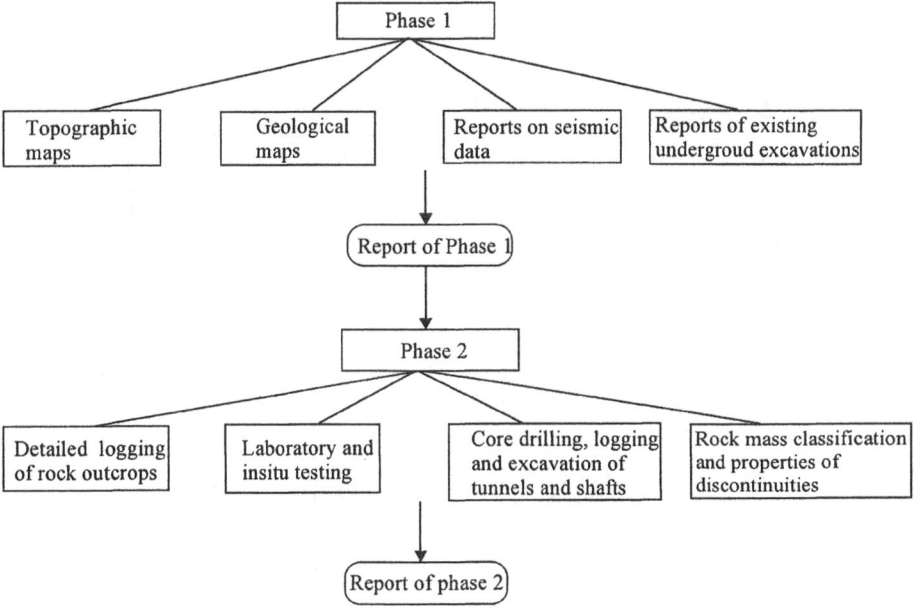

Figure 1.13. Detailed site investigation for geological structures.

methods are now popular because of the improved resolution of photographs, and also because of the large areas that can be surveyed and the data interpreted in a relatively short time, using modern computer technology. For the site investigation of nuclear power plants in USA, McEldowney and Pascucci (1979) discussed different techniques of remote sensing methods and their applicability to identify geological features such as faults. Near surface rock can be explored by observing natural and man-made structures such as landslides, open pit mines, and slopes. Exposed rock faces provide direct measurements of discontinuity properties over a large area, despite the fact that the rock face may be badly damaged due to induced stresses. In addition, close to the ground surface, rocks are usually subjected to weathering process, which results in increased discontinuity frequency, greater magnitude of apertures, or the infilling of joints with materials like clay and silt.

Basically, two different techniques, i.e. scanline method and window method are widely used to map joints on the face of the rock (ISRM, 1978; Pahl, 1981; Priest & Hudson, 1981). These two methods are more or less similar in the way of measurement, except in the scanline method, only the discontinuities that intersect the scanline are mapped, whereas, all discontinuities in the defined (predetermined) area are measured in the window method (Fig. 1.14). In the window

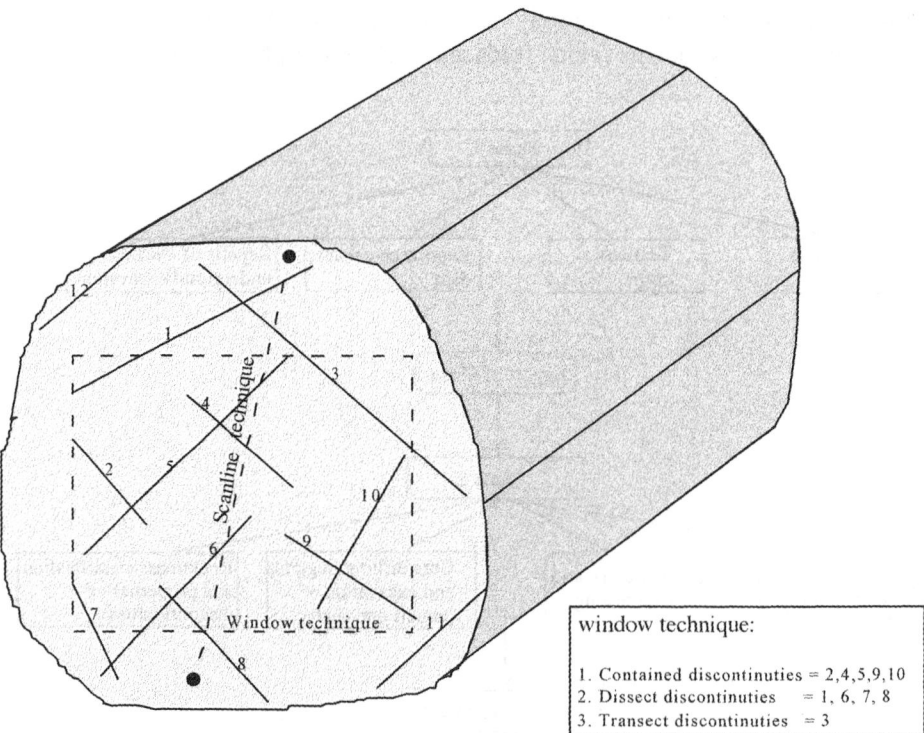

window technique:

1. Contained discontinuties = 2,4,5,9,10
2. Dissect discontinuties = 1, 6, 7, 8
3. Transect discontinuties = 3

Figure 1.14. Mapping techniques of discontinuities on an exposed rock surface.

technique, as described by Pahl (1981), the discontinuities within the window are classified into 3 classes: (a) contained, (b) dissect and (c) transect discontinuities. If both ends of the discontinuities are within the window, they are referred to as contained (see 2, 4, 5, 9 and 10 discontinues in Fig. 1.14), whereas when one end of the discontinuities is visible in the window, they are referred to as dissect (discontinues 1, 6, 7, 8 in Fig. 1.14). In some cases, discontinuities may terminate outside the window, and these are called transect discontinuities.

Although an extensive view of rock structures and discontinuities is provided by the exploration of open pit and shafts, however, these methods are not feasible at greater depths. At greater depths, the ground can be surveyed by other techniques such as the core-drilling method. Geophysical techniques (e.g., electrical, seismic and magnetic) are also widely used to map geological structures and to find properties of soil and rocks to a certain extent. The mechanism of seismic technique is based on the velocity of sound waves travelling through the rock media. Geophysical methods often detect underground cavities such as faults, abandoned mines caverns, and joints with large apertures.

1.5   PHYSICAL PROPERTIES OF DISCONTINUITIES

It is important to distinguish the common terms used to describe the planes of weakness developed in surrounding an excavation. Joints often threaten the civil or mining engineer, whereas in some occasions such as in oil recovery process, interconnected joints accelerates the recovery. Attewell and Woodman (1971), Priest (1993) and Goodman (1976) used the term discontinuities to describe a whole range of planes of weakness, such as joints, folds, faults, unconformities, outliers and inliers. According to the definition introduced by ISRM (1978), discontinuity is the general term for any mechanical fissure or joint in a rock mass having zero or low tensile strength. It is the collective term for most types of joints, bedding planes, foliations, weakness zones and faults.

The main focus of this section is to describe the geometrical and physical properties of joints rather than the properties of other structural features such as folds and faults. In order to characterise a discontinuity in the field, the following aspects of joints need to be considered:
(1) Length;
(2) Spacing;
(3) Orientation;
(4) Aperture;
(5) Surface geometry;
(6) Mode of origin and the presence of infill material;
(7) Wall strength;
(8) Number of discontinuity sets; and
(9) Block sizes.

Piteau (1970, 1973) and ISRM (1978) identified some of the above properties as influencing parameters during the design of engineering structures such as slopes, underground structures and foundations. In a rock mass, joints are usually subjected to gravitational and induced stresses associated with excavations and groundwater pressures. As a result, they undergo sliding, toppling or both, shear failure and large displacements. Aspects of discontinuity length, spacing, orientation and shape are discussed in the following sections, while the effects of surface geometry on fluid flow are discussed later in Ch. 3.

1.5.1   Discontinuity length

Length of a joint or the areal extent of a joint can be observed at the exposure of rock surface, such as at face of rock slope cut or tunnel roof or longwalls. The trace length or discontinuity length may vary from several centimetres to hundreds of metres. In two-dimensions, the trace length has two components, and they are the dip trace length and the strike trace length. Dip trace length is obtained by measuring length in the direction of the dip, whereas if the length measurement is done in the direction of strike, it is referred to as the strike trace length. Every effort should be made to measure both the dip and strike length, whenever possible. Some discontinuities may terminate when the existing weak planes meet, while the others may extend beyond them.

In various engineering applications such as dam foundations and tunnelling, it is important to study the degree of interconnectivity of discontinuities. The degree of interconnectivity of fractures varies depending upon their orientation and the trace lengths. The accurate mapping of trace length involves the measurement of the actual length of weak planes as well as recording the type of termination at both ends of the discontinuity. The length of individual discontinuity at rock surface can be measured using a measuring tape, and the measurements are usually done in the direction of dip and in the direction of strike. According to ISRM (1978), based on the magnitude of trace lengths of discontinuity sets, they can be grouped as shown in Table 1.1. As discussed by ISRM (1978), during the mapping process, it is important to note the type of joint termination as delineated below:
(1) Discontinuities which terminate in the rock exposure (r);
(2) Discontinuities which extend to the outside rock exposure (x); and

Table 1.1. Trace lengths of discontinuity sets (ISRM, 1978).

Discontinuity trace length (m)	Description
<1	Very low length
1–3	Low length
3–10	Medium length
10–20	High length
>20	Very high length

(3) Discontinuities which terminate against the other discontinuities in the rock exposure (*d*).

The ends of some discontinuities may not be visible because of excavation, or the presence of vegetation or features extending beyond the limits of the exposure. For example, when recording data, if the length of discontinuity is 12 m and one end terminates in the rock surface (*r*) and the other end of termination is outside the rock surface (*x*), the discontinuity is recorded as 12*rx*. It is also useful to calculate the termination index, which provides information such as the degree of block separation within the given area. Termination index for category (2) is represented as T_x and expressed by (ISRM, 1978)

$$T_x = \frac{\Sigma x}{(\Sigma(x + d + r))} \times 100\% \qquad (1.1)$$

A large value of T_x indicates that a large number of discontinuities terminate outside the rock surface. If T_x takes a larger value, it is then expected that the rock exposure contains a large number of discrete blocks, which suggest that the exposure has a well-interconnected fracture network. The category (1) listed above is usually smaller than the summation of category (2) and (3) in a rock mass (Piteau, 1973).

The mean discontinuity length (*l*) is computed as follows:

$$\bar{l} = \sum_{i=1}^{n} l_i / n \qquad (1.2)$$

where l_i = individual discontinuity length and *n* = number of discontinuities.

As discussed by Cruden (1977), Priest and Hudson (1981) and Kulatilake (1988), if one uses the scanline method, the mean discontinuity length has several drawbacks: (a) the measurements depend upon the orientation and position of the scanline, (b) accuracy of data are based on the exposed area (large discontinuities may terminate beyond the limits of the exposure, hence it is impossible to measure the full length) and (c) it is difficult to measure very small discontinuities. Because of these, one has to carry out statistical analysis to increase the reliability of the mapped data. In the past, probability distribution methods such as lognormal, hyperbolic and exponential have been used for quantifying the trace length (Cruden, 1977; Priest & Hudson, 1981). Detailed discussion on different distribution functions is not within the scope of this book, and the reader can obtain more information from Priest and Hudson (1981, 1983) and Lee and Farmer (1993).

1.5.2   Orientation of discontinuity

To define the orientation of a discontinuity on a rock surface, the dip angle and the dip direction at the point of intersection with scanline are needed. The dip is the steepest declination and measured from the horizontal, whereas the dip direction is measured clockwise from the true north. The dip angle is expressed in degrees.

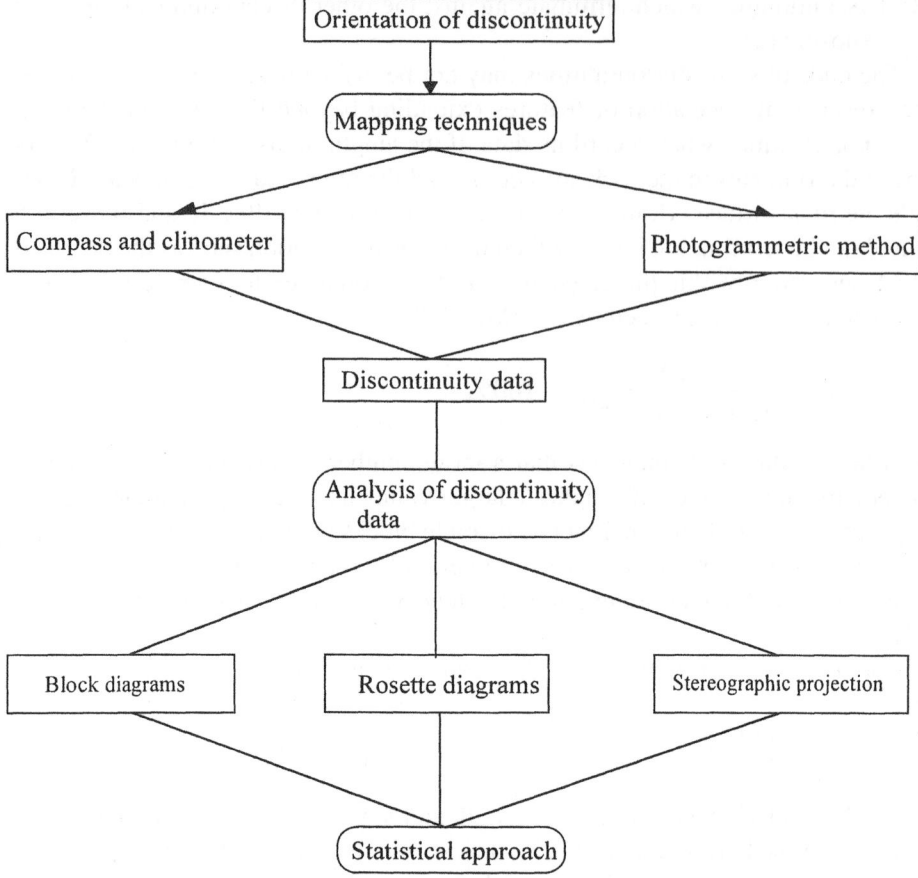

Figure 1.15. Detailed procedure for mapping and analysis of discontinuity orientations.

The overall strength and permeability of rock mass and the stability of engineering structures are influenced by joint orientation. Higher the interconnectivity of fractures depending upon orientation and lower the shear strength is, greater the risk of failure of rock mass. There are basically two methods suitable for measuring the orientation. They are (a) compass and clinometer technique and (b) photogrammetric technique. The detailed procedure involving mapping and analysing discontinuity orientation is given in Fig. 1.15 (Priest, 1993).

The compass and clinometer techniques are usually employed to conduct joint survey on exposed rock mass. The equipment such as the compass, clinometer, measuring tape and protractor are usually required to measure the joint orientation (ISRM, 1978). The photogrammetric method is usually employed for mapping of quite a large number of discontinuities, or in the case of inaccessible places and at areas affected by high magnetic fields, where a compass could indicate distorted readings. In order to determine the orientation of a given discontinuity, coordinates of at least 4 points are required. The procedures involved in a joint survey are described elsewhere (ISRM, 1978; Priest, 1993).

Oriented rock coring technique is also used to explore properties of rocks including discontinuity orientations at greater depth. However, during the drilling process, rock cores tend to rotate, therefore, special precautions should be taken to get the correct alignment of fractures. This problem can be overcome using non-rotating scribing core barrels (e.g., Craelius Core Barrel). In this technique, a vertical line is scratched on one side of the core closed to the drilling bit, which facilitates the alignment of the cores in the core box correctly. According to Christensen–Huegel method, orientation of core is measured by means of hardened steel groove scriber and compass photo device (ISRM, 1978). As discussed by Franklin and Dusseault (1989), an alternative technique developed by National Civil Engineering Laboratory in Portugal (Rocha & Barroso, 1971) is called the Integral Coring method, which is suitable even for sheared rocks. In this method, a small diameter is cored first and then a reinforcing rod is grouted to the broken core together.

It is important to note that the borehole method is relatively expensive, and the information produced from a small number of boreholes can be misleading. Therefore, one has to carefully plan and execute core-drilling exercises, including the use of hole inspection techniques such as borehole television cameras, photographic cameras and borehole periscopes. At greater depths, the use of borehole camera is limited because of high water pressures. For borehole lengths of up to 150 m, cameras can be used with high level of confidence (ISRM, 1978). Cameras are practical in large boreholes exceeding 76 mm diameter, whereas the periscope may be employed in smaller boreholes, subjected to limited depth of up to 30 m.

In order to examine the geological features, the bored core barrels should be kept on a core box with markers indicating the depths of geological horizons, and the start and end of each layer. Although quantifiable parameters such as Total Core Recovery (R), Rock Quality Designation (RQD) and Frequency (F) can be directly estimated, a closer look at the borehole wall is required for a quantitative description of orientation, infill, spacing and aperture. The dip angle (α) of individual discontinuities of cores can be measured relative to the core axis. Therefore, the true dip angle in degrees is calculated as 90-α. Using graphical techniques (e.g., stereographic projections), the dip angle and dip direction can also be obtained when data from non-parallel boreholes are available.

Once the preliminary data processing is completed, the next step is to analyse the data using statistical methods. The measured orientation data can be presented using (a) block diagrams, (b) rosette diagram and (c) stereographic projection methods. Block diagrams are usually employed for small number of discontinuities, and also when the orientations of discontinuities do not vary too much. Large numbers of discontinuity data are usually represented by graphical methods, such as the rosette diagram and stereographic projection. The simple technique 'rose diagram' is usually employed when the dip angles of most discontinuities are larger than 60° (Attewell & Farmer, 1976; Cawsey, 1977; Priest, 1993). In this method,

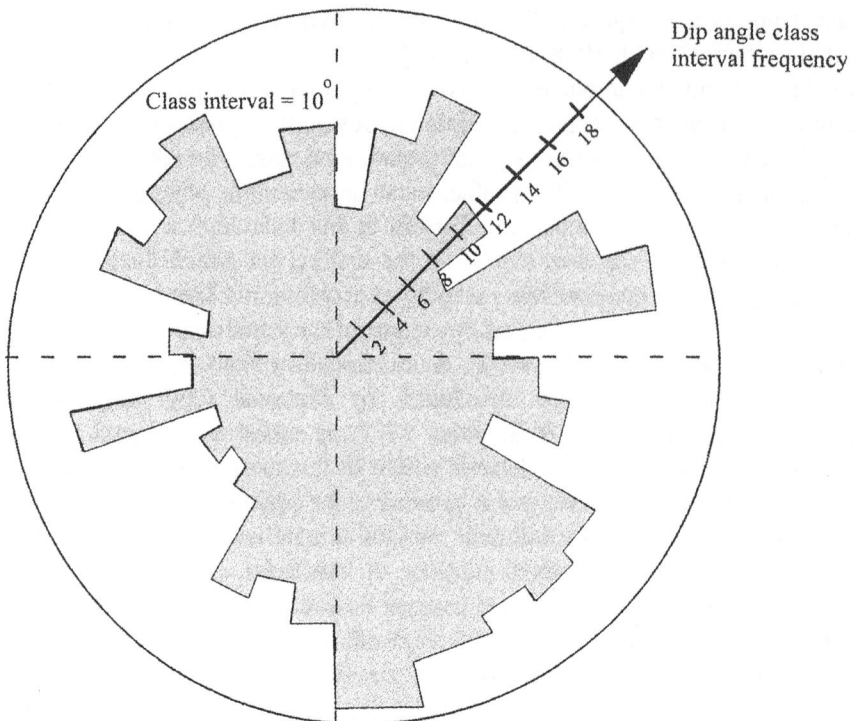

Figure 1.16. Typical rosette diagram to represent the orientation data.

data are plotted in a simple circular diagram in which the outward radius represents the frequency of the discontinuity orientation, whereas the horizontal angle represents the dip angle of each discontinuity. Before plotting the rose diagram (Fig. 1.16), it is recommended to plot the histogram for suitable class interval depending on dip angles. The histogram shows directly the dip angle, whereas the rose diagram does not. In order to represent the orientation data in 3-D, stereographic technique is widely used (Phillips, 1971; Duncan, 1981; Priest, 1993). In this method, the data is plotted on a special graph sheet with an equal angle (Fig. 1.17). The construction of the stereographic projections is described elsewhere (e.g., Priest, 1993).

1.5.3   Discontinuity spacing

Spacing between joints is the perpendicular distance between a pair of discontinuities, and for a discontinuity set, it is the mean perpendicular distance. Comprehensive definitions to discontinuity spacing are given by Priest (1993) as 'A "total spacing" is defined as the spacing between a pair of immediately adjacent discontinuities measured along a line at given location and orientation. A "set spacing" is defined as the spacing between a pair of immediately adjacent discontinuities from a particular discontinuity set, measured along a line at a given

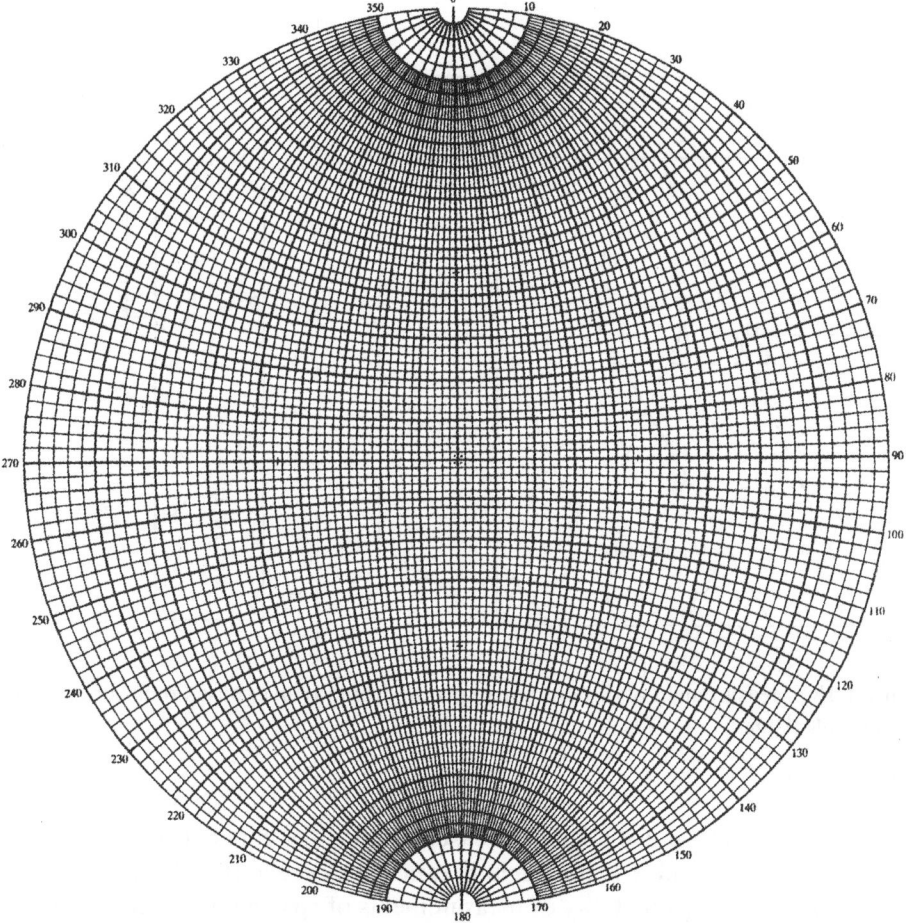

Figure 1.17. Typical stereographic projection: equatorial equal angle net.

orientation and location'. A 'normal set spacing' is defined as the set spacing when measured along a line that is parallel to the mean normal to the set. Discontinuity spacing largely influences the stability of the structure and the mass permeability and seepage characteristics of rock mass. For example, a closely spaced discontinuity reduces the cohesion between the blocks, whereas a widely spaced discontinuity consists of less reduced sliding force between blocks. The spacing measurements can be simply carried out using a measuring tape, a compass and a clinometer. For exposed rock surface, the perpendicular distance (x_i) between adjacent discontinuities may be calculated as follows:

$$x_i = d \sin \alpha \tag{1.3}$$

where d = measured distance along the tape or scanline and α = the angle between the scanline and the discontinuity.

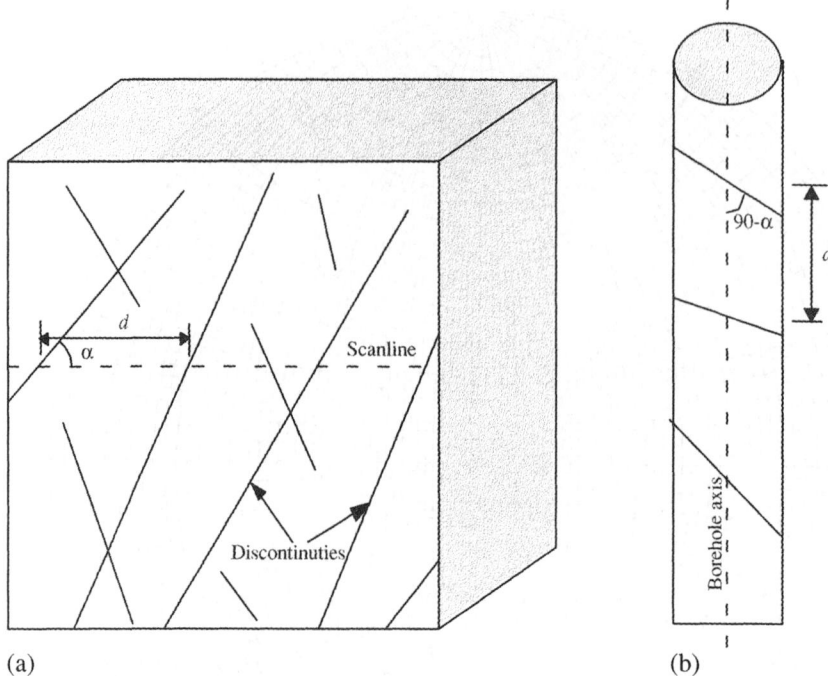

(a)   (b)

Figure 1.18. Measurement of discontinuity spacing from (a) an exposed rock surface and (b) borehole data.

For borehole data, d is the length measured along the core axis between adjacent discontinuities and α is the angle between the core axis and the individual joints (Fig. 1.18). The reliability of measurements of spacing in boreholes can be increased by an employing periscope or TV cameras.

The measured values depend on both scanline orientation and location, however, for parallel discontinuities, measured spacing varies only on the orientation of the scanline. For measured n number of discontinuities along the scanline length of X, the mean discontinuity spacing ($\bar{x}$) is given by

$$\bar{x} = \sum_{i=1}^{n} x_i/n \qquad (1.4)$$

where x_i = measured discontinuity spacing along the scanline.

In order to get a precise value for mean spacing, one needs to take a large number of measurements. The sample size can be calculated for a proportionate error at a certain level of confidence as follows (Priest & Hudson, 1981):

$$n = \left[\frac{z}{\varepsilon}\right]^2 \qquad (1.5)$$

where z = standard normal variable, ε = proportionate error and n = sample size.

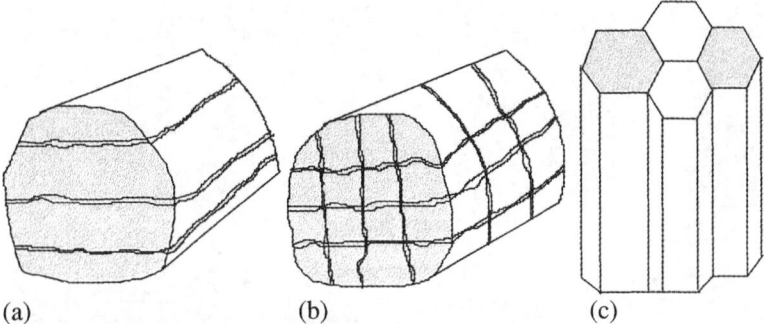

Figure 1.19. Simplified discontinuity shapes in a rock mass. (a) Rectangular joint, (b) square joints and (c) hexagonal joint.

For example, for a probable proportionate error of 5% at the 95% confidence level, the sample should contain at least 1530 discontinuities. The discontinuity spacing may also be expressed as a frequency, which equals the inverse of the mean spacing.

1.5.4   Discontinuity shape

The shape of discontinuity is important in determining the discontinuity size, which has not been thoroughly studied, because, it is not easy to measure the entire length of the fracture surface in the field. The geometry of a natural fracture is in the form of a complex polygon and is difficult to identify, as the joints are not completely exposed. A simplifying assumption used to idealise the discontinuity shape is the consideration of circular, rectangular, square or elliptic shapes (Robertson, 1970; Baecher et al., 1977; Rasmussen et al., 1985). In some rocks such as basalt, joints have clearly visible hexagonal shapes. In orthogonal joint models, joint shapes are assumed as rectangular if joints terminate at the intersection of the other joint plane. As shown in Fig. 1.19, if joints form two/three orthogonal sets of parallel joint network, it can be modelled as an orthogonal model for flow deformation characteristics (Snow, 1969). Discontinuities may also be assumed as circular or elliptical discs in shape (Baecher et al., 1977; Long & Witherspoon, 1985; Warburton, 1980). In this approach, a given discontinuity set is represented by a set of parallel discs whose centres are in space, and the disc radii may be assumed to take a lognormal distribution.

1.5.5.   Infill material in discontinuities

Filled material within discontinuities can influence the strength, deformation and permeability characteristics of jointed rock mass. The filled rock-joint behaviour is governed by material properties, water content, permeability, and the thickness of infill. Figure 1.20 shows a horizontal joint filled with some clay and silt. The

Figure 1.20. A horizontal joint filled with other materials, Southern Freeway, Bulli, Australia.

infill material can typically be clay, silt, fault gouge, calcite or chlorite. During the mapping of joints, it is important to collect some infill material for testing at a later stage.

1.6   PHYSICAL AND MECHANICAL PROPERTIES

The rock-mass behaviour is a combination of both rock-joint properties and intact material properties. The rock-joint properties were discussed in Section 1.5. In this section, properties of rock mass are briefly described, and the reader is advised to refer to other texts for detailed information including test procedures (ISRM, 1978; ASTM, 1995). Figure 1.21 summarises some typical physical properties of intact rock material and rock mass which need to be considered in engineering projects.

1.6.1   Water content and porosity

Water content in a rock specimen is directly related to the degree of saturation and the porosity. In order to carry out non-destructive and rapid method to determine the water content of rock at shallow depths, the nuclear method as described in ASTM (1995) can be carried out. Unlike in destructive tests, the nuclear method

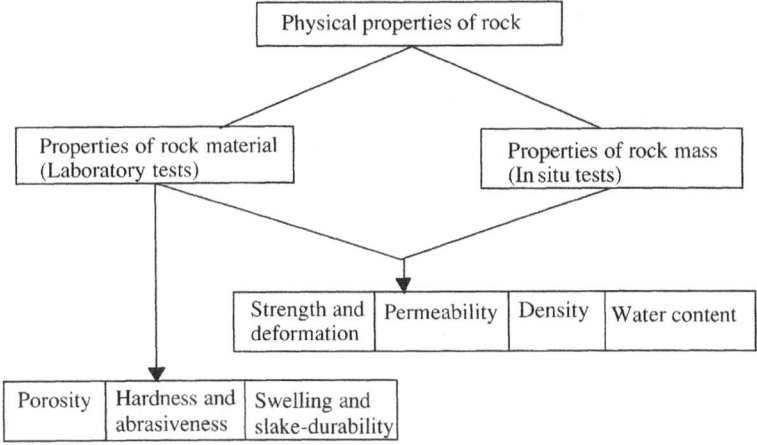

Figure 1.21. Important physical properties of rocks.

could be executed from a single location for a number of tests, in order to obtain a series of test data, facilitating statistical analysis. The water content determined by this method depends on the chemical composition of the rock, sample heterogeneity, material density and the surface texture of rocks.

Porosity, defined as the ratio of pore volume to the total volume, governs the strength, deformability and permeability characteristics of rocks. For example, uniaxial compressive strength decreases with the increase in porosity, whereas the permeability is expected to increase. Depending on the formation (geological process), the porosity of rocks varies from one rock to another. It is usually found that the porosity of igneous rocks (e.g., granite, basalt) does not often exceed 5%. The total pore volume can be readily determined by the weight difference between the fully water-saturated and the oven-dried specimen.

1.6.2 Elastic modulus

In the laboratory, the uniaxial compression apparatus is often used to determine the elastic modulus (e.g., Young's modulus, shear modulus and bulk modulus). However, in order to simulate the correct confining stress in the field, it is better to use the triaxial test method, in which the specimen is subjected to a given confining pressure, while applying the axial stress. The uniaxial compression test gives reliable values for fairly isotropic rock specimens. For a given loading rate, the change of axial and diametric deformation of the specimens is measured using a dial gauge and a clip gauge, respectively. For given test data, the Young's (E), shear (G) and bulk moduli (K) as well as Poisson's ratio (v) are calculated using the following equations:

The axial strain (ε_a),

$$\varepsilon_a = \Delta a / L \qquad\qquad (1.6a)$$

The diametric strain (ε_d),

$$\varepsilon_d = \Delta d/d \tag{1.6b}$$

where Δa and Δd are axial and diametric deformation, and L and D are length and the diameter of the specimen, respectively. The shear modulus (G) and bulk modulus (K) are calculated using the following elastic formulae:

$$G = \frac{E}{2(1 + \nu)} \tag{1.7a}$$

$$K = \frac{E}{3(1 - 2\nu)} \tag{1.7b}$$

Depending on the engineering requirements, the deformation modulus can be calculated either as a secant modulus or as a tangent modulus from the stress-strain plot. Typical strength and deformation properties for intact rock specimens are given in Table 1.2. The presence of joints leads to a decrease in the modulus and strength of rocks. Build-up of water pressure in joints further reduces the compressive and shear strength.

Using the rigid or flexible loading method, the in-situ modulus of a rock mass can also be determined in underground cavities. The load is usually applied to a steel plate on a levelled area on the rock surface, and the deformation is observed using displacement transducers. The modulus (E) is calculated for a rigid plate as follows (ASTM, 1995-D4394):

$$E = \frac{(1 - \nu^2)P}{2W_a R} \tag{1.8}$$

where P = total load on the plate, W_a = average deflection of the plate, R = radius of the rigid plate.

Alternatively, a radial jacking test can also be used to determine the deformation modulus and the anisotropic behaviour of rocks (ASTM, 1995). This technique

Table 1.2. Typical strength and deformation moduli for various rocks.

Rock type	Young's modulus (GPa)	Poisson's ratio	Bulk modulus (GPa)	Shear modulus (GPa)	Friction angle (Deg.)	Cohesion (MPa)
Granite	73.8	0.22	43.9	30.2	51	55.1
Quartzite					48	70.6
Basalt					31	66.2
Shale	11.1	0.29	8.8	4.3	14.4	38.4
Marble	55.8	0.25	37.2	22.3	–	–
Sandstone	19.3	0.38	26.8	7.0	27.8	27.2
Siltstone	26.3	0.22	15.6	10.8	32.1	34.7
Limestone	28.5	0.29	22.6	11.1	42	6.72

Table 1.3. Tensile strength of various rocks.

Rock type	Tensile strength, (MPa)	Compressive strength, (MPa)	Reference
Pikes Peak Granite	11.89	226	Miller, 1965
Nevada Test Site Granite	11.75	141	Stowe, 1969
Nevada Test Site Basalt	13.0	148	Stowe, 1969
John Day Basalt	14.5	355	Miller, 1965
Dworshak Dam Gneiss	6.9	162	Miller, 1965
Micaceous shale	2.1	75.2	Blair, 1956
Flaming Gorge Shale	0.2	35.2	Brandon, 1974
Tavernalle Limestone	3.9	97.9	Miller, 1965
Bedford Limestone	1.6	51	Miller, 1965

gives more reliable test data than the other two techniques discussed above. This is because, in the field, a larger volume of rock is affected, so that the influence of discontinuities is taken into account. In-situ uniaxial compressive tests are also used to measure the deformability and the strength of rock mass with closely spaced joints, approximately 30–500 mm.

1.6.3   Tensile strength

The direct estimation of tensile strength of rocks using the uniaxial tensile test is not an easy task because of the difficulty involved in sample preparation. The Brazilian test seems to be desirable because of its simplicity and low cost. The description of Brazilian test procedures can be found in ISRM, (1978) and ASTM (1995). The tensile strength (σ_t) is calculated using the following formula:

$$\sigma_t = 2P/\pi DL \tag{1.9}$$

where P = failure load, D = diameter and L = thickness of the specimen.

The tensile strength decreases with the increase in water content, the porosity and interconnectivity of joints. For a single fractured specimen, the influence of fracture on tensile strength is minimal when the fracture orientation is parallel to the loading direction. Typical values of tensile strengths of rocks are given in Table 1.3.

1.7   ROCK-MASS PERMEABILITY

1.7.1   Introduction

The term rock-mass permeability often appears in discussions on groundwater flow and is of fundamental importance in rock engineering problems. The concept of permeability defines the resistance to flow under an applied hydraulic gradient. The average permeability of a rock mass depends on the porosity, the

Table 1.4. Typical permeability of porous rocks (Brace et al., 1968).

Rock type	Permeability, nanodarcy	Reference
Fine-grained Dolomite, Tennessee	80	Ohle, 1951
Fine-grained Limestone, Tennessee	30	Ohle, 1951
Coarse-grained Dolomite, Tennessee	6000	Ohle, 1951
Granite, Barriefield, Ontario	50	Ohle, 1951
Granite, Quincy	4600	Ohle, 1951
Diabase, Hudson, New York	0.8	Ohle, 1951

number of fractures present and their interconnectivity, fracture aperture, hydraulic head, properties of fluid itself and ground stresses.

The permeability is more correctly known as the 'intrinsic permeability' and it is independent of the properties of the fluid, having units of (m^2). In a strict sense hydraulic conductivity should be used in the place of permeability since water is, in most cases, is the permeant. Hydraulic conductivity (K) relates the intrinsic permeability to fluid dynamic viscosity (μ) and is a function of both permeant properties and pore geometry:

$$K = \frac{kg\rho}{\mu} = \frac{kg}{\nu} \tag{1.10}$$

where K has units of ms^{-1}, ρ = density of the fluid, k = intrinsic permeability and ν = kinematic viscosity.

The energy head causing water flow is calculated from Bernoulli's equation. In groundwater problems where the seepage velocity (V) is small, the effect of kinematic head can be ignored (i.e. assume that $V^2/2g \approx 0$). Thus, head (or total head, as it is sometimes described) is defined as the sum of pressure head and elevation head above a given datum:

$$h = \frac{p}{\gamma_f} + z \tag{1.11}$$

where h is head (m), p = pressure (kPa), γ_f = unit weight of fluid (kN/m^3) and z is elevation above datum (m).

1.7.2   Primary and secondary permeability

Primary permeability is directly related to the material porosity, which varies depending upon factors such as the rock type, geological history, and in-situ stress conditions. Porosity can vary between less than 1% for shales and granites up to 50% or more for some clays and sandstones. Secondary permeability is a function of the fracture aperture geometry and connectivity.

Water flow occurs in rocks through a combination of intergranular pores and interconnected fractures. Rock-mass conductivity (K) is, therefore, the sum of the

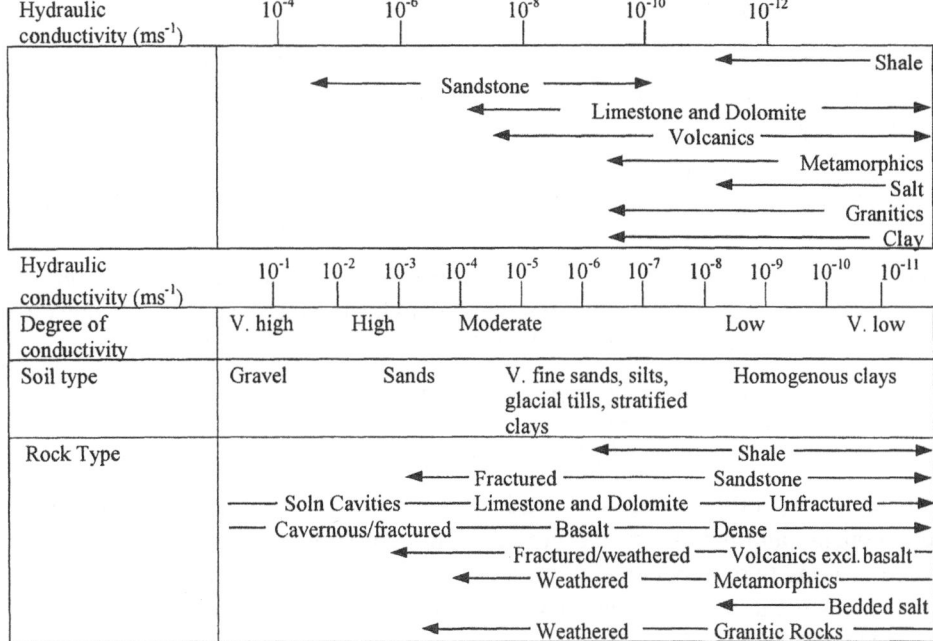

Figure 1.22. Typical values of primary and secondary hydraulic conductivity (after Isherwood, 1979) from laboratory and field tests.

matrix or primary conductivity (K_m) and the fracture or secondary conductivity (K_f). Both flow mechanisms occur in most rocks but it is commonly accepted that due to the low primary porosity of many rocks, fracture flow is often the dominant mechanism (Fig. 1.22):

$$K = K_m + K_f \qquad (1.12)$$

Consideration of Eqn. 1.12 indicates that in cases where the primary permeability is about 2 orders of magnitude less than the secondary permeability, the effect of the primary permeability can be ignored for most practical engineering purposes. Practical implications of Eqns. 1.10–1.12 are elucidated in Ch. 3.

The typical variation in hydraulic conductivity for different rock types is often obtained by comparison of laboratory tests and field testing. This variation is a function of the size of the sample in relation to the scale of rock-mass variability in the field.

The Representative Elemental Volume (REV) is defined as a notional volume of rock mass that if tested would contain the intrinsic features of the rock mass. Theoretically, this would allow a rock mass comprising elements of primary and secondary conductivity to be considered as an equivalent porous media. Although a useful concept, in reality, the actual volume varies with the rock type and location, and in certain circumstances may be indefinable.

1.7.3   Flow in porous media

Under conditions commonly encountered in engineering projects, seepage is observed to be intergranular, laminar and viscous allowing seepage calculation using Darcy's Law:

$$Q = KA\frac{\Delta h}{L}$$

(1.13)

where Q = flow rate (m^{3}s^{-1}), K = hydraulic conductivity (ms^{-1}), A = cross-sectional area of flow (m^2) and Δh = change in head measured over a distance L.
    Darcy's law can be more formally written as

$$q_i = -\frac{K_{ij}}{\gamma_f}\frac{\partial P}{\partial x_j} = -\frac{k_{ij}}{\mu_f}\frac{\partial P}{\partial x_j}$$

(1.14)

where q_i = specific discharge, k_{ij} and K_{ij} = permeability and conductivity tensor components and $\partial P/\partial x_j$ = the pressure gradient causing flow; i and j are unit vectors along two perpendicular axes.
    If a gas is the permeant the matrix coefficient of air permeability is estimated according to Eqn. 1.15:

$$k = \frac{2qp_e\mu L}{(p_i^2 - p_e^2)A}$$

(1.15)

where q = gas flow rate, μ = dynamic viscosity of the gas, L = length of the sample, D = sample diameter, A = cross-sectional area of sample, p = gas pressure measured at the inlet and exit; the subscripts i and e represent inlet and exit, respectively.
    The matrix permeability coefficient for transient method (decay of pressure is observed) is given by Kranz et al. (1979):

$$k_m = \frac{\alpha\beta\mu LV_1V_2}{A(V_1 + V_2)}$$

(1.16)

where β = isothermal compressibility of fluid, A = cross-sectional area, V_1 and V_2 = volume of the pore fluid at the top and the bottom of the sample, respectively, L = length of the specimen and α is an empirical constant for a given initial pressure.

1.7.4   Flow in discontinuous media

Flow is often approximated to occur between smooth parallel plates, assuming that flow is laminar and viscous. In such a case, the conductivity of a single fracture is given by the 'cubic law':

$$K_f = \frac{ge^3}{12vb}$$

(1.17)

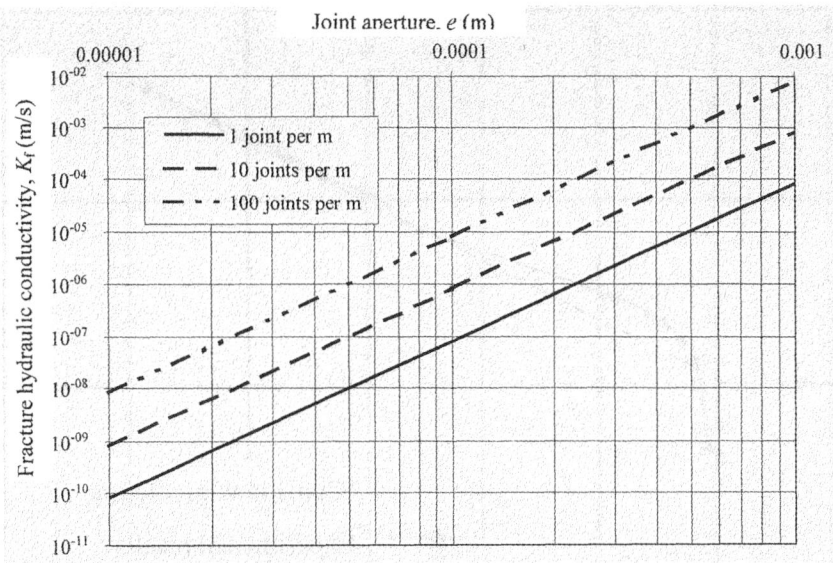

Figure 1.23. Influence of joint aperture (*e*) and joint spacing on hydraulic conductivity (K_f) in the direction of a set smooth joints (after Hoek and Bray, 1981).

where K_f = fracture conductivity (ms^{-1}), e = hydraulic aperture (m), g = acceleration due to gravity (ms^{-2}), v = kinematic viscosity, which is 1.01×10^{-6} (m^{2}s^{-1}) for pure water at 20°C, and b is the spacing between fractures (m). The sensitivity of fracture conductivity to aperture and fracture frequency is well illustrated in Fig. 1.23.

The use of the cubic law is valid for most engineering calculations, except where fractures are significantly rough or pressure gradients are large (Fig. 1.24). In these circumstances, laminar flow behaviour can be replaced by transitional or turbulent flows. Early investigations (Louis, 1968) applied the concepts of pipe hydraulics to fracture flow incorporating the pressure drop coefficient λ and the Reynolds number Re.

Reynolds number (Re) is defined in terms of the aperture:

$$Re = \frac{VD_h}{v} \tag{1.18}$$

where V = average velocity, D_h is the hydraulic diameter of a fracture of aperture e and v is the kinematic viscosity of the permeant.

Louis (1968) developed relationships between pressure drop coefficient, relative roughness, Reynolds number and unit flow rate for a laminar, transitional and turbulent flow. Louis (1968) quantified roughness using an empirical relative roughness coefficient defined as k/D_h where k is the absolute roughness (i.e. the maximum amplitude of the asperities measured parallel to the plane of the fracture, Wittke, 1990).

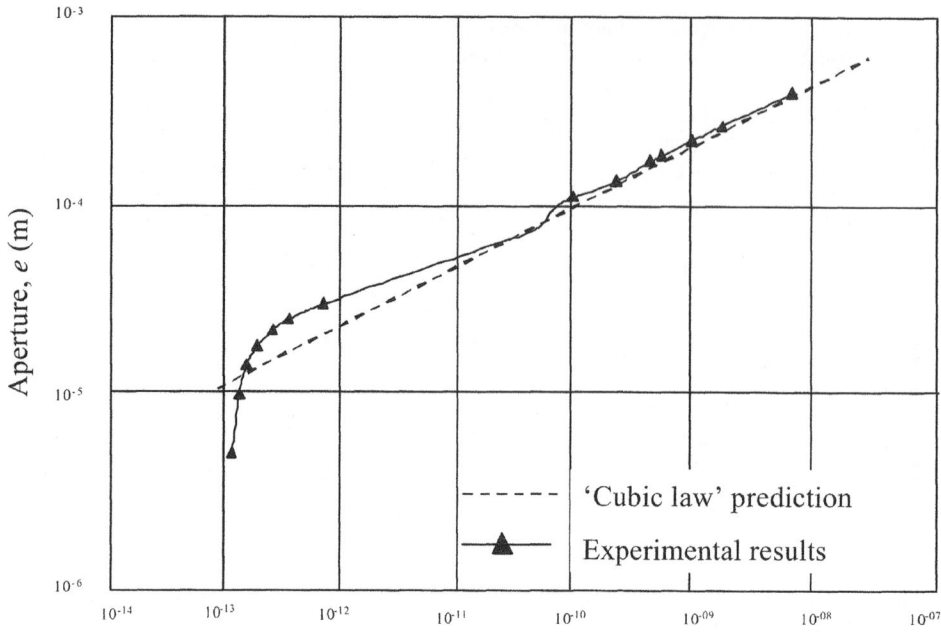

Flow rate per unit pressure gradient (m⁴/ Pa s)

Figure 1.24. Sample plot showing the range of validity of the 'cubic law' after Long et al. (1996) and Cook et al. (1990).

Louis (1968) presented the relationship between the total energy gradient J_c and the pressure drop coefficient λ. Where the velocity V is constant along the flow path, $J_c = J$ thus

$$J = \frac{1}{D_h}\frac{V^2}{2g}\lambda \tag{1.19}$$

The relationships of the pressure drop coefficient (λ) to Reynolds Number (Re) for the range of flow types are given in Eqns. 1.20–1.22. Laminar flow was found to extend to a critical Reynolds number (Re_k) of 2300 for hydraulically smooth surfaces ($k/D \leq 2300$). The onset of turbulent conditions occurred at Re_k as low as 100 with increased fracture roughness.

$$\lambda = f(Re) \text{ hydraulically smooth flow} \tag{1.20}$$

$$\lambda = f\left(Re, \frac{k}{D_h}\right) \text{ transition flow} \tag{1.21}$$

$$\lambda = f\left(\frac{k}{D_h}\right) \text{ completely rough flow} \tag{1.22}$$

The flow type regions defined by Louis (1976) in Fig. 1.25 are labelled as 1, 2 and 3 corresponding to hydraulically smooth, intermediate and completely rough,

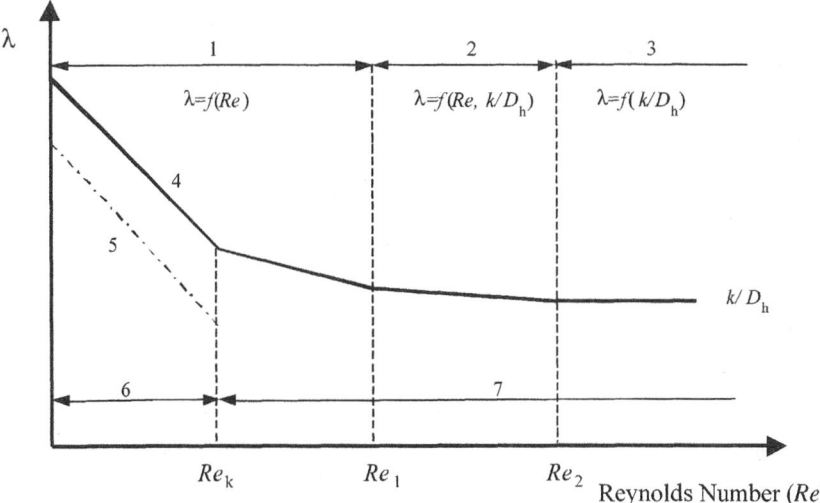

Figure 1.25. Plot of Pressure Drop Coefficient (λ) versus Reynold's Number (Re) after Louis (1976).

respectively. The range of laminar and turbulent flow is also indicated as domains 6 and 7. Figure 1.25 also shows the similarity in flow response between the rock joints and conduits, as represented by regions 4 and 5, respectively.

The algebraic expressions from Fig. 1.26 allow the construction of relationships between relative roughness and pressure drop with Reynolds number for the range of identified flow types. These functions are plotted in Figs. 1.27 and 1.28.

1.7.5   In-Situ testing

The rock-mass hydraulic conductivity is routinely measured as part of the geotechnical field investigations. Common industry methods include borehole tests (constant or falling head), or packer tests where an isolated section of a borehole is subjected to pressurised pumping. When interpreting test results it should be remembered that field trials can be easily affected by smearing of clay-rich material at the borehole wall, resulting in clogging of pores and fractures. Care should be taken in the selection of the drilling method and the preparation of the test section to reduce as far as possible this effect. Evaluation of design parameters should be based upon a number of tests, and combined with careful borehole logging so that the variability in conductivity can be assessed and analysed.

Constant or falling head tests
The advantages of these simple tests are that they are quick and easy to perform and require no specialised drilling equipment. A thorough discussion of head tests is given by Hvorslev (1951), and a summary of the main points is produced herein.

	Flow type	Pressure drop coefficient (λ) and associated author	Unit flow rate (q)
Relative roughness $k/D_h \le 0.033$ (parallel flow)	Laminar	**I**　$\lambda = \dfrac{96}{Re}$ Poiseuille	$q = \dfrac{g}{12v} e_i^3 J_i$
	Turbulent	**II**　$\lambda = 0.316\,Re^{-0.25}$ Blasius	$q = \left[\dfrac{g}{0.079}\left(\dfrac{2}{v}\right)^{0.25} e_i^3 J_i\right]^{4/7}$
		III　$\dfrac{1}{\sqrt{\lambda}} = -2\log\dfrac{k}{3.7D_h}$ Nikuradse	$q = 4\sqrt{g}\left(\log\dfrac{3.7D_h}{k}\right)e_i^{1.5}\sqrt{J_i}$
Relative roughness $k/D_h > 0.033$ (non-parallel flow)	Laminar	**IV**　$\lambda = \dfrac{96}{Re}\left[1 + 8.8\left(\dfrac{k}{D_h}\right)^{1.5}\right]$ Louis	$q = -\dfrac{ge_i^3 J_i}{12v\left[1 + 8.8\left(\dfrac{k}{D_h}\right)^{1.5}\right]}$
	Turbulent	**V**　$\dfrac{1}{\sqrt{\lambda}} = -2\log\dfrac{k}{1.9D_h}$ Louis	$q = 4\sqrt{g}\left(\log\dfrac{1.9D_h}{k}\right)e_i^{1.5}\sqrt{J_i}$

Figure 1.26. Pressure drop coefficients and the unit flow rate for a single joint, from Louis (1976).

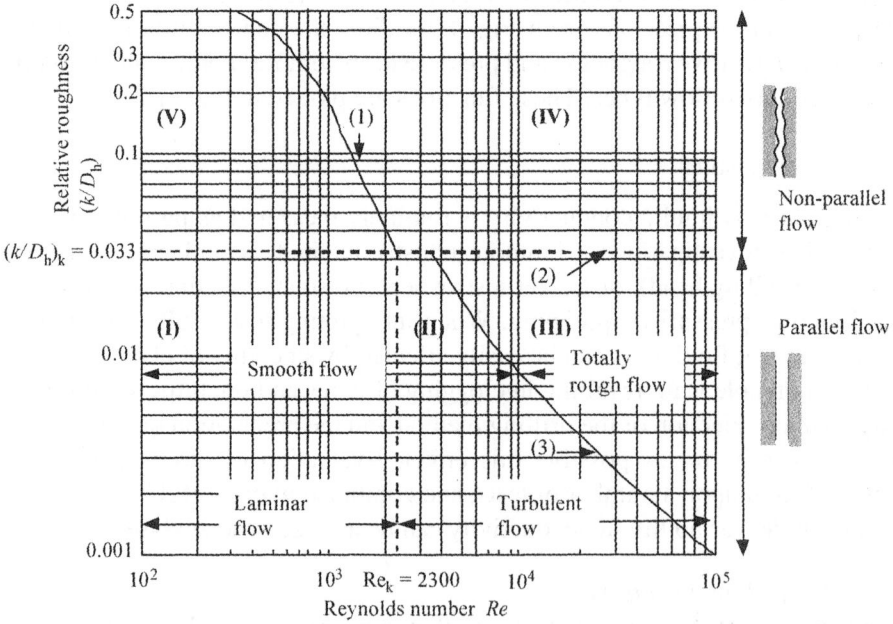

Figure 1.27. Relationship of flow behaviour to relative roughness and Reynolds Number, as defined by Louis (1968) and Thiel (1989).

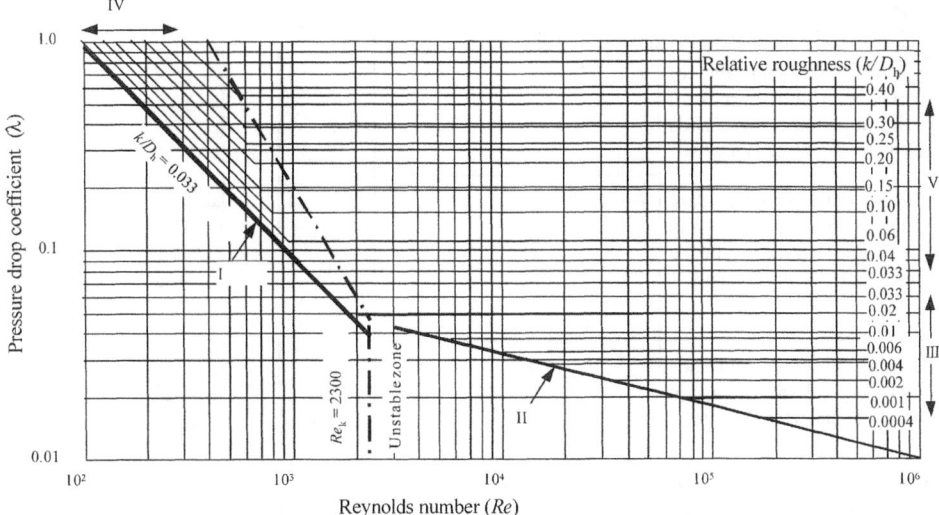

Figure 1.28. Joint flow relationships based on pressure drop coefficient, Reynolds number and relative roughness, after Louis (1976).

The test methodologies apply to saturated ground conditions and involve the following techniques:
(1) Maintaining a constant head of water in a borehole (using a constant inflow or extraction rate) over time; or
(2) Adding or extracting a known quantity of water from the borehole, and monitoring the recovery of the borehole water level with time.

Knowing the borehole geometry and the geological environment of the test section enables the conductivity to be determined, as explained below.

Rock-mass conductivity (K_m) is calculated from falling head tests using:

$$K_m = \frac{A}{F(t_2 - t_1)} \log_e \frac{H_1}{H_2} \tag{1.23}$$

or from constant head tests, using:

$$K_m = \frac{q}{FH_c} \tag{1.24}$$

where A = cross-sectional area of the borehole (m^2), F = is Hvorslev's dimensionless shape factor for different test section conditions (Fig. 1.29), H is the head (m) measured at time t (s) from the standing water level in the borehole, q = constant flow rate (m^{3}s^{-1}), H_c is the head maintained during the constant head test (m).

Packer tests
Packer tests are commonly known as pressure or Lugeon tests, which comprise pumping water into the test sections at staged known pressures until the flow rate

End conditions	Shape Factor F	
Casing flush with end of borehole in soil or rock of uniform permeability. Inside diameter of casing is d cms.	$F = 2.75\,d$	
Casing flush with boundary between impermeable and permeable strata. Inside diameter of casing is d cms.	$F = 2.0\,d$	
Borehole extended a distance L cm beyond end of casing. Borehole diameter D in cms.	$F = \dfrac{2\pi L}{\text{Log}_e\left(2L/D\right)}$ for $L > 4D$.	
Borehole extended a distance L beyond the end of the casing in a stratified soil or rock with different horizontal and vertical permeability. Borehole diameter is D.	$F = \dfrac{2\pi L}{\text{Log}_e\left(2mL/D\right)}$ where $m = \left(k_h/k_v\right)^{0.5}$ for $L > 4D$.	
Borehole extended a distance L beyond the end of the casing which is flush with an impermeable boundary. Borehole diameter is D.	$F = \dfrac{2\pi L}{\text{Log}_e\left(4L/D\right)}$ for $L > 4D$.	

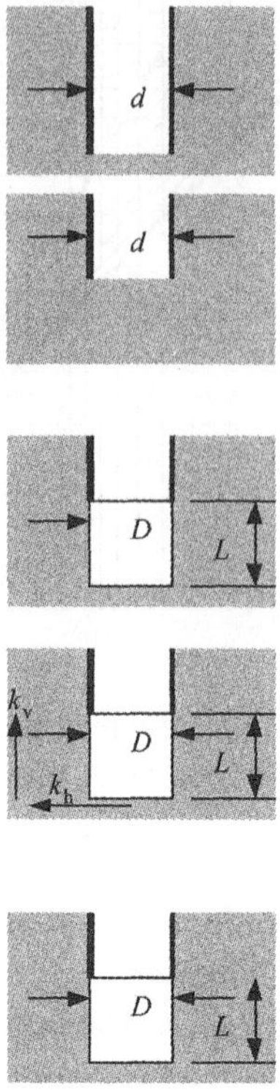

Figure 1.29. Details of shape factor calculation for falling and constant head test, after Hoek and Bray (1981).

becomes constant. The test section is isolated from the rest of the borehole using inflatable packers. The test results are described in Lugeon units (uL), where a Lugeon is defined as the water loss of 1 l/min per 1 m of test section at an effective pressure of 1 MPa. Moye (1967) and Houlsby (1971) have described the use and interpretation of the Packer test. A comprehensive discussion of the Packer test, the related equipment and the analysis is provided by Fell et al. (1992).

Various relationships between the Lugeon value flow rate and an equivalent conductivity (K_e) have been suggested. For instance, Hoek and Bray (1981) suggested relationships based upon the availability of an observation borehole. It is often the case that only a single borehole is available, in which case the Hvorslev (1951) method for a constant head test using a shape factor applicable to a stratified rock mass provides an approximate solution:

$$K_e = \frac{q\text{Log}_e(2mL/D)}{2\pi LH_c} \tag{1.25}$$

where q is the pumping rate needed to maintain a constant pressure over the test section, $m = (K_e/K_p)^{0.5}$, K_e is the conductivity perpendicular to the borehole, K_p is the conductivity in the plane of the borehole, the test section has length L and diameter D and the constant head used for the test is H_c. The mathematical formulation assumes that $K_e > K_p$.

REFERENCES

ASTM Standards (American Society of Testing Materials), 1995. Section 4, Vol. 04.08, Soil and Rock (1) (ASTM D420-D4914).

Attewell, P.B. and Farmer, I.W. 1976. *Principles of Engineering Geology.* Chapman & Hall, London.

Attewell, P.B. and Woodman, J.P. 1971. Stability of discontinuities rock masses under polyaxial stress system. *Stability of Rock Slopes, Proc. 13th U.S. Symp. on Rock Mechanics,* ASCE, New York, pp. 665–683.

Baecher, G.B., Lanney, N.A. and Einstein, H.H. 1977. Statistical description of rock properties and sampling. *Proc. 18th U.S. Symp. on Rock Mechanics,* pp. 1–8.

Baron, M., Bleich, H. and Weidlinger, P. 1960. Theoretical studies on ground shock phenomena. The Mitre Corp. SR-19, Bedford, Mass.

Biot, M.A. 1957. Folding instability of a layered visco-elastic medium under compression. *Proc. Ser. Assoc.,* Roy. Soc. London, 242 p.

Biot, M.A. 1959. The influence of gravity on the folding of a layered visco-elastic medium under compression. *J. Franklin Inst.,* 211–228.

Biot, M.A. 1961. Theory of folding of stratified visco-elastic media and its implications in tectonic and orogenesis. *Geol. Soc. Am. Bull.,* 72: 1595–1620.

Blair, B.E. 1956. Physical properties of mine rock, Part IV. U.S. Bureau of Mines, Inv. 5244.

Brace, W.F., Walsh, J.B. and Frangos, W.T. 1968. Permeability of granite under high pressure. *J. Geophys. Res.* 73(6): 2225–2236.

Brandon, T.R. 1974. Rock mechanic properties of typical foundation rocks. U.S. Bureau of Reclamation Report REC-ERC 74-10.

Cawsey, D.C. 1977. The measurement of fracture patterns in the Chalk of southern England. *Eng. Geol.,* 11: 210–215.

Coates, D.F. 1964. Classification of rocks for rock mechanics. *Int. J. Rock Mech. Min. Sci. Geomech. Abstr.,* 1(1): 421–429.

Cook, A.M., Myer, L.R., Cook, N.G.W. and Doyle, F.M. 1990. The effect of tortuosity on flow through a natural fracture. In *Rock Mechanics Contributions and Challenges,* Balkema, Rotterdam, pp. 371–378. Proc. 31st U.S. Symp. on Rock Mechanics.

Cruden, D.M. 1977. Describing the size of discontinuities. *Int. J. Rock Mech. Min. Sci. Geomech. Abstr.,* 14: 133–137.

Duncan, A.C. 1981. A review of Cartesian coordinate construction from a sphere, for generation of two-dimensional geological net projections. *Comput. Geosci.,* 7(4): 367–385.

Fanklin, J.A. and Dusseault, M.B. 1989. *Rock Engineering.* McGraw-Hill, New York, 600 p.

Fell, R., Stapledon, D. and Macgregor P. 1992. *Geotechnical Engineering of Embankments Dams.* Balkema, Rotterdam, 675 p.

Goodman, R.E. 1976. *Methods of Geological Engineering in Discontinuities Rocks.* West Publishing, St Paul, Minnestota.

Hobbs, B.E., Means, W.D. and Williams, P.F. 1976. *An Outline of Structural Geology.* Wiley, New York, 571 p.

Hoek, E. and Bray, J.W. 1981. *Rock Slope Engineering.* Revised 3rd ed., IMM, London, 358 p.

Houlsby, A.C. 1971. Routine interpretation of the Lugeon water test. *Q. J. Eng. Geol.,* 9: 303–312.

Hvorlsev, M.S. 1951. Time lag and soil permeability in groundwater measurements. U.S. Corps of Engineers Waterways Experiment Station, Bulletin No. 36, 50 p.

Isherwood, D. 1979. Geoscience Data Base Handbook for Modelling Nuclear Waste Repository. Vol. 1 NUREG/CR-0912 VL, UCRL-52719, V1.

ISRM 1978. International Society for Rock Mechanics Commission on Standardisation of Laboratory and Field Tests suggested methods for the quantitative description of discontinuities in rock masses. *Int. J. Rock Mech. Min. Sci. Geomech. Abstr.,* 15: 319–368.

ISRM 1981. Basic geotechnical description of rock masses. ISRM commission on classification of rocks and rock masses, M. Rocha, coordinator. *Int. J. Rock Mech. Min. Sci. Geomech. Abstr.,* 18(1): 85–110.

Kranz, R.L., Frankel, A.D., Engelder, T. and Scholz, C.H. 1979. The permeability of Whole and Jointed Barre Granite. *Int. J. Rock. Mech. Min. Sci. Geomech. Abstr.,* 16: 225–334.

Kulatilake, P.H.S. 1988. State-of the–art in stochastic joint geometry modeling. *Proc. 29th U.S. Symp. on Rock Mechanics,* Minneapolis, pp. 215–229.

Lee, C.H. and Farmer, I. 1993. *Fluid Flow in Discontinuous Rocks.* Chapman & Hall, London, 169 p.

Long, J.C.S. 1996. *Rock Fractures and Fluid Flow, Contemporary Understanding and Applications.* Committee on Fracture Characterisation and Fluid Flow, US National Committees on Rock Mechanics, National Academy Press, 551 p.

Long, J.C.S. and Witherspoon, P.A. 1985. The relationship of the degree of interconnection to permeability of fracture networks. *J. Geophys. Res.,* 90(B4): 3087–3098.

Louis, C. 1968. Etudes des écoulements d'eau dans les roches fissures et de leurs influences sure la stabilité des massifs rocheux. *Bull. De la Direction des Etud. et Rech. EDF, sér. A,* 3, T2-F.

Louis, C. 1976. *Introduction à l' hydraulique des roches.* PhD thesis, Paris.

McEldowney, R.C. and Pascucci, R.F. 1979. Application of remote sensing data to Nuclear Power plant site investigation. *Rev. Eng. Geol.* (American Geological Society), 4: 121–139.

Miller, R.P. 1965. *Engineering Classification and Index Properties for Intact Rock.* PhD thesis, University of Illinois.

Moye, D.K. 1967. Diamond drilling for foundation exploration. *J. I. E. Australia,* Civil Eng. Transactions.

Nockolds, S.R., Knox, R.W.O'B. and Chinner, G.A. 1978. *Petrology for Students.* Cambridge University Press, Cambridge, UK.

Ohle, E.L. 1951. The influence of permeability on ore distribution in limestone and dolomite. *Econ. Geol.,* 46: 667.

Pahl, P.J. 1981. Estimating the mean length of discontinuity traces. *Int. J. Rock Mech. Min. Sci. Geomech. Abstr.,* 18: 221–228.

Pettijohn, F.J., Potter, P.E. and Siever, R. 1972. *Sand and Sandstone.* Springer-Verlag, Berlin, 618 p.

Phillips, F.C. 1971. *The Use of Stereographic Projection in Structural Geology.* 3rd ed., Edward Arnold, London.

Piteau, D.R. 1970. Geological factors significant of the stability of slopes cut in rock. *I Symp. on Planning Open Pit Mines,* South African Institute of Mining and Metallurgy, Johannesburg, pp. 33–53.

Piteau, D.R. 1973. Characterizing and extrapolating rock joint properties in engineering practice. *Rock Mech. Suppl.,* 2: 5–31.

Price, N.J. 1966. *Fault and Joint Development in Brittle and Semi-Brittle Rock.* Pergamon Press, London, 176 p.

Priest, S.D. 1993. *Discontinuity Analysis for Rock Engineering.* Chapman & Hall, London, 473 p.

Priest, S.D. and Hudson, J.A. 1981. Estimation of discontinuity spacing and trace length using scanline surveys. *Int. J. Rock Mech. Min. Sci. Geomech. Abstr.,* 18: 183–197.

Priest, S.D. and Hudson, J.A. 1983. Discontinuity frequency in rock masses. *Int. J. Rock Mech. Min. Sci. Geomech. Abstr.,* 20: 73–89.

Rasmussen, T.C., Huang, C.H. and Evans, D.D. 1985. Numerical experiments on artificially-generated, three-dimensional fracture networks and examination of scale and aggregation effects. *Hydrogeology of rocks of low permeability Int. Assoc. of Hydrologists, Mem.,* (Vol. 17), pp. 676–680.

Robertson, A. 1970. In The interpretation of geological factors for use in slope theory. In *Planning Open Pit Mines.,* van Rensburg, P.W.J., ed., Balkema, Cape Town, South Africa, pp. 55–71.

Rocha, M. and Barroso, M. 1971. Some applications of the new integral sampling method in rock masses. *Proc. Int. Symp. on Rock Fracture,* Nancy, France, pp. 1–12.

Santosh, K.G. 1991. *Geology – The Science of the Earth.* Khanna Publishers, India, 439 p.

Snow, D.T. 1969. Anisotropic permeability of fractured media. *Water Resources Research,* 5: 1273–1289.

Smith, H.S. and Erlank, A.J. 1982. Geochemistry and petrogenesis of komatites from the Barberton greenstone belt, South Africa. In *Komatiites,* Arndt, N.T. and Nisbet, E.G., eds., George Allen and Unwin, London, pp. 347–397.

Spencer, E.W. 1969. *Introduction to the Structure of the Earth.* McGraw-Hill, New York.

Stowe, R.L. 1969. Strength and deformation properties of granite, basalt, limestone, and tuff. US Army Corps Engineers, WES Misc. Paper C-69-1.

Terzaghi, K. 1946. In Introduction to Tunnel Geology. In *Rock Tunneling with Steel Supports.* Proctor, R. and White, T. eds., Youngstown Printing, Ohio.

Thiel, K. 1989. *Rock Mechanics in Hydroengineering, Developments in Geotechnical Engineering, Vol. 51.* Elsevier, Amsterdam, 408 p.

Tucker, M.E. 1993. *Sedimentary Rocks in the Field.* John Wiley & Sons, New York, 153 p.

US Bureau Mines I.C. 1962. A study of *mining examination* techniques for detecting and identifying underground nuclear explosions. *US Bureau Mines I.C.*

Warburton, P.M. 1980. A stereological interpretation of joint trace data. *Int. J. Rock Mech. Min. Sci. Geomech. Abstr.,* 17: 181–190.

Wittke, W. 1990. *Rock Mechanics Theory and applications with Case Histories.* Springer Velag, Berlin, 1073 p.

CHAPTER 2

Specialized triaxial equipment and laboratory procedures

2.1 INTRODUCTION

The material properties as well as stress–strain behaviour of engineering materials including soil and rock are very useful for the design of various engineering structures on the surface and underground. For geological materials such as soil and rocks, there is a high demand for laboratory testing techniques. In order to provide meaningful data from laboratory testing, the apparatus must be truly capable of simulating existing insitu field conditions, including the stress–strain behaviour and permeability characteristics. For this purpose, various types of triaxial apparatus have been developed during the past 5–6 decades (Handin, 1953; Anderson & Simons, 1960; Hoek & Franklin, 1968; Hambly & Reik, 1969; Dusseault, 1981; Smart, 1995; Indraratna & Haque, 1999). Triaxial apparatus may be classified depending on the (a) capacity of the triaxial cell (i.e. high-pressure, or low-pressure), (b) loading system (i.e. plain strain, quasi-static stress, polyaxial stress) and (c) use of single-phase flow or multiphase flows (Table 2.1 and Fig. 2.1).

The triaxial testing of cylindrical soil specimens and solids started in 1930s and the main aim of this equipment was limited to studying the stress–strain behaviour. Subsequently, the triaxial equipment was advanced to test harder materials like rocks. Bishop and Henkel (1969) have extensively discussed the use of triaxial apparatus to study the properties of soil for different boundary conditions. A rock or soil element below the ground surface is normally subjected to three principal ground stresses apart from the fluid pressures, as shown in Fig. 2.2. In order to represent realistic boundary conditions, the testing apparatus must be capable of applying the ground stress independently to the additional fluid pressures. However, most types of triaxial cells have been designed to evaluate basic rock strength and deformation properties. For instance, the conventional triaxial equipment has the ability to use cylindrical specimens subjected to two distinct stress fields (i.e. major and minor principal stresses, σ_1 and σ_3) and the intermediate stress (σ_2) is assumed to be equal to σ_3, by the application of all round fluid pressures.

Table 2.1. Classification of triaxial apparatus.

Classification	Applications	References
Strength of the pressure cell		
Low-pressure triaxial apparatus	Soil and soft rock	Anderson & Simons (1960) Bishop & Henkel, 1953; Hon-Yam & Ronald, 1967;
High-pressure triaxial apparatus	For most type of rocks including hard rocks	Handin, 1953; Meier et al., 1985; Paterson, 1970
Stress state		
Quasi-static triaxial state	Two stresses are equal $\sigma_1 \neq \sigma_2 = \sigma_3$	Saada & Baah, 1967; Hoek & Franklin, 1968
Poly-triaxial state	Principle stress are not equal $\sigma_1 \neq \sigma_2 \neq \sigma_3$	Sture & Desai, 1979; Crawford et al., 1995; Hambly & Reik, 1969
Shape of specimen		
Solid-cylindrical	Common	Paterson, 1970; Shibuya & Mitachi, 1997; Smart, 1995
Solid-cubical	Not common	Michelis, 1988; Airey & Wood, 1988; Amadei & Robison, 1986
Hollow	Not common	Saada & Baah, 1967; Dusseault, 1981; Hight et al., 1983
Rectangle	Not common	Reik & Zacas, 1978; Hambly & Reik, 1969
Type of flowing fluids		
Single-phase flow	Fluid flow through the specimen is single flow (e.g. either water or gas)	Most work on carried out in triaxial test is based on single-phase flow
Two-phase flow	Two fluids or more flow through the specimen (e.g. water + gas or water + gas + oil)	Very few two-phase or multiphase triaxial apparatus are available. Authors have designed a two-phase high-pressure triaxial apparatus.

Before describing the appropriate triaxial equipment for single or two-phase flow through soil or rock specimens, it is important to appreciate the historical development of various forms of triaxial facilities. The following part of this chapter describes the main features of triaxial equipment developed for both soil and rock testing, indicating the main requirements governing their design.

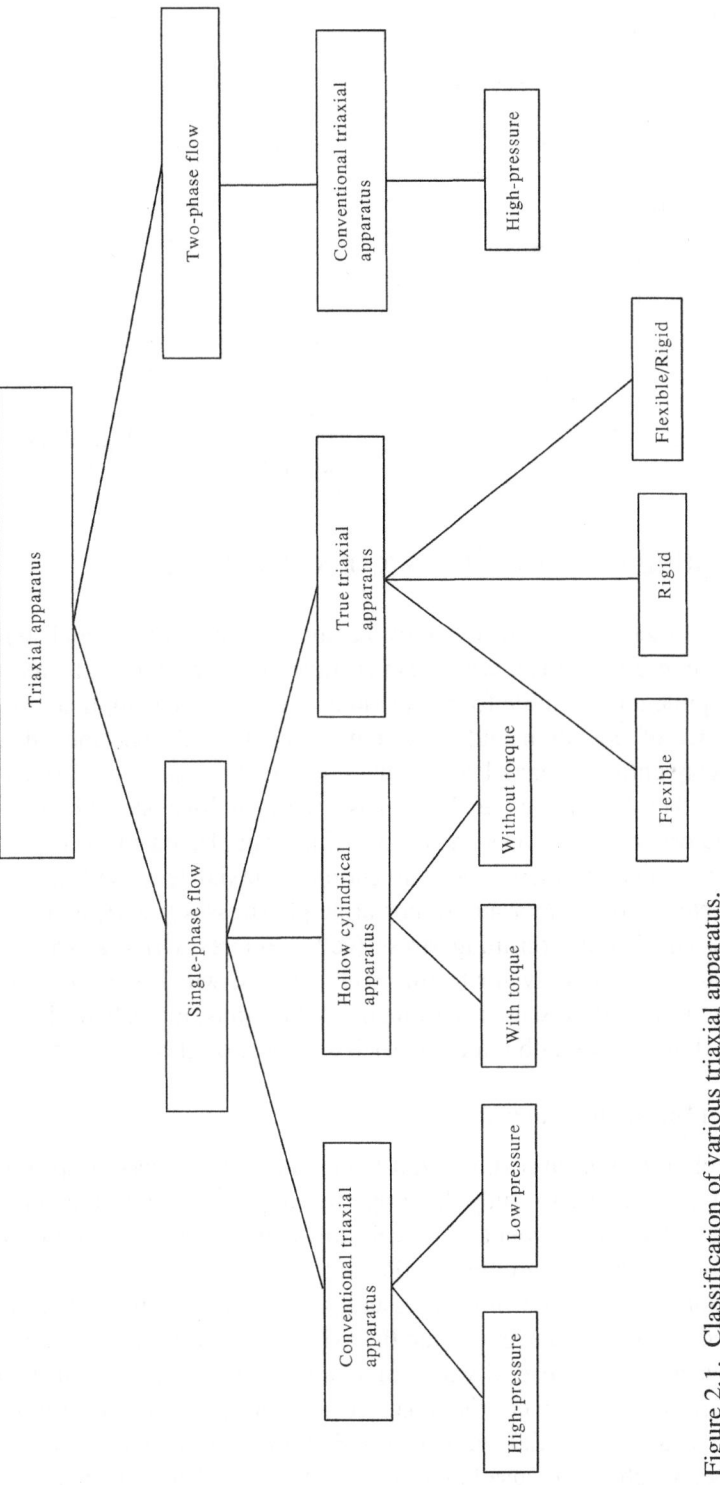

Figure 2.1. Classification of various triaxial apparatus.

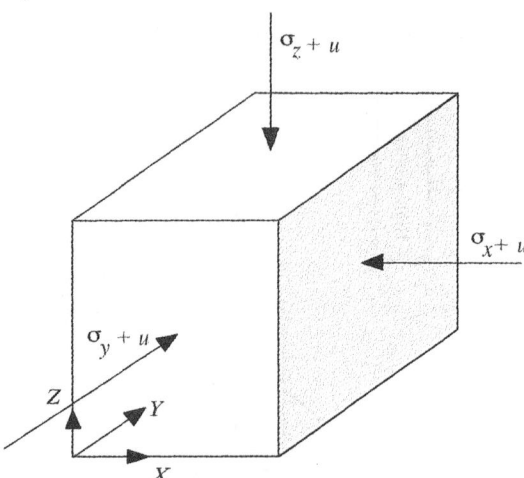

Figure 2.2. Rock element subjected to ground and fluid stresses where, σ is the geological stress and u is the fluid stress.

2.2   LOW-PRESSURE TRIAXIAL CELLS (CONVENTIONAL)

For laboratory investigation of soil and rock behaviour, the conventional axisymmetric triaxial compression test which can sufficiently represent field stresses, has become popular and versatile for many kinds of geotechnical applications, due to its simplicity of operation and provision of reliable results. One other advantage of the conventional triaxial cell is that no special preparation of specimens is necessary other than the two ends, because, the cylindrical shape of samples used in laboratory testing is the same as that of the samples taken from field boreholes. As mentioned earlier, in the conventional triaxial tests, cylindrical samples are subjected to two major and minor principle stresses (i.e. σ_1 and σ_3), i.e. axial pressure and lateral confining stress (cell pressure). Original triaxial facilities were constructed mainly for testing soil samples, which were usually subjected to small all round pressures (less than 1 MPa), thus, the cell wall was usually constructed of low strength materials such as plastic or glass.

2.2.1   Bishop and Henkel triaxial cell

Bishop and Henkel (1969) discussed in detail the features of the triaxial apparatus, application of triaxial tests to study the properties of soil as well as the benefits and limitations of the equipment. This cell was capable of withstanding a maximum confining pressure of around 1 MPa and was designed to test 38 mm diameter soil specimens. The confining pressure was applied by the cell fluid (water) pressurised around the specimen, and the axial load was applied to the top of the specimen via a shaft. Instead of water, oil could also be used as a cell fluid because of its high viscosity, hence it's enhanced resistance to leakage through the membrane. To measure the axial deformation, Bishop and Henkel (1969) used a vernier telescope, which was focused on top of the steel ball. The soil specimen was confined within a rubber membrane having a thickness of 0.26 mm. The commercially available transducers can measure pressure levels accurately, up to several decimal places. However, during 1950s and 1960s pressure transducers were not properly developed, hence other techniques were employed to measure cell pressure and pore pressure in the triaxial tests. For example, Bishop and Henkel (1953) developed a self-compensating mercury control technique for measuring cell pressures. Pore pressure was recorded using a null indicator. The equipment could conduct both drained and undrained tests for given radial and axial pressures. The equipment was not strong enough for rock testing, but this concept was the beginning of testing of water-saturated (single-phase) specimens.

2.2.2 NGI triaxial testing apparatus

Berre (1982) reported of two types of triaxial cells, one for static loading and the other for cyclic loading which were designed by the Norwegian Geotechnical Institute (NGI) for testing of soil. Both cells were identical except the cyclic loading cell had an improved axial loading system. Basically, they are similar to the Bishop and Henkel (1969) triaxial cell, but equipped with traditional measuring devices for pressure measurements as well as electronic devices such as pressure transducers and linear variable differential transformers (LVDTs) for automatic datalogging. The cell could accommodate specimens having diameters of 54 and 80 mm, with a maximum cell pressure of 2 MPa. The triaxial cell was usually filled with liquid paraffin in order to reduce the friction between the piston through the cell and the top seating. Moreover, paraffin would minimise the leakage problems through the membrane due to the high viscosity. When the sample diameter is not exactly the same as the membrane diameter, reducing the sample size to fit into the available membrane is not an easy task, particularly for very soft soil, which can develop fine cracks upon disturbance. To overcome these difficulties, Iversen and Moum (1974) and Berre (1982) proposed the use of paraffin around the specimen instead of the membrane.

2.3 TRUE TRIAXIAL CELLS

In the true triaxial tests, the samples are generally subjected to three different stresses (i.e. σ_1, σ_2 and σ_3), and the typical stress field on a specimen is shown in Fig. 2.2. The main significance of the true triaxial concept is the recognition of intermediate stress (σ_2) which is assumed to be the same as σ_3 in the conventional triaxial apparatus. Depending on the boundary conditions, there are three distinct design features of the true triaxial apparatus:
(1) All rigid boundaries;
(2) All flexible boundaries; and
(3) Combination of rigid and flexible boundaries.

In the last two decades, several researches including Meier et al. (1985), Hambly and Reik (1969), Smart (1995), Crawford et al. (1995), Amadei and Robinson (1986), Hight et al. (1983), Hon-Yam and Ronald (1967), Michelis (1988), and Airey and Wood (1988), and have discussed the various forms of development of true triaxial testing. Majority of the existing true triaxial devices can either be stress

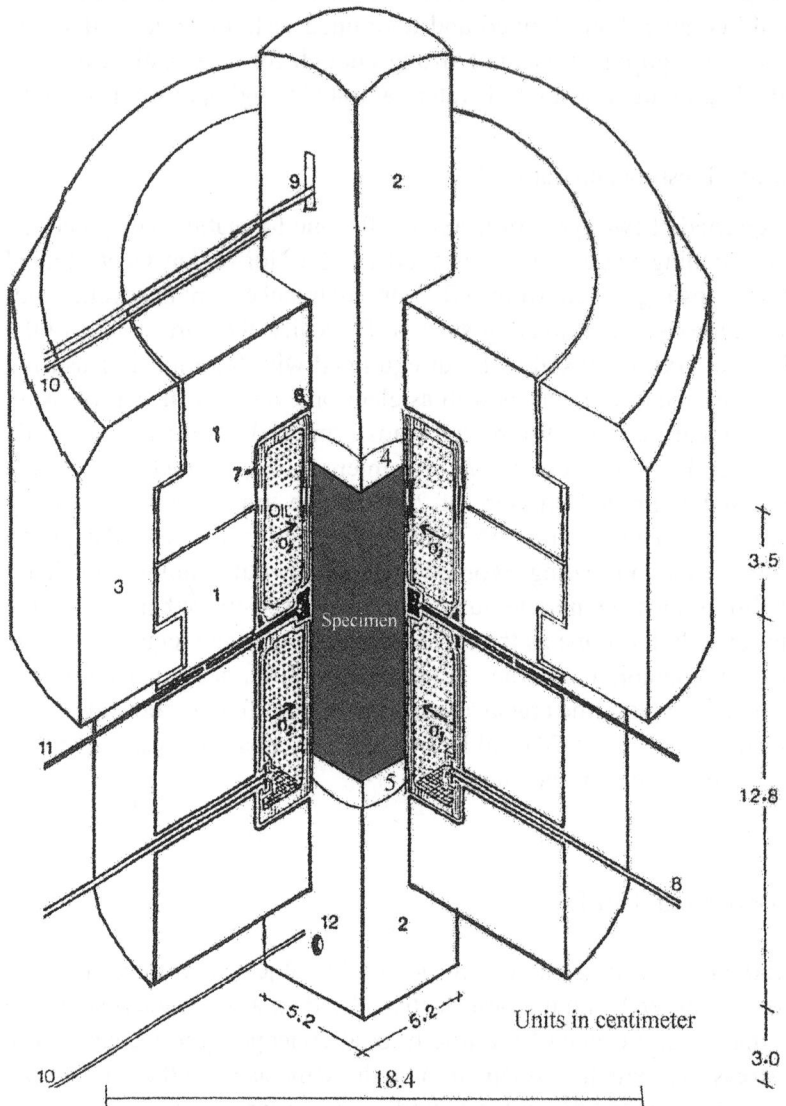

Figure 2.3. True triaxial cell for soil and rock (modified after Michelis, 1988). 1 = Cylindrical body, 2 = piston, 3 = locking segment, 4 = spherical seat and load cell, 5 = spherical seat, 6 = very thin copper plate, 7 = PVC fluid cushion, 8 = high-pressure tube, 9 = partial axial deformation rod, 10 = total axial deformation rod, 11 = lateral deformation rod, 12 = exit pore water and strain gauge cables.

controlled or strain-controlled (Lade, 1973; Michelis, 1988; Smart, 1995). In Sections 2.3.1–2.3.3 three kinds of true triaxial apparatus based on stress-controlled, strain-controlled and both stress–strain controlled are elucidated.

2.3.1   True triaxial cell for rock testing

Another version of the true triaxial cell is illustrated in Fig. 2.3 (Michelis, 1988). The cell pressure is applied to the specimen through oil filled flexible membranes, and the axial load is applied via a rigid piston. The cell can withstand up to a 250 MPa confining pressure and up to a 1500 MPa axial stress. Lateral and axial deformations are recorded using LVDTs, as discussed later in Section 2.8. The major advantage of this application is that both rocks and soil specimens can be tested at low to high pressures. One major drawback of this equipment is that pore water pressure could not be measured, hence only a total stress analysis could be performed. Therefore, this equipment was not appropriate for testing fractured or porous rocks with internal pore pressures.

2.3.2   Poly-axial cell for rock testing

Most true triaxial testing cells have been designed for testing cubical or rectangular specimens. A novel cell capable of testing cylindrical rock core plugs under realistic poly-axial stress state has been presented by Smart (1995). As in a conventional triaxial cell, confining pressure and axial stress are applied via a rubber membrane and rigid plate, respectively. The two distinct horizontal stresses (σ_2 and σ_3) are obtained by arranging 24 PVC tubes around the specimen as shown in Fig. 2.4. The axial cross section of the cell is also shown in Fig. 2.5. The tubes are

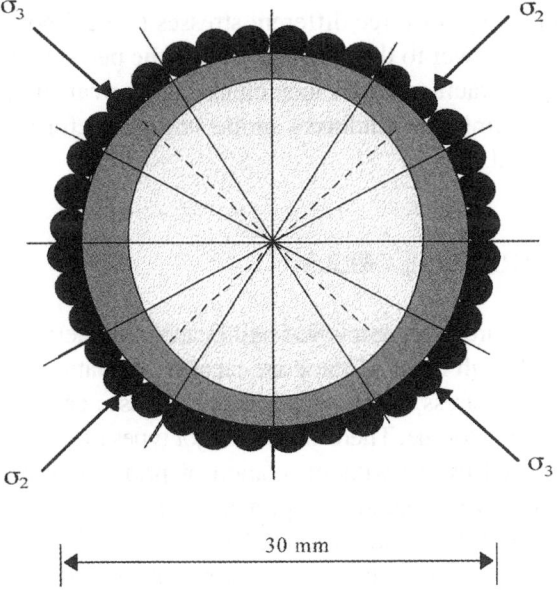

30 mm

Figure 2.4. Stress state on true triaxial cell (from Smart, 1995).

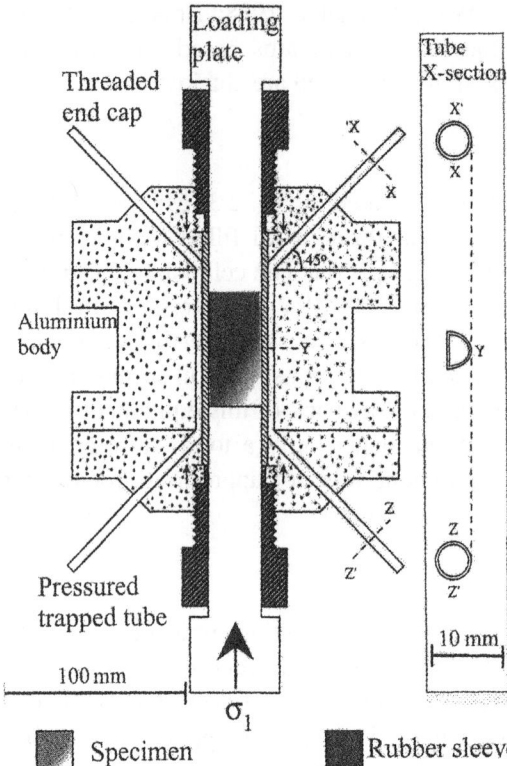

Figure 2.5. Axial cross section of the true polyaxial cell (after Smart, 1995).

encased in individual compartments machined in the body of the cell. Once pressurised, these tubes transmit load to the specimen via the rubber liner. Although the new apparatus ensures the application of three different stresses to a cylindrical core plug, it does not provide the answer to the question of how the permeability characteristics of fractured and unfractured rock specimens can be evaluated. In addition, the effect of increased membrane thickness on the strength-deformation behaviour is not clearly addressed.

2.4   HOLLOW CYLINDRICAL TRIAXIAL CELLS

The conventional and true triaxial cells are often employed significantly to determine a wide variety of stress paths, even though none of these are capable of rotating the principal stresses. Under the same field stress conditions, principal stresses can still rotate especially upon the application of a torque. There are two major types of hollow cylinder triaxial apparatus characterised by (a) without rotation of principle stress directions (Dusseault, 1981) and (b) with rotation of principal stress directions (Broms & Casbarian, 1965; Hight et al., 1983; Saada & Baah, 1967; Lade, 1973). The objective of hollow triaxial devices is to obtain three different principal stresses or to

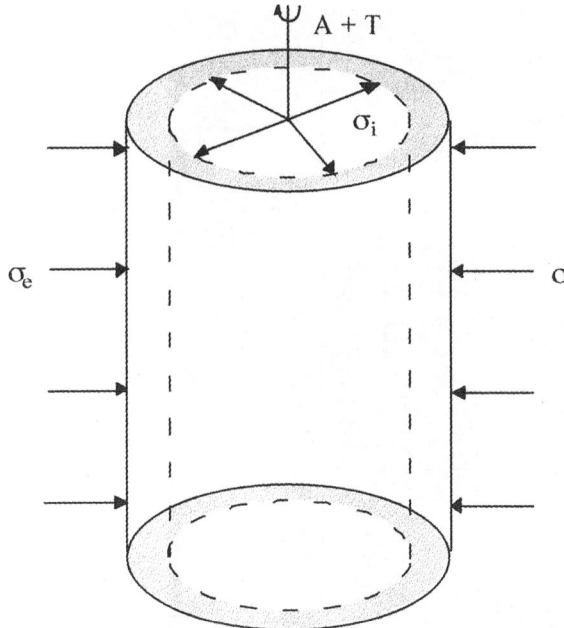

Figure 2.6. Hollow cylindrical specimen subjected to torque, axial and radial stresses.

rotate the principal stresses. Figure 2.6 shows a hollow cylindrical sample subjected to torque, axial and radial pressures. The torque, T, is applied about the vertical axis, and the external (σ_e) and internal (σ_i) radial confining stresses are applied to the specimen. The experimental procedure is a somewhat tedious task due to the need for preparing hollow specimens, practical difficulty of testing fractured rock specimens, as well as the possible development of new cracks during the specimen preparation.

The structural design of a typical hollow cylinder apparatus (Hight et al., 1983) is briefly discussed here. In this hollow cylinder device (Fig. 2.7), a large hollow sample is subjected to combined axial, internal and external radial pressures plus a torque. Moreover, the magnitude and the direction of minor and major stresses can be controlled with the magnitude of the intermediate stress. Radial stresses are directly applied to the inner and outer cylindrical surfaces of the specimen by fluid pressure working through flexible membranes. Proximity transducers are fixed on inner and outer surfaces of the specimen to measure the radial deformations. A complete description of the apparatus is given by Hight et al. (1983).

2.5 HIGH-PRESSURE TRIAXIAL FACILITIES

In order to perform tests on hard material, such as concrete and rocks, as well as to achieve a better understanding of fractured and intact rocks at higher loading conditions, various researches have acknowledged the importance of high-pressure triaxial equipment. Permeability characteristics of fractured and intact rocks under

Figure 2.7. Hollow cylindrical triaxial cell (from Hight et al., 1983; reproduced with kind permission from Geotechnique).

elevated compressive stress are important in many applications, such as, in the mining and petroleum industry.

A simple and inexpensive triaxial cell for testing rock core plugs at high confining pressure was designed by Hoek and Franklin (1968). The section of the

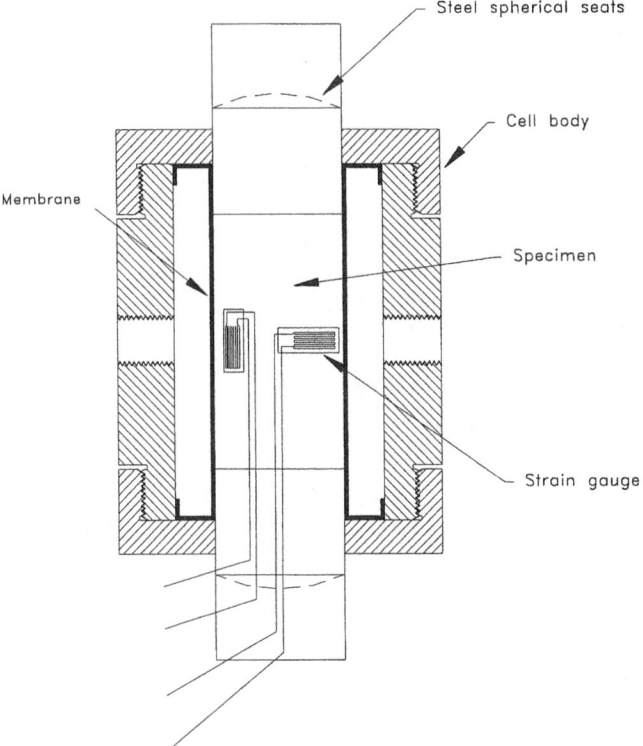

Figure 2.8. High-pressure triaxial cell (modified after Hoek and Franklin, 1968).

apparatus is shown in Fig. 2.8. Because of the simplicity of the device, the tests can be carried out in the field as well as in the laboratory at a maximum cell pressure of 70 MPa. The cell pressure is provided by a hydraulic pump connected to an oil inlet in the cell. Deformation is recorded using strain gauges attached to the surface of the rock. Rock specimens are usually covered with 1.6 mm thick rubber sealing sleeves. Although the equipment provides stress–strain behaviour of rock for preliminary investigation, the permeability characteristics of rocks cannot be measured using this facility.

For rock specimens, the provisions for measuring the permeability together with strength parameters under drained and undrained conditions associated with volume changes and pore pressures have been fully accommodated in the high-pressure triaxial apparatus developed at the University of Wollongong (Indraratna & Haque, 1999). The high-pressure triaxial system (Fig. 2.9) described herein comprises of five major components, namely

(1) High pressure cell assembly;
(2) Volume change device;
(3) Pore pressure measurement system;
(4) Axial loading device; and
(5) Digital display unit.

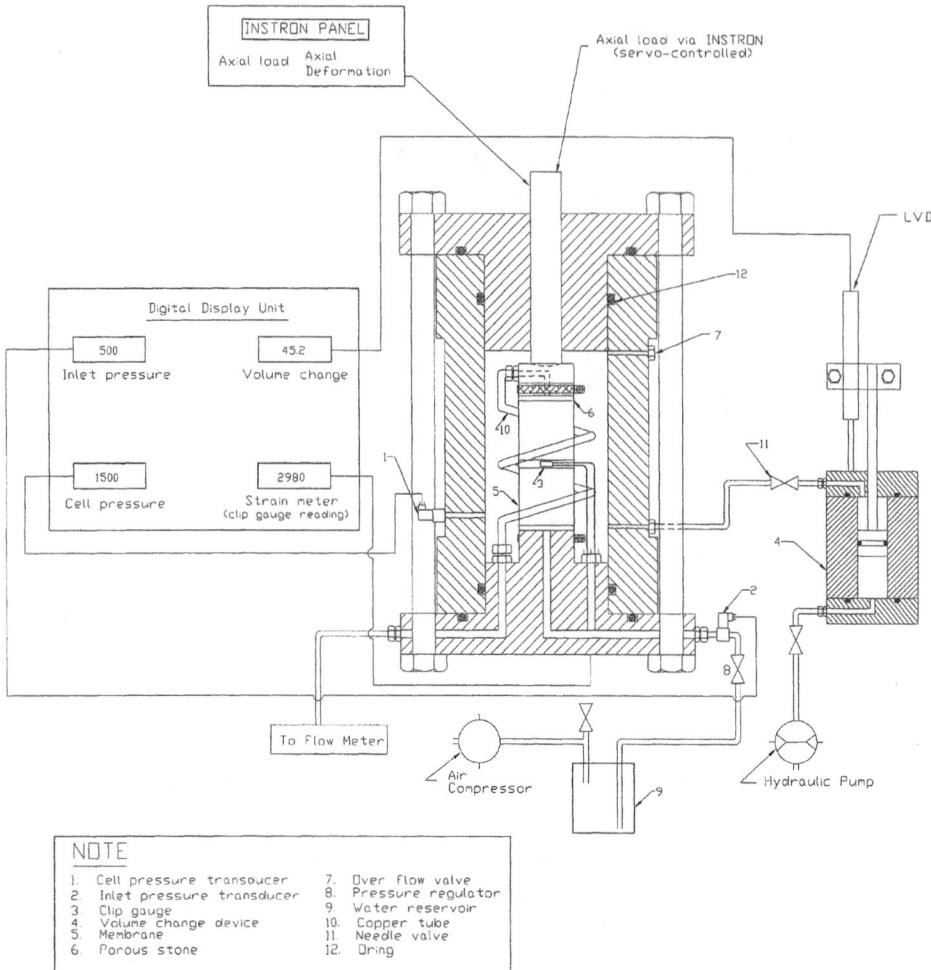

Figure 2.9. High-pressure triaxial cell (Indraratna & Haque, 1999).

The cell is made from high-yield steel having a 100 mm internal diameter and a 120 mm height. The cell walls can withstand a maximum pressure of 150 MPa with a factor of safety 2.0. Sample sizes up to 54 mm in diameter (NX core size) and 120 mm in height can be tested. The cell is confined at the top and bottom by a thick, stepped steel plate, which is firmly held in place by six steel bolts. The water pressure inlet and outlet valves as well as the strain meter connections (i.e. for clip gauge reading) are attached to the bottom plate. The outlet at the bottom of the cell wall is connected to a hydraulic jack for pumping oil prior to testing. The overflow valve for expelling air from the cell is located at the top plate of the cell. Once the specimen is set up inside the cell, oil is manually poured from top of the cell up to the level of the overflow valve. The top plate is then mounted and all the bolts are tightened. Subsequently, using the hydraulic jack, oil is further pumped into the cell via the cell outlet (Fig. 2.9) until all entrapped air is expelled through the overflow valve, which is then closed. Two transducers, one at the inlet and the other fixed to the cell wall are provided to measure the pore pressure and confining pressure, respectively. The volume change device and the clip gauge transducer are discussed in Section 2.8.

2.6   TWO-PHASE FLOWS IN SOIL AND ROCK

In order to study stress–strain behaviour and permeability characteristics of rock or soil, one has to incorporate actual fluid flow in the testing equipment. Specimens can be either fully saturated or unsaturated. Fully saturated soil or rock carries a single fluid, in which case, a single-phase flow analysis can be carried out. If two or more fluids are found, then the specimen is considered to be in an unsaturated state. Under these circumstances, a two-phase flow analysis should be carried out. Figure 2.10 clearly shows whether the single- or multiphase flow analysis is carried out depending on the number of fluids present in the media. Stress states on unsaturated and fully saturated rock/soil element are shown in Fig. 2.11. For an example, in two-phase flow of water–gas, capillary pressure (i.e. $P_w - P_a$) acts in addition to the ground stresses. However, if $P_a > P_w$, then suction pressure (i.e. $P_a - P_w$) acts on the solid element to increase the apparent strength.

2.6.1   Apparatus for unsaturated (two-phase) soils

For unsaturated soils, various laboratory methods (Klute, 1965; Barden et al., 1969) have been used to investigate permeability characteristics under steady- and unsteady-state conditions. According to the approach used by Klute (1965), water is supplied to the specimen from an overhead water tank, and the pressure is measured by two-tensiometers as shown in Fig. 2.12. The constant air pressure is measured using a manometer. The tests have been carried out for different suction pressures ($P_a - P_w$), where P_a is air pressure and P_w is water pressure. In Klute's apparatus, the change of permeability with respect to the change of confining pressure or axial stress has not been addressed.

An advanced triaxial equipment was developed to measure both water and air permeability of soil (Barden et al., 1969), in which the effect of lateral pressure was incorporated. In this apparatus (Fig. 2.13), air and water pressures are applied to the specimen from top and bottom respectively, for a given lateral pressure. The fluid flow direction is not properly modelled in the apparatus, because fluid flow in reality is not just in one direction. Moreover, this equipment is not capable

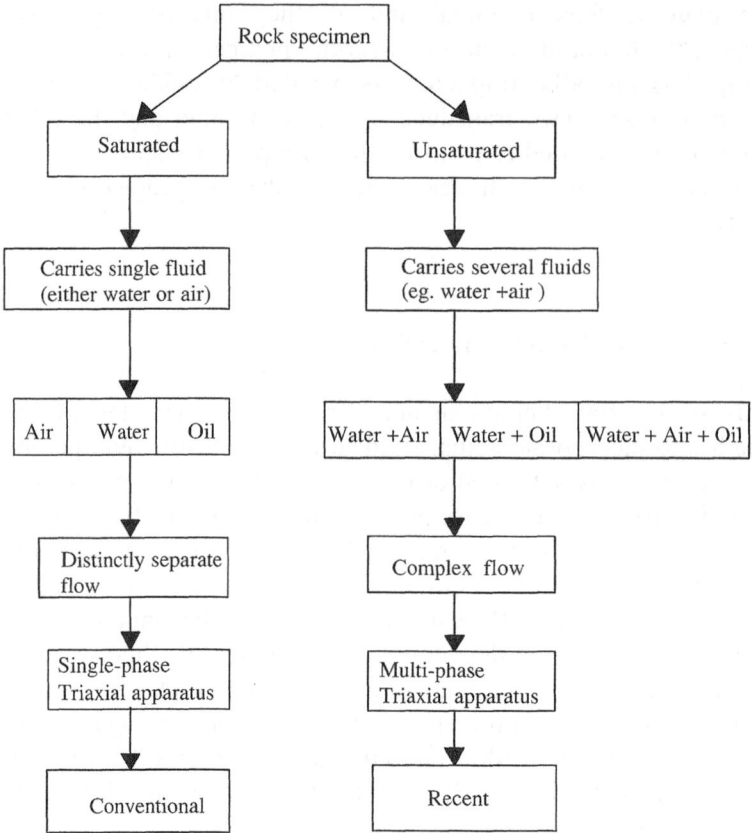

Figure 2.10. State of fluid flow in a rock specimen.

of applying an axial stress, and also, no strain measurement devices are attached to the specimen. Therefore, stress-dependant permeability cannot be evaluated correctly.

2.6.2 Two-phase triaxial equipment for rocks

Although much experimentation has been carried out to understand the complete two-phase (air–water) flow behaviour in the field of chemical and mechanical engineering, the proper understanding of two-phase flow behaviour in jointed rocks still remains at infancy because of its complexity. As discussed earlier, some triaxial facilities are capable of measuring either the pore water pressure or pore air pressure or both within a fractured rock, but are not capable of measuring the relative permeability (air or water) of a fractured specimen. However, a fractured rock mass is generally associated with a multi-phase flow system, such as, water + air, water + air + solid and water + air + oil. In order to conduct an experimental study of two-phase flows through fractured rock specimens, the

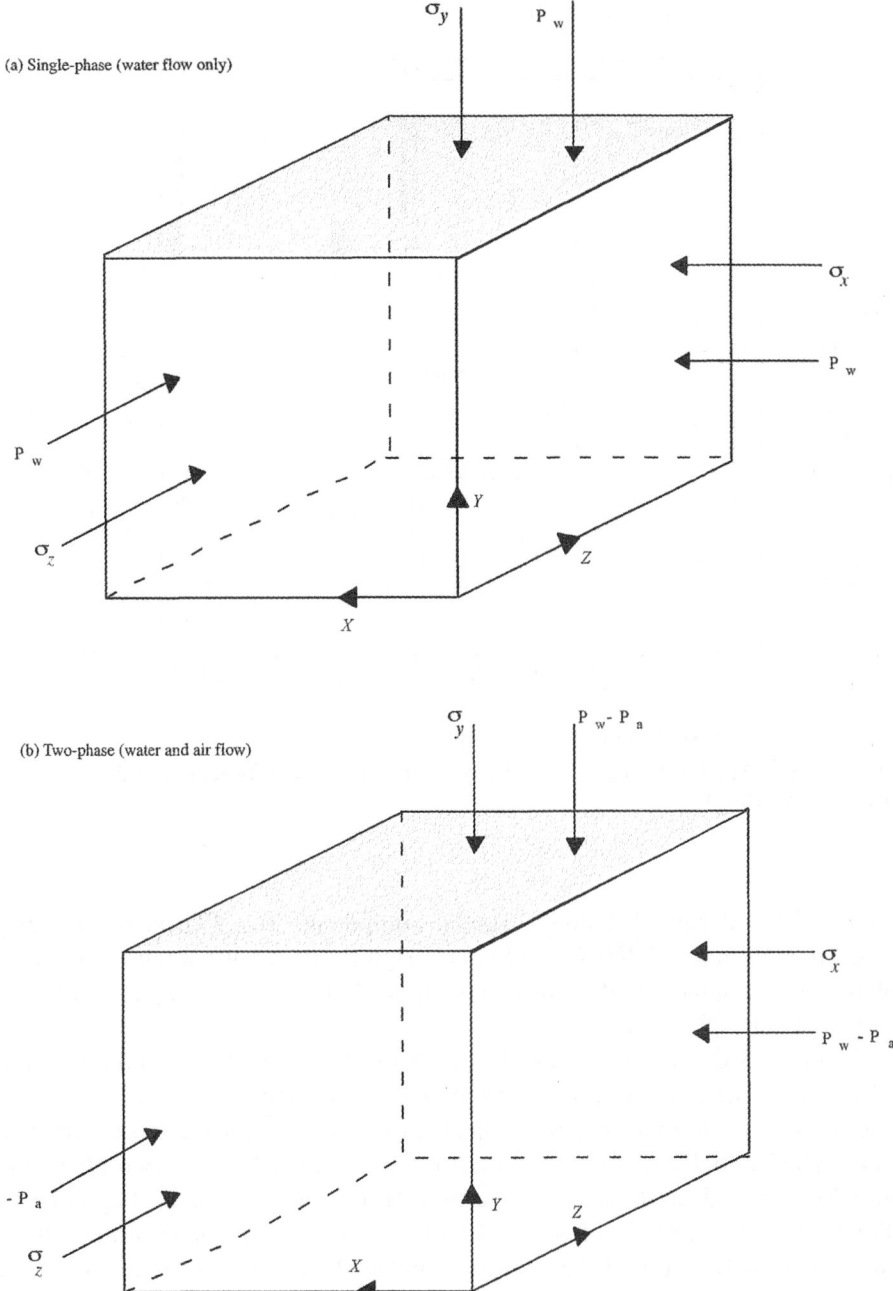

Figure 2.11. Comparison of stress states under single-phase and two-phase flow conditions.

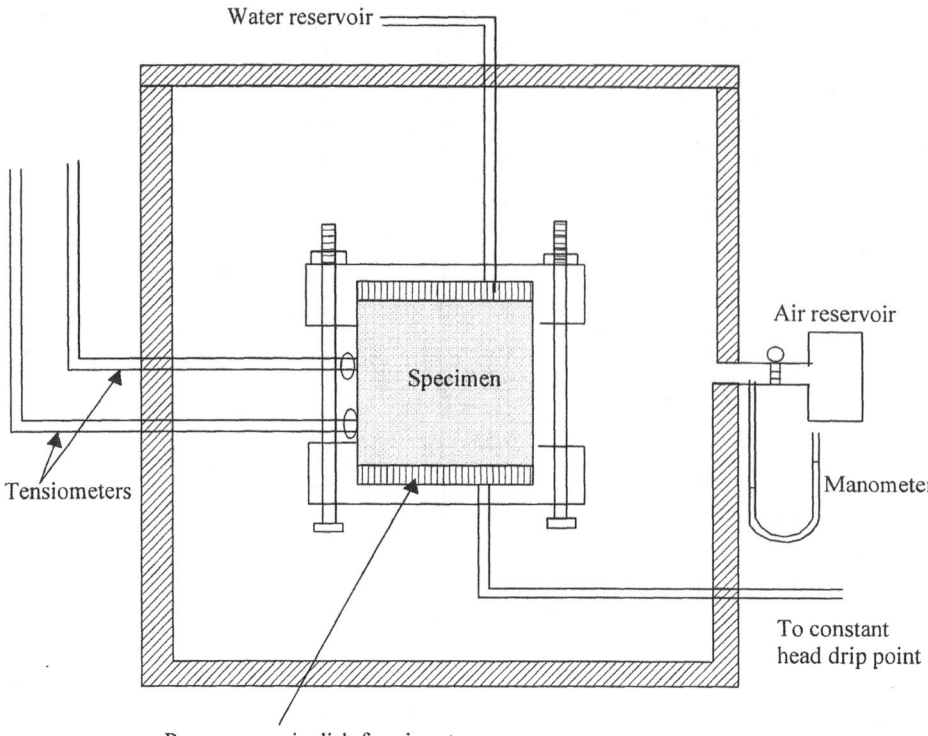

Water reservoir

Air reservoir

Specimen

Tensiometers

Manometer

To constant
head drip point

Porous ceramic disk for air entry

Figure 2.12. Apparatus used to measure permeability coefficient of unsaturated soil (from Klute, 1965).

authors have designed a novel triaxial equipment: *Two-Phase, High-Pressure Triaxial Apparatus (TPHPTA)*. This is a significant modification of the single-phase triaxial apparatus designed previously (Indraratna & Haque, 1999) at the University of Wollongong.

Section 2.7 describes the salient features of the two-phase triaxial apparatus, which can measure relative permeability characteristics, as well as the stress–strain behaviour of rocks subjected to axial and confining pressure conditions. The TPHPTA is capable of carrying out both single and two-phase flows through soft and hard rocks. A photograph and the schematic diagram of the TPHPTA are illustrated in Figs. 2.14 and 2.15. The cell is made from high-yield steel having a 100 mm internal diameter and a 120 mm height. The cell walls can withstand a maximum pressure of 150 MPa with a factor of safety of 2.0. The cell is confined at the top and bottom by a thick, stepped steel plate which, is firmly held in place by six steel bolts. The modified cell can accommodate a range of specimens from 45 to 60 mm in diameter. In two-phase flow, both water and air simultaneously flow through the specimen. In this equipment, water and air phases are carried by two separate lines to the bottom end of the specimen. In

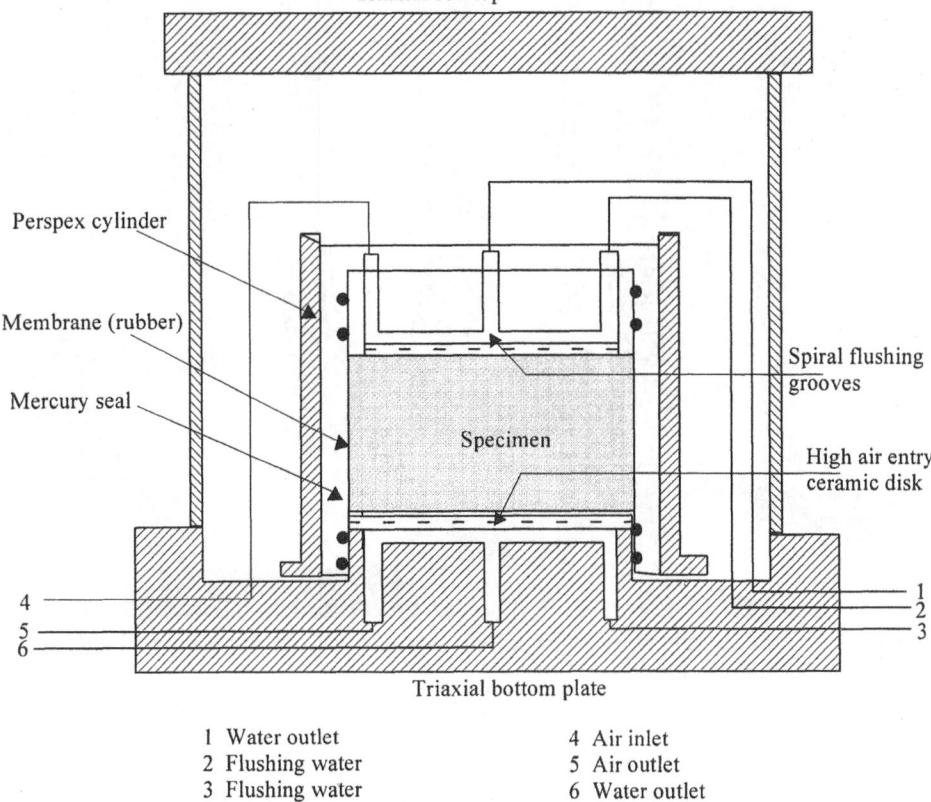

Figure 2.13. Triaxial apparatus for unsaturated soil (from Barden et al., 1969).

1 Water outlet	4 Air inlet
2 Flushing water	5 Air outlet
3 Flushing water	6 Water outlet

order to prevent water and air interaction before entering the specimen, the separate lines which carry water and air are integrated with several on/off valves and check valves as shown in the Fig. 2.16. These valves have been attached to the bottom plate to ensure that there is no back flow of one phase through the line of the other phase. In order to measure the pressure of each phase, a pressure transducer is attached to each phase line.

The outlet at the bottom of the cell wall is connected to a hydraulic jack for pumping oil prior to testing. The overflow valve for expelling air from the cell is located at the top plate of the cell. Once the specimen is set up inside the cell, oil is manually poured from top of the cell up to the level of the overflow valve. The top plate is then mounted and all the bolts are tightened. Subsequently, using the hydraulic jack, oil is further pumped into the cell via the cell outlet until all entrapped air is expelled through the overflow valve, which is then closed.

The readings of the inlet water and pressure transducers, outlet air and water pressure transducers, volume change device, axial and lateral deformations are monitored continuously and displayed digitally on the instrumentation

Figure 2.14. Photograph of two-phase high-pressure triaxial apparatus (TPHPTA) at University of Wollongong.

display unit. In order to measure the quantity of each phase, the mixture from the specimen is separated into water and gas using the gravity separation technique. This permits phase separation at the outlet using a dreschel bottle as shown in Fig. 2.17. Different separation techniques for gas–liquid, liquid–liquid, liquid–solid are discussed in Section 2.7. The water-flow measurement is recorded using an electronic weighing scale, and an electronic film flow meter is used to monitor the continuous air flow. Details of different air-flow meters including the film flow are discussed in Section 2.10.

In two-phase flows, the time taken to reach the steady-state condition is much longer than the single-phase condition for given boundary conditions. Therefore, a datalogger is employed for the acquisition of measurements from all transducers, flow measurement devices, strain meters and LVDTs. When output voltage of some transducers is not large enough, it is first amplified by an amplifier before sending to the datalogger.

This test procedure developed by the authors is relevant to the cylindrical rock specimens tested in TPHPTA. After smoothing both ends of the rock specimen, it is then covered by a polyurethane membrane, and subsequently placed on the bottom seat of the triaxial cell. In order to measure the lateral deformation of the rock specimen, two specially designed clip gauges are mounted at 1/3 length of the specimen on the membrane. Using two horseshoe clamps, the membrane is tightened to the top and bottom seating so that no fluid flow

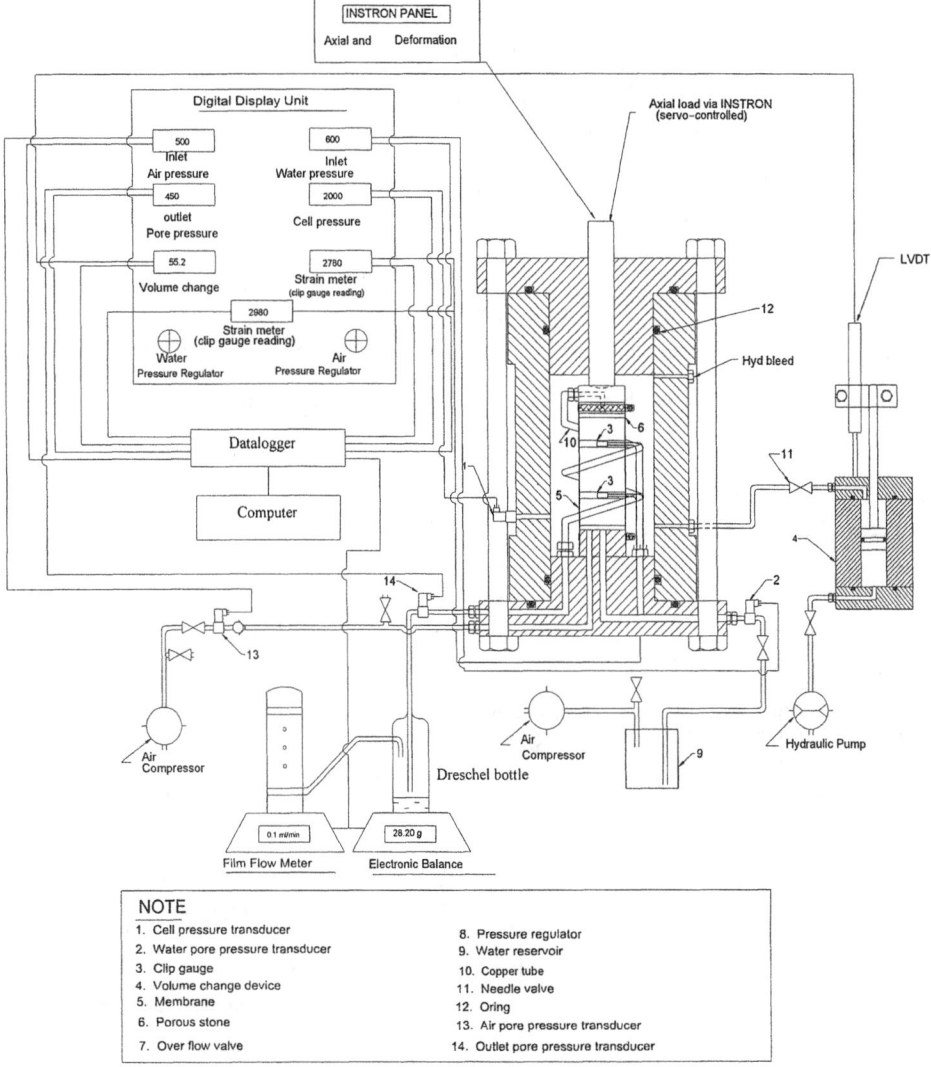

Figure 2.15. Schematic diagram of the two-phase, high-pressure triaxial apparatus (TPHPTA).

through the membrane and the specimen takes place. The spiral tube (see Fig. 2.15) is fixed to carry fluid flow from the specimen to the outlet. Oil is filled inside the cell from the top to the air bleeding hole, and then the top and bottom bases are tightened by six bolts. In order to ensure that there is no trapped air inside the cell, oil is further pumped in to the cell using a hydraulic jack. The specimen is first saturated with one fluid phase, and then the second phase is forced through the specimen. The readings of the inlet air and water pressure transducers, outlet air and water pressure transducers, cell pressure transducer, volume change device, axial and lateral deformations are monitored

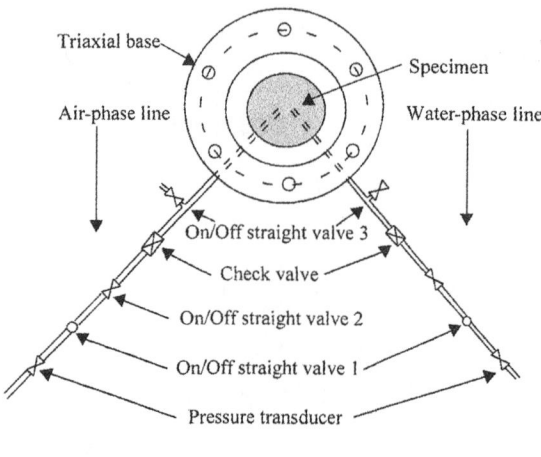

Figure 2.16. Schematic diagram of the triaxial base with water and air phase lines.

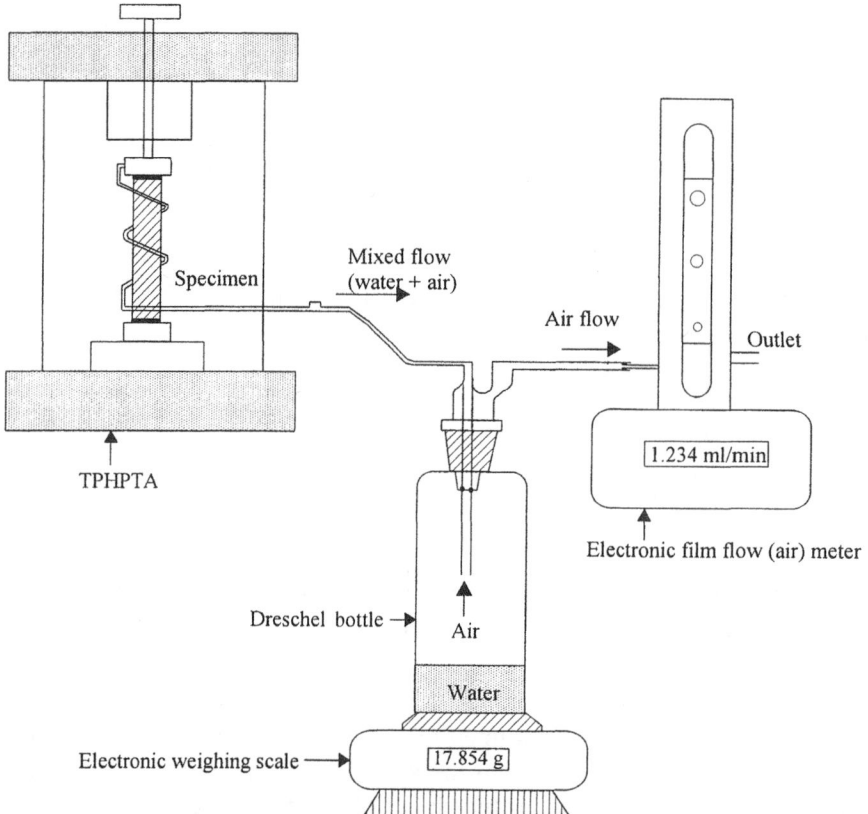

Figure 2.17. Separation of mixed flow (i.e. water + air).

continuously, and displayed digitally on the instrumentation display unit prior to recording by the datalogger. Once the water and air mixture passes through the dreschel bottle, air-flow rates and water flow quantities are recorded by the film flow meter and electronic weighing scale, respectively. Separation techniques of mixtures are discussed in Section 2.7. Table 2.2 summarizes the typical tests that can be conducted using TPHPTA.

2.7 SEPARATION TECHNIQUES

The most common types of fluid phases that are encountered in hydro-mechanics are (a) gas–liquid, (b) liquid–liquid, (c) gas–solid and (d) liquid–solid. The separation techniques of these mixtures are listed in Fig. 2.18. However, no attempt is made here to describe all the separation techniques, except for gas–liquid mixtures. The gas in liquid can be generally of two types which are difficult to separate:
(1) Unstable bubbles
(2) Stable gas bubbles
The unstable gases in water are normally separated by the action of gravity. The water-gas flow is allowed to move in a laminar flow path until the gas bubbles come to the surface. Depending on the gas-flow rates, different sizes of separators (e.g. static tank and continuous flow tanks) can be used. The gravity action may not be enough for gases in high viscosity fluids, and in this case, a centrifugal action (e.g. using Versator machine by Cornell Machine Co.) may be employed. Techniques such as, thermal, electrical and mechanical methods are applied for stable gas in liquids. In the TPHPTA, the mixture contains both gas and water, where gas is in unstable form. The fluid mixture from the triaxial specimen is allowed to flow through a dreschel bottle, as shown earlier in Fig. 2.17. The unstable gas in the water separates from the bottle and passes to the film flow meter where the water stays in the dreschel bottle, which sits on an electronic balance.

2.8 VOLUMETRIC AND LATERAL DEFORMATION MEASUREMENT

In the triaxial test, the volume change or deformation state is normally expressed by axial, lateral (diametric) and volumetric strains. There are basically three categories of deformation measurement devices:
(1) Contact measurements-strain gauges;
(2) Linear Variable Differential Transformers (LVDTs); and
(3) Non-contact devices such as, optical and inductive instruments.
   It is common practice to record axial deformation using dial gauges or LVDTs, and volumetric changes using volume change devices, such as burettes. However, in some cases, it is important to measure the diametric deformations. For example, in fractured rock media, permeability is a function of the joint apertures, thus the change in joint apertures in lateral and axial directions of fractured specimens in triaxial tests is important to assess. Various techniques,

Table 2.2. Testing procedures for two-phase flow measurement.

Stages		Cell pressure	Axial stress	Fluid pressures		Comments
				Water (p_w)	Air (p_a)	
Stage 1	a	Constant	Constant	Constant	No air flow	Steady-state water flow is observed (joint is initially saturated with water)
	b	Constant	Constant	Constant	Variable	Air pressure is gradually increased from zero. For ($p_w > p_a$), steady-state is observed for both water and air
	c	Constant	Constant	Constant	Variable	Air pressure is gradually increased until air pressure equals water pressure. Capillary pressure becomes zero ($p_w = p_a$). Steady-state two-phase flow is observed
	d	Constant	Constant	Constant	Variable	Air pressure is gradually increased until no water flow is observed from the specimen. For different suction pressures ($p_a > p_w$), steady-state two-phase flow is observed
Stage 2		Constant	Constant	Variable	Constant	Same as stages 1a, 1b, 1c and 1d
Stage 3		Variable	Constant	Constant	Constant	Cell pressure is changed at zero capillary pressure, ($p_w = p_a$)
Stage 4		Constant	Variable	Constant	Constant	Axial stress is changed at zero capillary pressure, ($p_w = p_a$)

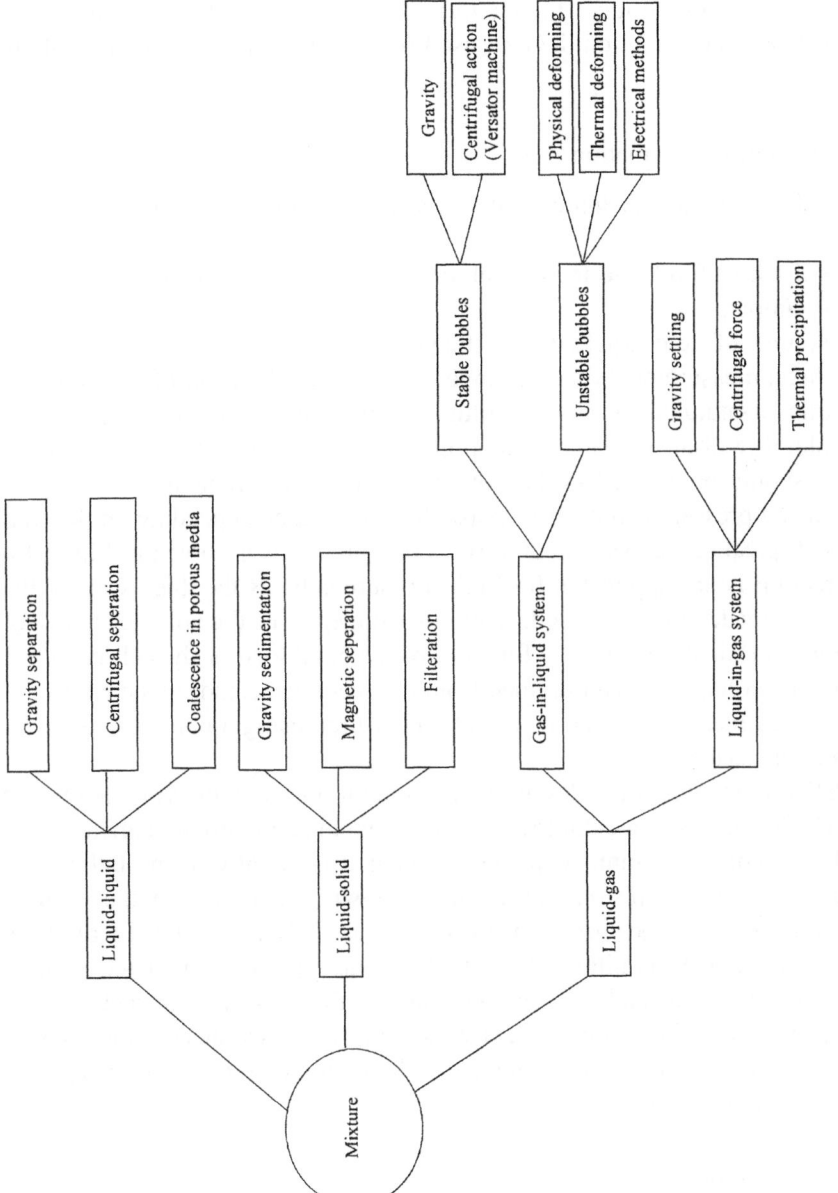

Figure 2.18. Separation techniques for different mixtures.

such as, fixing strain gauges directly on specimens, indirect method of using volume change (Wawersik, 1975), cantilever devices (Hobbs, 1970), winding a peripheral wire around the specimen (Attinger & Köppel, 1983) and clip gauge transducers have been employed to measure lateral deformations in the triaxial and uniaxial tests conditions. In Sections 2.8.1 and 2.8.2 distinct volumetric as well as lateral deformation devices used for testing of both rock and soil are discussed.

2.8.1   Volumetric deformation

Basically, there are three techniques of measuring the volume change in the triaxial tests:

(1) The volume of fluid entering the cell to compensate for the change in volume of the sample;
(2) The volume of fluid expelled from the pore space of the soil; and
(3) The direct measurement of the change in length and diameter of the sample.

For partially saturated specimens, volume changes are due to the compressibility and solubility of air in water in the pore spaces. It is important to note that a correction should be applied to the volume change measurements, because the increase in cell pressure can itself increase the cell volume depending on the wall thickness. Typical cell expansion at various cell pressures is shown in Fig. 2.19. This correction is not applicable for high-pressure cells as the thick wall of the cell is usually made of high strength steel. For drained tests, the volume change of the saturated specimen is purely a function of the axial load or the cell pressure. Therefore, the volume of water drained from the specimen is indeed the change of volume of specimen, which can be directly measured using a burette or an electronic weighing scale.

In TPHPTA, an alternative design is presented to measure the volume change device, which consists of a cylindrical chamber having an internal diameter of 25 mm and a height of 90 mm. A piston is attached co-axially to the cylindrical chamber, in which the piston moves up or down depending on the volume increase or decrease of the specimen (Figs. 2.20 and 2.21). The movement of the piston is continuously monitored by a LVDT. The top chamber of the volume change device is connected to a hydraulic jack and the bottom is connected to the cell. Once the cell is filled with oil, the hydraulic jack is disconnected and the volume change device is connected to the cell. The required cell pressure is applied by another hydraulic jack.

2.8.2   Lateral deformation

An indirect method of estimating the radial strain based on the volume change of the cell fluid was proposed by Wawersik (1975). In the conventional triaxial test, the change of the volume of the specimen is replaced by an equal volume of the cell fluid. Using this concept, Wawersik (1975) used an experimental setup

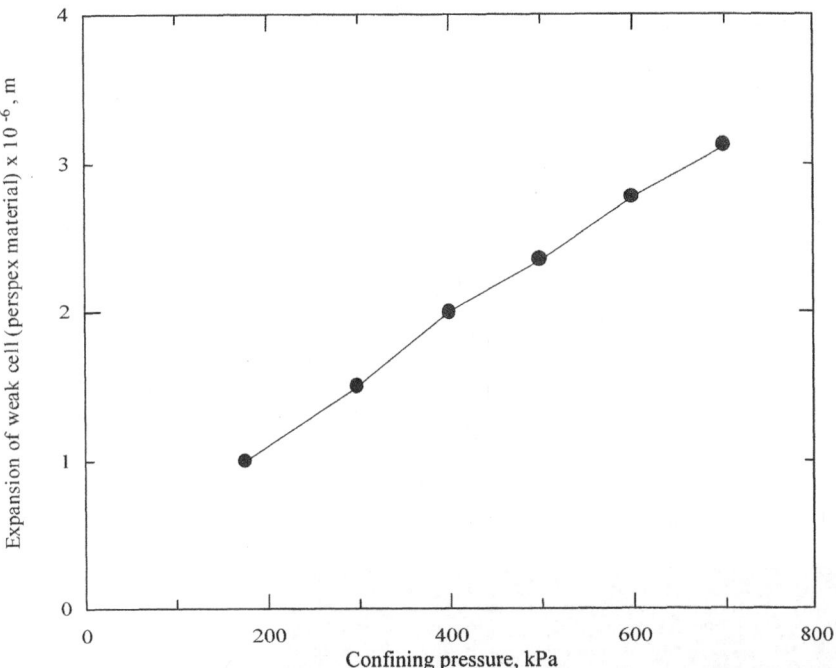

Figure 2.19. Typical expansion of cell for different confining pressures.

(Fig. 2.22) with the intention of measuring the volume of the cell fluid displaced, as a function of specimen deformation. For a homogeneous sample, the radial strain is given by the following expression:

$$\varepsilon_2 = \varepsilon_3 = \frac{1}{1-\varepsilon_1}\left[\frac{\Sigma \Delta V}{C_1} - (C_2\varepsilon_1 + C_3)F + C_4(\varepsilon_1 + C_5F)\right] \qquad (2.1)$$

where $C_1 = 2A_tL_t$, $C_2 = \dfrac{v_s}{E_sA_t}$, $C_3 = \dfrac{v_s}{E_sA_t}\left(\dfrac{L_t}{L_p - L_t}\right)$,

$$C_4 = \left(\frac{A_t - A_s}{2A_t}\right) \text{ and } C_5 = \frac{L_c}{E_sA_tL_t}.$$

A_t = cross-sectional area of test sample end caps that are commonly placed between the specimen ends and the loading pistons, L_t = Length of test sample, L_e = combined length of end caps, E_s, v_s = elastic constants of loading piston and end-cap material, A_s = cross-sectional area of loading pistons, L_p = effective internal length of pressure vessel, F = axial force and ($\Sigma \Delta V$ = cumulative, incremental volume adjustments of confining pressure medium.

Accuracy of calculated strain values entirely depends on the measured volume change of the specimen. Wawersik (1975) discussed the disadvantages of strain gauges for direct strain measurements. It is possible to eliminate such problems

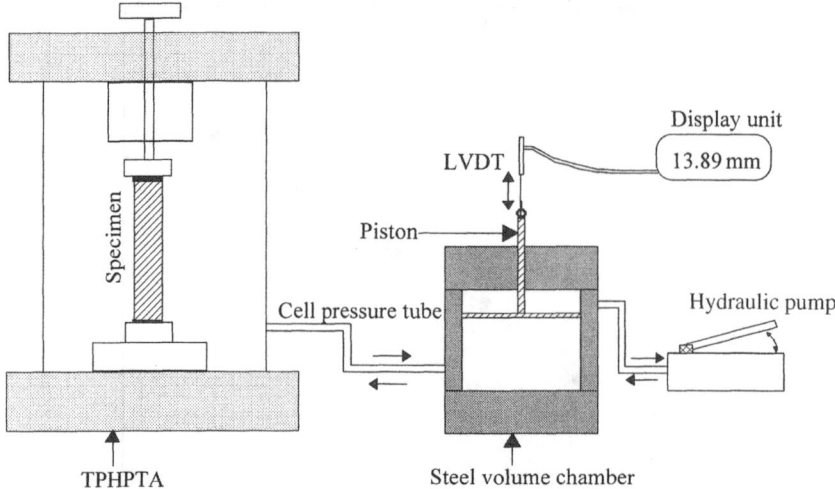

Figure 2.20. Volume change device based on the flow of cell fluid.

Figure 2.21. Volume-change measurement device installed in TPHPTA (University of Wollongong)

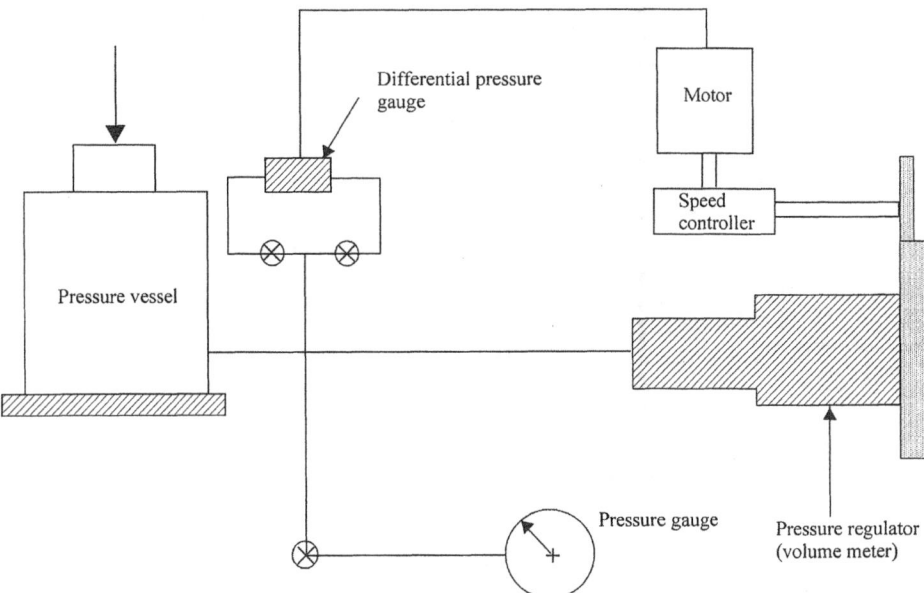

Figure 2.22. Schematic diagram of the strain device (Wawersik, 1975).

using proper strain gauges with appropriate fixing materials, such as compatible strain glues and clip gauges. However, Eqn. (2.1) can still be used to estimate lateral strains in order to perform direct and indirect comparisons. The demerits of Wawersik (1975) method include the inability to obtain volume change measurements at varying changing cell pressures.

Attinger and Koppel (1983) used a technique, in which a wire was wound three times around the specimen with a prestress of about 100 MPa. The strain measurements are based on the principle of the change of electrical resistance of a stretched resistance wire. The electric current is applied to the resistance wire via two copper conductors. The resistance wires and the copper wires are glued to the specimen as shown in Fig. 2.23. The actual change of resistance is determined by the Wheatstone bridge principle. An amplifier is used to amplify the output signal, which is proportional to the lateral strain. The method described here gives the circumference strain rather than the local strain, which is usually obtained by gluing a strain gauge to the specimen. The application of this method to triaxial apparatus is not an easy task, because leakage can occur due to the two copper conducting wires at the one end of the specimen. The uncertainty of measuring large deformations is also regarded as a problem.

In a true triaxial cell (Michelis, 1988), displacement of all four sides of the cubical specimen can be recorded using thin rods and Linear Variable Differential Transformers (LVDTs) (Fig. 2.24). The cumulative deformation is transmitted through rods to two pistons, and then through tubes (filled with mercury) to one piston, where the LVDT is connected. The strain measured represents local strain

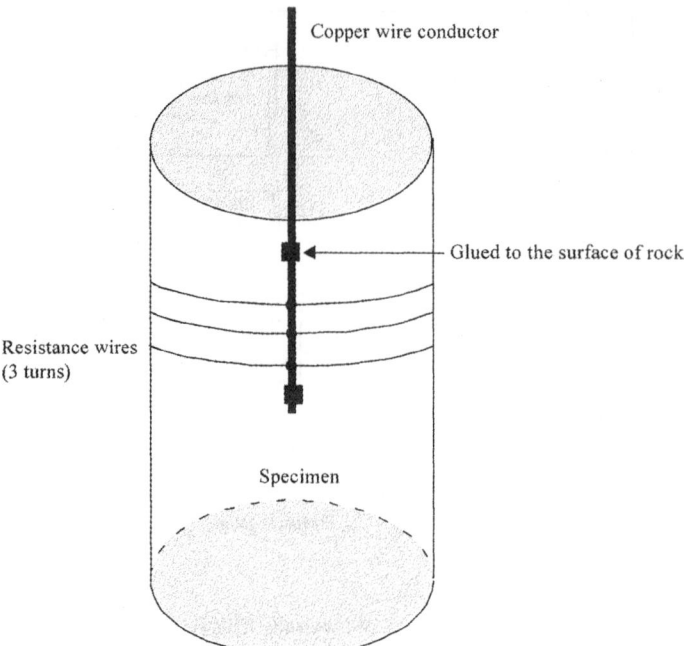

Figure 2.23. Lateral deformation of rocks using resistance-wire method (modified after Attinger and Köppel, 1983).

in two directions. In order to obtain an average strain, one has to install at least three strain devices to one side of the specimen. However, in this technique, practical difficulties arise due to the presence of several pistons and LVDTs.

Lateral deformation can be measured using two micrometers mounted on the exterior of the cell chamber as shown in Fig. 2.25 (Silvestri et al., 1988). Accuracy of the local strain entirely depends on the accuracy of micrometer. Mochizuki et al. (1988) designed a new technique to measure deformation using no-contact gap sensors as well as evaluating the shape of the deformed rectangular specimen. Figure 2.26 shows a rectangular specimen with markers of aluminum foil (10 × 20 × 0.1 mm) on its sides together with gap sensors set in position for testing. The no-contact gap sensors induce the variation of outlet voltage when conductors such as iron or aluminum change the distance to the markers on the specimen sides. The main advantage of this method is that the deformation pattern of each side can be mapped. Although it is suitable for uniaxial conditions, this approach is not practical for triaxial situations.

The opening or closure of a regular rock fracture can be measured by a specially designed, sensitive clip gauge (Fig. 2.27), which can be fitted circumferentially to the specimen (Fig. 2.28). The clip gauge transducer is calibrated using an internal micrometer and a strain meter. The internal micrometer is usually placed along the diameter of the clip gauge, and its diameter can be adjusted to touch the clip gauge

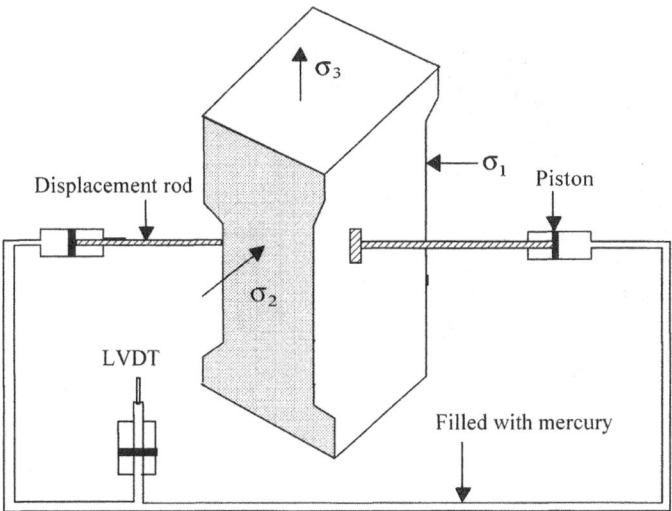

Figure 2.24. Deformation measuring device for a specimen subjected to tension (Michelis, 1988).

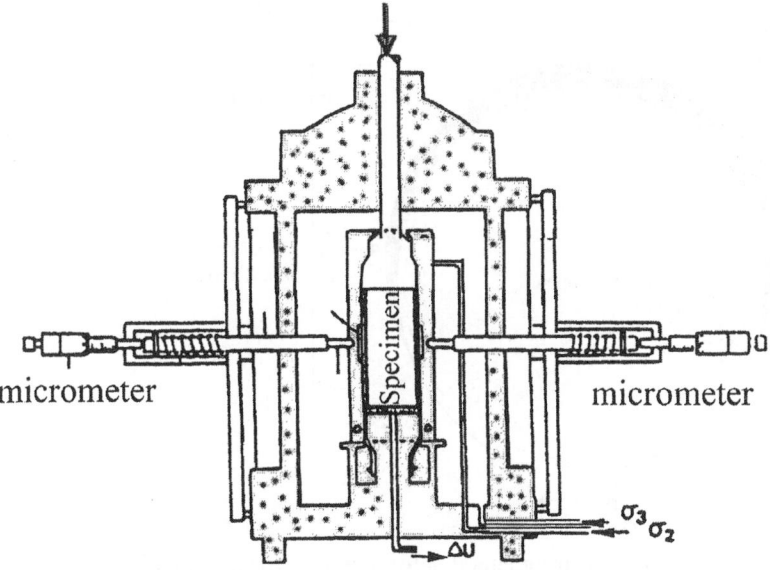

Figure 2.25. Lateral deformation technique used by Silvestri et al. (1988) (modified).

at both ends of it. The corresponding strain meter reading with a gauge factor of say 2.14 would be recorded. Figure 2.29 indicates that the clip gauge transducer produces a linear relationship between the diameter change and the strain meter readings. The relationship between the strain meter readings with deformation is thereby established, which is ultimately used for the back calculation of aperture

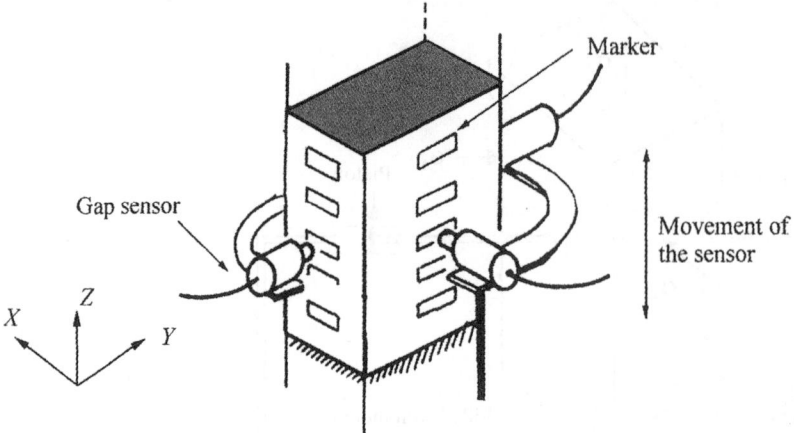

Figure 2.26. Lateral deformation of rock using no-contact gap sensors (after Mochizuki, 1988).

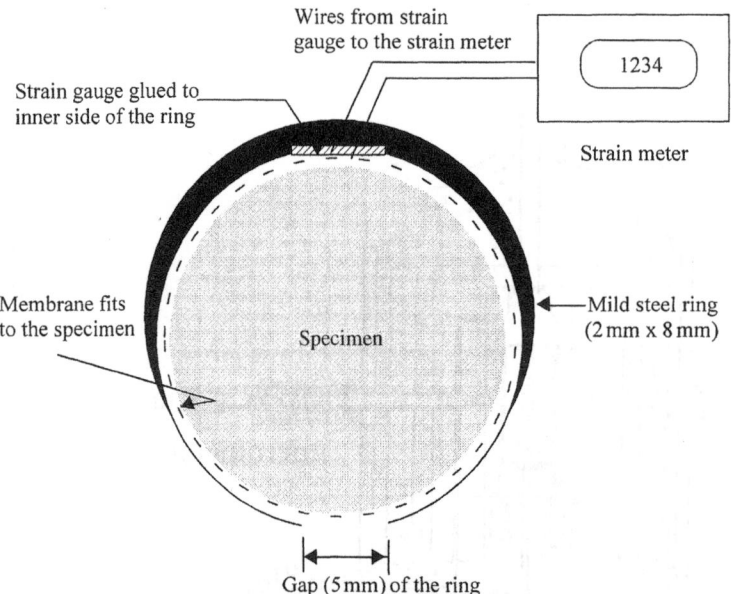

Figure 2.27. Clip gauge for measuring lateral deformation of rocks (TPHPTA).

changes in a regular joint under various confining and driving pressures. Advantages of clip gauges over other methods are listed below:

(1) No difficulty involved in mounting the transducer on a metal ring. In most other devices, the transducers are mounted on the surface of the rock specimen, which becomes difficult on coarse grained, porous, fractured and saturated rock;

Figure 2.28. Clip gauges mounted to the specimen in TPHPTA (University of Wollongong)

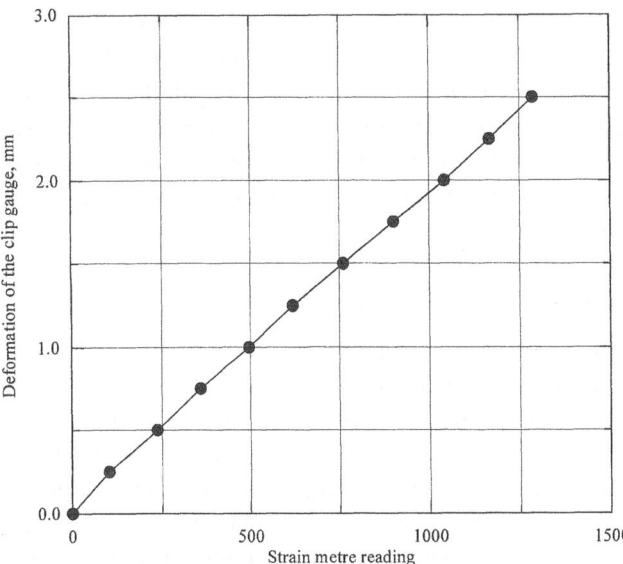

Figure 2.29. Calibration chart for clip gauge.

(2) Measures circumferencial strain rather than local strains;
(3) Accuracy, simplicity and re-usability;
(4) Possibility of using several clip gauges around a single specimen;
(5) Water proofing of the strain gauge is guaranteed;
(6) Applicable for uniaxial, triaxial and polyaxial stress states.

2.9   MEMBRANES AS SPECIMEN JACKETS

This is certainly one of the most critical aspects in conventional triaxial testing because the role of membrane on sample behaviour cannot be ignored. If confining pressure is increased significantly, then the membrane can penetrate into the sample surface. The membrane penetration is significant for coarse-grained soil or fractured rock specimens. Therefore, the thickness, type of material and easy manufacturing process are essential requirements in the design of membranes. In the past, various materials such as natural and synthetic rubber and annealed copper jackets of thickness ranging from 0.2 to 3.0 mm have been used for manufacturing membranes. Magnitude of cell pressures, surface of specimen (i.e. fractured/coarse grained), shape and size of the specimen, change of temperature, duration of tests and types of cell fluid govern the required thickness and type of membrane. Hoek and Franklin (1968) used two kinds of synthetic rubber for manufacturing membranes:
(1) Silicone rubber – for low cell pressures;
(2) Urethane rubber – to withstand high pressures.

A 1.6 mm thick silicone membrane and urethane membrane can withstand up to 70 MPa confining pressure. Bicycle inner tubes, which come in a variety of diameters, are often recommended for membranes when they are subjected to low cell pressures. According to Dusseault (1981), 1.5 mm thick latex rubber can withstand up to 6 MPa cell pressure, and 3 mm thick Neoprene membrane can withstand even greater pressures. For testing granite specimens under high pressure, Brace et al. (1968) used a 3 mm thick polyurethane rubber jacket, which was strengthened by clamping with several loops of steel wire. The drawbacks of thick membranes are that they provide excessive confinement which leads to increased compressive strength, as well as influencing the failure mode.

2.9.1   Membranes for high-pressure triaxial systems

For the TPHPTA the authors have carefully designed a series of moulds using perspex material to accommodate different specimen sizes (field cores) as shown in Fig. 2.30. Having considered the important role of the membrane, authors have selected polyurethane as membrane material for TPHPTA. The membrane should not be too thin or soft as it can get damaged either at high lateral pressures or during extended testing periods. On the other hand, harder material influences the stress–strain behaviour of rocks. Polyurethane has a wider range of hardness than other materials such as, rubber and plastic (Fig. 2.31). Under triaxial test conditions,

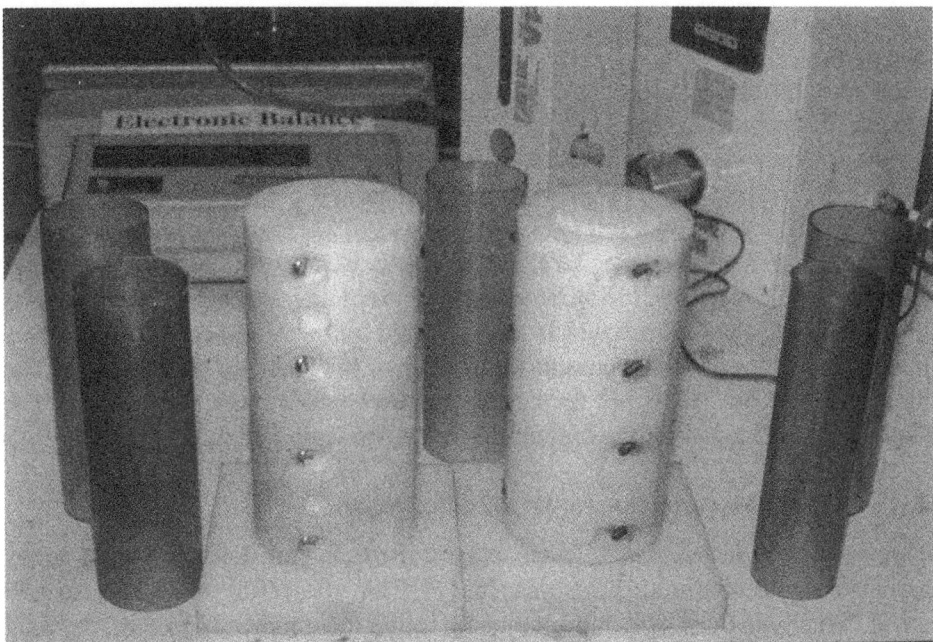

Figure 2.30. Moulds manufactured using perspex material.

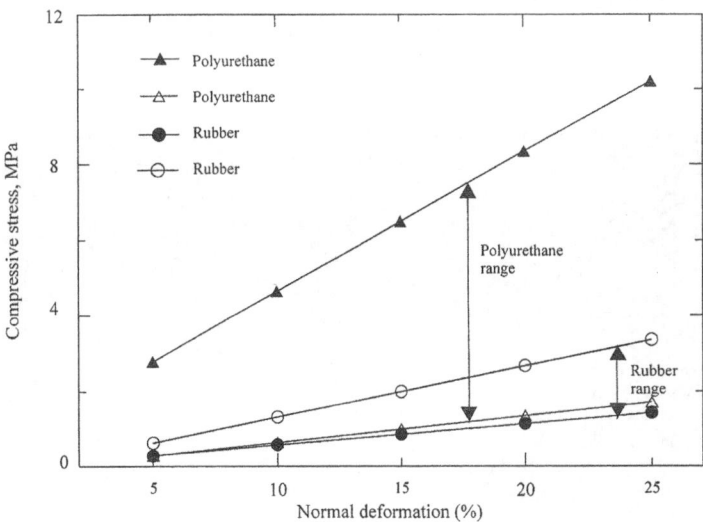

Figure 2.31. Comparison of hardness of polyurethane with rubber.

the membrane is often surrounded by an oil medium and the inner surface is in contact with gases or water. It is of importance to note that the membrane material should be water resistant and it should not react with chemicals such oil and kerosene. The water absorption of polyurethane is as low as 0.3% by weight, and swelling is negligible. Also, polyurethane does not react with oil or kerosene.

Era Polymers Pty Ltd (1998) supplies polyurethane (TU 801 and TU 901) as two parts:

(1) Part A-resin;
(2) Part B-hardener.

The corresponding mixing ratios by weight are 100/51 for TU 801 and 100/44 for TU 901, respectively. The mixing could be carried out using a special type of resin gun (dispenser), dual foil pack (two tubes) and 12″ long nozzles with internal static mixer, as shown in Fig. 2.32. The dispenser enables two parts of material A and one part of material B to flow into the 12″ long nozzle, before the material is thoroughly mixed. This mixture is then pumped into a mould (pre-coated with a release agent) from the bottom until it overflows, to ensure no entrapped air. The mould is kept for 24–48 hrs for curing purposes under room temperature. The stress–strain behaviour of the cast membrane is shown in Fig. 2.33.

2.9.2 Membrane correction for lateral deformation

This correction is necessary when recording the lateral deformation of joints associated with confining pressures. In TPHPTA, the deformation is measured using two clip gauges, which are mounted to the membrane. Therefore, the total deformation indicated by the strain meters include the deformation of both the membrane and the specimen. The following procedure can be undertaken to estimate the membrane deformation only:

(1) Consider different specimens (i.e. 44, 51 and 54 ϕ mm) made of high strength steel;

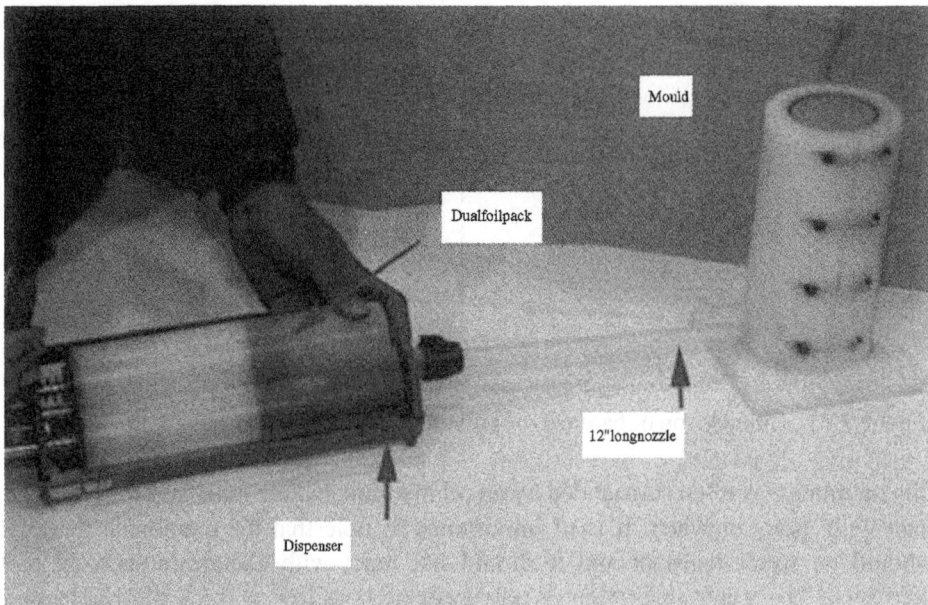

Figure 2.32 Casting of membranes for TPHPTA.

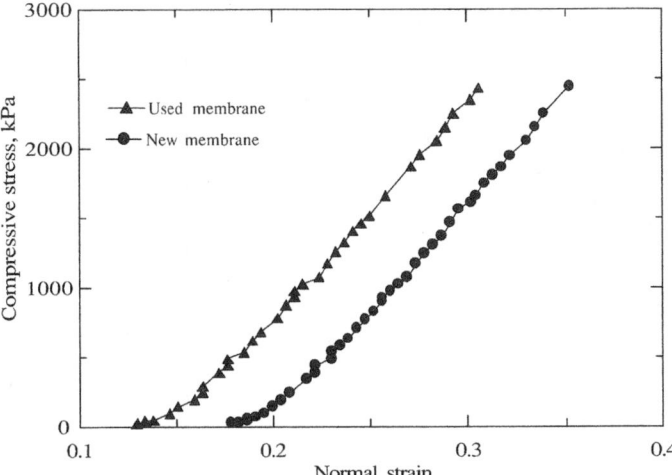

Figure 2.33. Stress–strain behaviour of new and used membranes.

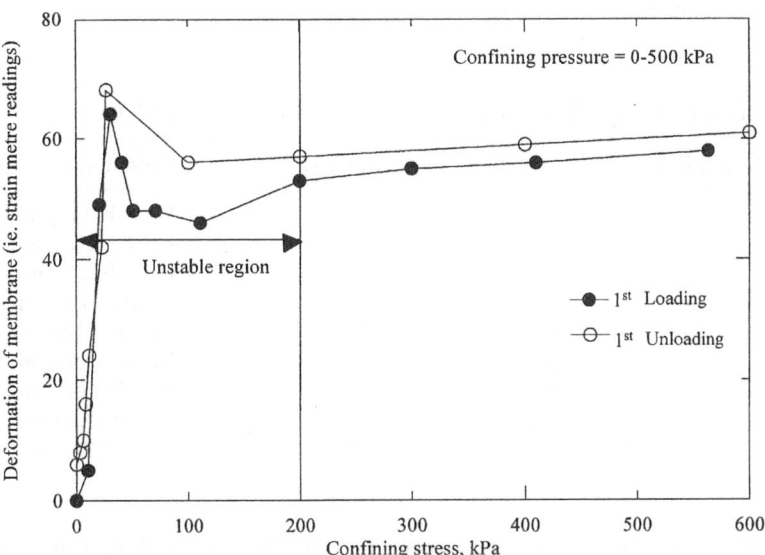

Figure 2.34. Deformation of membrane at varying cell pressure (0–500 kPa).

(2) Each steel specimen to be covered with the membrane and two clip gauges attached (mounted to the membrane);

(3) The clip gauges are then connected to the two strain meters;

(4) The triaxial cell is filled with oil, and a small cell pressure (10 kPa) is applied to record the initial strain meter readings;

(5) For all specimens at different cell pressures, the strain meter readings are recorded (Figs. 2.34 and 2.35). According to Fig. 2.35, almost 60% of membrane deformation is attained below a confining pressure of 500 kPa;

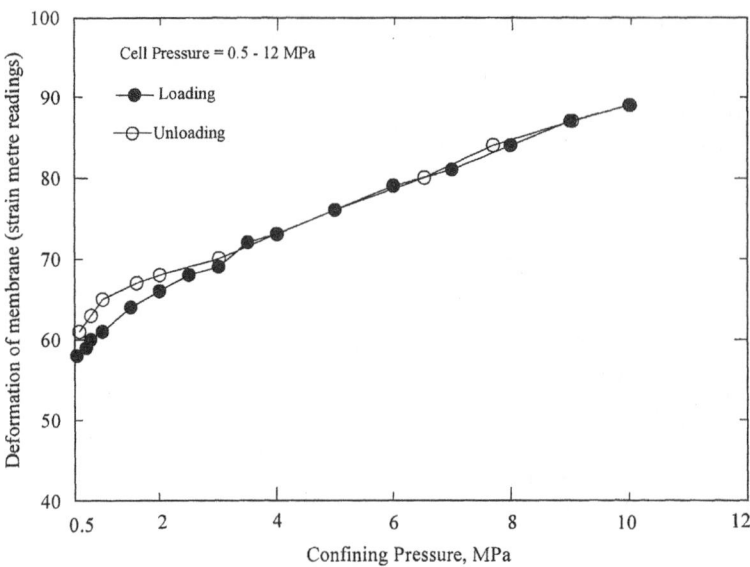

Figure 2.35. Deformation of membrane at different cell pressures.

Table 2.3. Membrane correction factors for different cell pressures for TPHPTA.

Cell pressure ranges (kPa)	Correction factors (strain rate)	Comments
0–50		Unstable region
50–200		Unstable region
200–600	1.5	Unstable region
600–1500	1.0	Nearly constant
1500–4000	2.0	Nearly constant
4000–8000	2.0	Nearly constant
8000–12000	1.0	Nearly constant
12000–15000	1.0	Nearly constant

(6) The following average correction factors are then estimated for different ranges of cell pressures base on the tests conducted on all specimens (Table 2.3).

2.9.3 Determination of cracking pressure

Cracking pressure can be defined as the minimum inlet fluid pressure, which permits flow between the specimen wall and the membrane for a given confining pressure. In the past, various researches have estimated the difference between the fluid pressure and the cell pressure in the range of 100–200 kPa (i.e. within this range, no fluid flow takes place between the specimen and membrane). However, cracking pressure depends on the confining stress, fluid viscosity, the material properties of membrane (e.g. thickness, hardness) as well as their end conditions, including the effect of clamps or 'o'-rings. The following procedure is suggested for estimating the cracking pressure:

(1) For example, consider a high strength steel specimen, 54 mm in diameter. To ensure 'zero' permeability, the specimen is to be coated with a paint;

(2) After mounting the steel specimen in the TPHPTA, the fluid pressure is varied for a constant cell pressure. For instance, at 250 kPa cell pressure, observe whether any flow takes place at an inlet water pressure of 75 kPa;

(3) If not, increase the inlet water pressure gradually (cell pressure kept constant) until some flow is observed at the outlet;

(4) This inlet water pressure causing water flow between the sample and the membrane is now defined as the cracking pressure at the corresponding cell pressure;

(5) Similarly, for both water and air, the cracking pressure is observed for different cell pressures.

Figure 2.36 shows the normalized inlet fluid pressure against the confining pressures. The normalized fluid pressure is the inlet pressure divided by the confining pressure. The points on the graph represent the minimum normalized inlet pressure causing flow between the specimen and the membrane. At low confining pressures (<500 kPa), the normalized cracking pressure for air is between 0.6 and 0.8, and for water it is between 0.8 and 0.9. Clearly, the cracking pressure for low

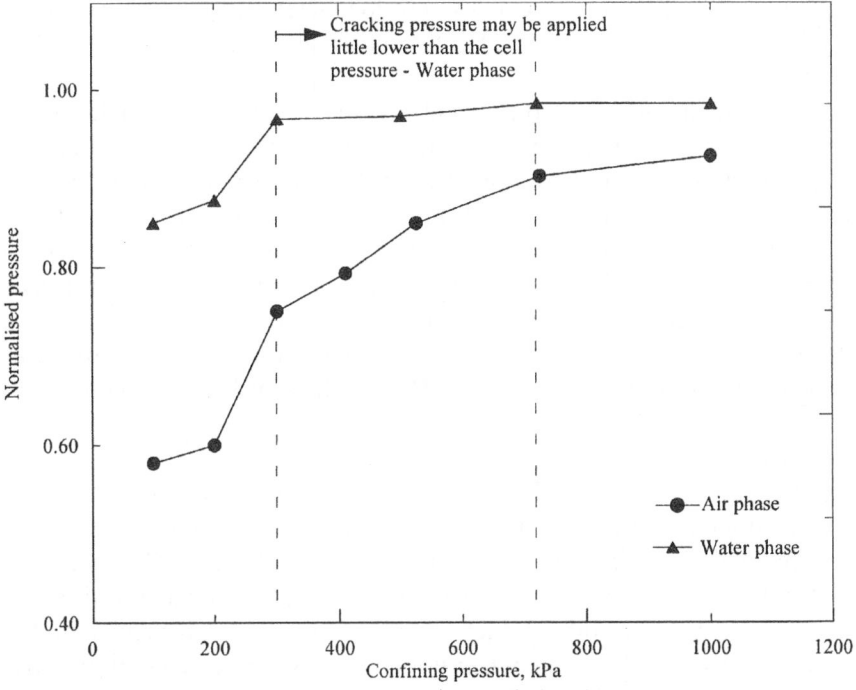

Figure 2.36. Cracking pressure–cell pressures relationship.

viscosity fluids, such as air should be smaller than for fluids such as water and oil. It is expected that at elevated confining pressures, the cracking pressure for high viscosity fluids can reach the magnitude of the applied confining pressure.

2.10 FLOW METERS

For routine practical flow measurements, there exits a number of flow meters and flow measuring techniques. Flow meters are grouped into displacement type and velocity type. Venturi-meters, orifices, nozzles and elbow devices are the most common velocity-type instruments. The velocity-type flow meters are normally employed to measure large quantities of flow, such as in large pipes, rivers and streams. Displacement-type flow devices indicate flow rate directly, by recording the volume and the time. Weighing meters and rotary types are examples of displacement-type flow meters. These are normally employed to record small flow rates, and mainly used in the laboratory environment and domestic water distribution systems.

Apart from the cost, selection of flow meters for a particular application depends on the type of flow (e.g. gas, liquid or multiphase), special fluid constraints (e.g. corrosive), design constraints (e.g. precision and flow range) and environmental considerations (e.g. humidity and temperature). Generally, orifice, venturi-meters and vortex devices are used for both gas-and water-flow measurements. Thermal flow devices, shielded microflowmeter and bubble meters are used only for gas-flow measurements. For two-phase flows, electromagnetic flow meters may be useful.

Flow measurements through soil or rock usually involve very small quantities of gas or water or both. At high lateral pressures, flow rate can be smaller than say 0.2 ml/min. In general, for small flow volumes of air, shielded microflowmeters or bubble-meters can be employed, whereas for small flow rates of water, a precision weighing scale is usually sufficient. For TPHPTA, a typical bubble meter, *Electronic Film Flowmeters* (STEC, 1998) are most appropriate, and they are designed for automatic measurement with high accuracy. Such electronic flow meters are also equipped with a high precision sensor to incorporate any atmospheric pressure change. This equipment basically consists of two parts:

(1) Measuring unit; and
(2) Measuring tubes, for typical flow rates of 0.2 ml/min to 10 l/min.

The operating principle of the film flow meter is shown in Fig. 2.37.

When non-soluble gas in water, such as, N_2, air, O_2, H_2, CO_2, CH_4 and Ar, a soap water film formed at the mouth of the measuring tube by a film ring will move up in the measuring tube, which is calibrated to a high precision. The equipment calculates the flow rate when it measures the time taken by the soap film to move between the 'start' and 'stop' detection points. The microcomputer

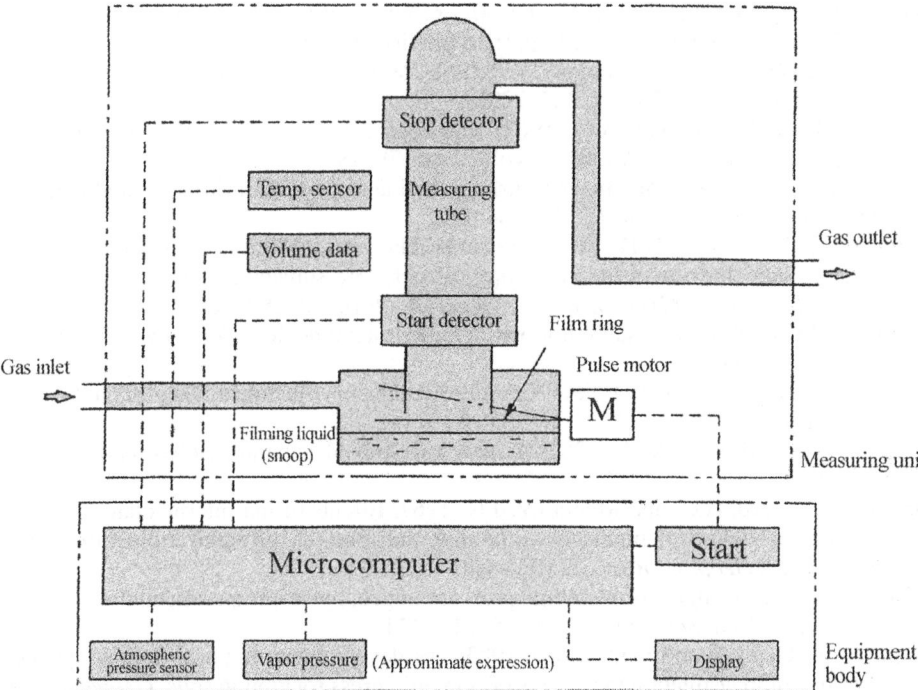

Figure 2.37. Operating principle of the film flow meter (STEC, 1998).

processes the temperature and the atmospheric pressure to correct the measured flow rate.

REFERENCES

Airey, D.W. and Wood, D.M. 1988. The Cambridge true triaxial apparatus. In *Advanced Triaxial Testing of Soil and Rock, ASTM, STP 977,* Donaghe, R. T., Chaney, R. C. and Silver, M. L., eds., American Society of Testing and Materials, Philadelphia, pp. 796–805.

Amadei, B. and Robison, J. 1986. Strength of rock in multi-axial loading conditions. *Proc. 27th U.S. Rock Mechanics,* Tuscaloosa, Alabama, pp. 47–55.

Anderson, A. and Simons, N.E. 1960. Norwegian triaxial equipment and technique. *Research Conference on Shear Strength of Cohesive Soils,* American Society of Civil Engineers, New York, pp. 695–709.

Attinger, R.O. and Köppel, J. 1983. A new method to measure lateral strain in uniaxial and triaxial compression tests. *Rock Mech.,* 16: 7.

Barden, L., Madedor, A.O. and Sides, G.R. 1969. The flow of air and water in partly saturated clay soil. *Int. Symp. Fundamentals of Transport Phenomena in Flow through Porous Media,* Haifa, Israel.

Berre, T. 1982. Triaxial testing at the Norwegian Geotechnical Institute. *Geotech. Test. J.,* 5: 3–17.

Bishop, A.W. and Henkel, D.J. 1953. Pore pressure changes during shear in two undisturbed clays. *Proc. 3rd Int. Conf. on Soil Mechanics,* pp. 94–99.

Bishop, A.W. and Henkel, D.J. 1969. *The Measurement of Soil Properties in the Triaxial Test*. Edward Arnold, London, 228 p. (first published in 1957).

Brace, W.F., Walsh, J.B. and Frangos, W.T. 1968. Permeability of granite under high pressure. *J. Geophys. Res.,* 73(6): 2225–2236.

Broms, B.B. and Casbarian, A.O. 1965. *Proc. 6th Int. Conf. on Soil Mechanics and Foundation Engineering,* Montreal, vol. 1, pp. 179–183.

Cole, D.M. 1978. A technique for measuring radial deformation during repeated load triaxial testing. *Can. Geotech. J.,* 15: 426–429.

Crawford, B.R., Smart, B.G.D., Main, I.G. and Liakopoulou-Morris, 1995. Strength characteristics and shear acoustic anisotropy of rock core subjected to true triaxial compression. *Int. J. Rock Mech. Min. Sci. Geomech. Abstr.,* 32: 189–200.

Dusseault, M.B. 1981. A versatile hollow cylinder triaxial device. *Can. Geotech. Test. J.,* vol. 18, No. 1, pp. 1–7.

Era Polymers Pty. Ltd. 1998. A New Era in Polyurethane – Technical Data Manual. NSW 2019, Australia.

Hambly, E.C. and Reik, M.A. 1969. A new true triaxial apparatus. *Geotechnique,* 19: 307–309.

Handian, J., Heard, H.C. and Magouirk, J.N. 1967. Effects of the intermediate principal stress on the failure of limestone, dolomite and glass at different temperatures and strain rates. *J. Geophys. Res.,* 72: 611–640.

Handin, J. 1953. An application on high pressure in geophysics: Experimental rock deformation. *Trans. Am. Soc. Mech. Eng.,* 75: 315–324.

Hight, D.W., Gens, A. and Symes, M.J. 1983. The development of a new hollow cylinder apparatus for investigation the effects of principle stress rotation in soils. *Geotechnique,* 33(4): 355–383.

Hobbs, D.W. 1970. Stress – strain time behaviors of a number of coals measure rocks. *Int. J. Rock Mech. Min. Sci.* 7: 149–170.

Hoek, E. and Franklin, J.A. 1968. Simple triaxial cell for field or laboratory testing of rock. *Trans. Inst. Min. Metall.,* A 22–A26.

Holubec, I. and Finn, P.J. 1969. A lateral deformation transducer for triaxial testing. *Can. Geotech. J.,* 6: 353–356.

Hon-Yam, K. and Ronald, F.S. 1967. A new testing apparatus. *Geotechnique,* 17: 40–57.

Indraratna, B. and Haque, A. 1999. Triaxial equipment for measuring the permeability and strength of intact and fractured rocks. *Geotechnique,* 49(4): 515–521.

Iversen, K. and Moum, J. 1974. The paraffin method; Triaxial testing without a rubber membrane. *Geotechnique,* 24(4): 665–670.

Klute, A. 1965. Laboratory measurement of hydraulic conductivity of unsaturated soil. In *Methods of Soil Analysis,* Black, C.A., Evans, D.D., White, J.L., Ensminger, L.E. and Clark, F.E., eds., American Society of Agronomy, Madison, WI 9(1): 253–261.

Lade, P.V. 1973. Torsion shear test on cohesionless soil. *Proc. 5th Pan-Am. Conf. on Soil Mechanics and Foundations Engineering,* Buenos Aires.

Meier, R.W., Ko, H.Y. and Sture, S. 1985. A direct tensile loading apparatus combined with a cubical test cell for testing rocks and concrete. *Geotech. Test. J.,* 8(2): 71–78.

Michelis, P. 1988. A true triaxial cell for soil and rock. In *Advanced Triaxial Testing of Soil and Rock, ASTM, STP 977,* Donaghe, R.T., Chaney, R.C. and Silver, M.L. eds., American Society of Testing and Materials, Philadelphia, pp. 806–818.

Mochizuki, A., Mikasa, M. and Takahashi, S. 1988. A new independent principal stress control apparatus. In *Advanced Triaxial Testing of Soil and Rock, ASTM, STP 977,* Donaghe, R.T. Chaney, R.C. and Silver, M.L. eds., American Society of Testing and Materials, Philadelphia, pp. 844–858.

Paterson, M.S. 1970. A high-pressure, high-temperature apparatus for rock deformation. *Int. J. Rock. Mech. Min. Sci.,* 7: 517–526.

Reik, G. and Zacas, M. 1978. Strength and deformation characteristics of jointed media in true triaxial compression. *Int. J. Rock Mech. Min. Sci. Geomech. Abstr.,* 15: 295–303.

Saada, A.S. and Baah, A.K. 1967. *Proc. 3rd Pan-American Conf. on Soil Mechanics and Foundations Engineering,* Caracas, vol. 1, pp. 67–88.

Shibuya, S. and Mitachi, T. 1997. Development of a fully digitized triaxial apparatus for testing soils and soft rocks. *J. Southeast Asian Geotech. Soc.* (Asian Institute of Technology).

Silvestri, V. Yong, R.N. and Mohamed, A.M.O. 1988. A true triaxial testing cell. In *Advanced Triaxial Testing of Soil and Rock, ASTM, STP 977,* Donaghe, R.T., Chaney, R.C. and Silver, M.L. eds., American Society of Testing and Materials, Philadelphia, pp. 819–833.

Smart, B.G.D. 1995. A true triaxial cell for testing cylindrical rock specimen. *Int. J. Rock Mech. Min. Sci. Geomech. Abstr.,* 3(32): 269–275.

STEC, 1998. Film Flow Meter – Instruction Manual, Kyoto, 601–8116, Japan.

Sture, S. and Desai, C. 1979. Fluid cushion truly triaxial or multiaxial testing device. *Geotech. Test. J.,* 2: 20–33.

Wawersik, W.R. 1975. Technique and apparatus for strain measurements on rock in constant confining pressure experiments. *Rock Mech.,* 7: 231–241.

CHAPTER 3

Fully saturated fluid flow through a jointed rock mass

3.1 INTRODUCTION

Fluid-flow analysis plays a major role in various geotechnical applications (e.g., mining and petroleum industry and nuclear waste storage plants), and the understanding of flow mechanisms is essential for the development of a hydromechanical flow model suitable for underground excavations in rock. Accurate prediction of mine water inflow to a tunnel is one of the essential tasks of underground works during the design and construction stages. Under the current increased environmental and regulatory controls, the evaluation of the quantity and quality of total inflow to excavations and the procedures for discharging polluted mine water are significant factors in the development and operational stages of underground mining. Moreover, in nuclear wastage storage plants, special attention should be given to prevent any radioactive contamination of groundwater. The accurate prediction of inflow volumes is expected to minimise most environmental hazards, damage to mine equipment and to reduce the time delay associated with dewatering.

Adverse situations and the risk of inundation can only be mitigated by correct evaluation of the protection measures during the design stages. Therefore, planning decisions concerning groundwater control measures such as grouting and dewatering should be implemented in advance, so that the whole operation system would contribute more efficiently towards greater economies of scale within a safer work environment. For rational analysis and design, it is essential to understand the hydraulic and mechanical behavior of rock mass, the response of the natural joint system, and how the groundwater table responses to changes induced by proposed excavation sequence.

In a comprehensive study of flow analysis, one has to consider an array of geo-hydrological factors as illustrated in Fig. 3.1. If the rock mass is saturated with a single fluid, then a single-phase flow analysis must be carried out. Fully saturated flow techniques have been well established during the past four decades. However, less attempts have been made on multiphase flow analysis in rock media, as discussed later in Ch. 4. Depending on the availability of geological

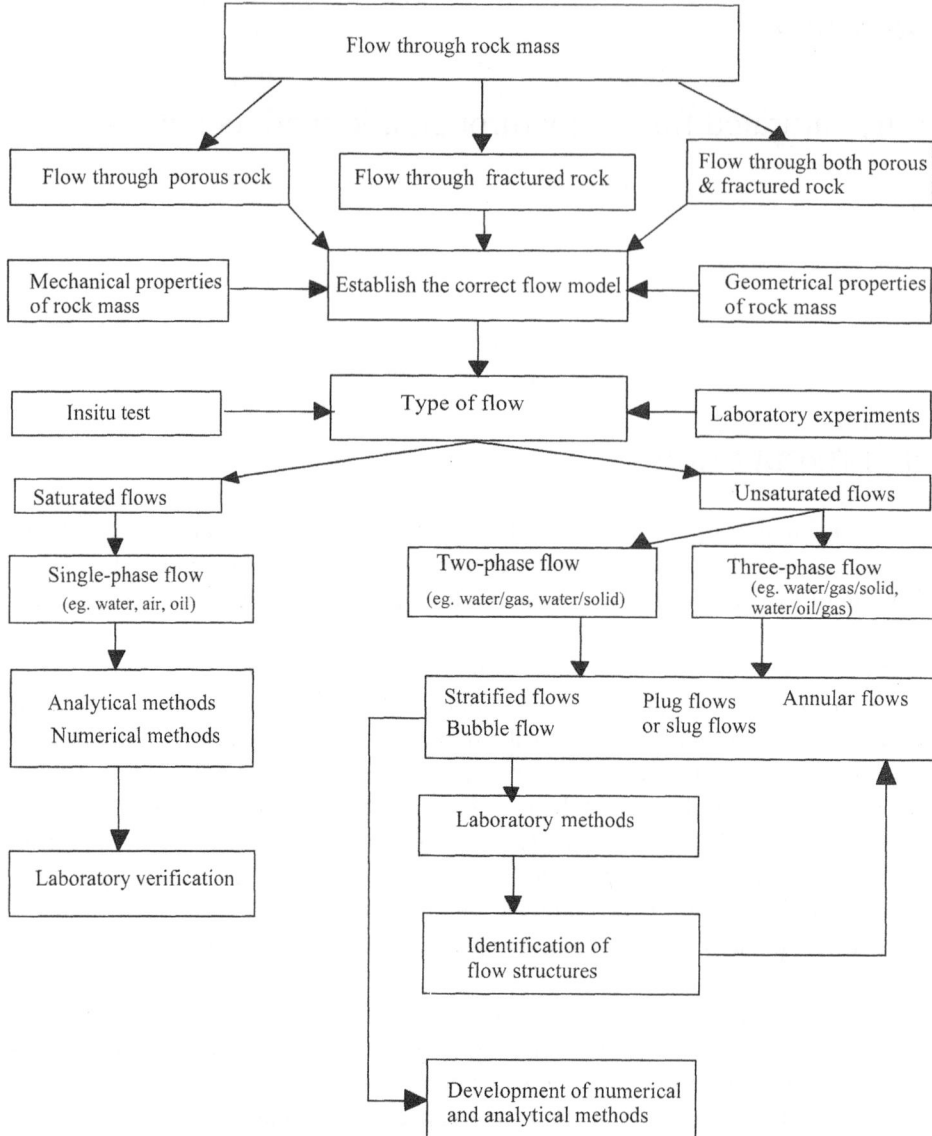

Figure 3.1. Flow analysis through a rock mass.

data and the required accuracy of flow estimation corresponding to the availability of time, numerical techniques and computer resources, the most appropriate flow approach can be selected for a partic ular case (i.e. discrete, continuum or combination of both). The use of a discrete method is more realistic, if fluid flow takes place mainly through a network of fractures (Fig. 3.2). In contrast, a continuum approach can be employed for relatively high porosity rocks (e.g., sandstone and limestone) or if the fracture density is extremely high (Fig. 3.2E), in which

Figure 3.2. Application of fractured and matrix permeability in different locations (modified after Brady & Brown, 1994). A = Intact rock (continuum flow) – Matrix permeability, B = Single discontinuity (discrete fracture flow) – Fractured permeability, C = Two discontinuities (discrete fracture flow) – Fractured permeability, D = Several discontinuities (discrete fracture flow) – Fractured permeability, and E = Rock mass (continuum flow) – Matrix permeability.

case, the media is assumed to behave as a porous medium. Comprehensive description of fluid flow models is given later in Ch. 5.

Permeability is simply the ability to conduct fluids, such as water, gas or multiphase flows (e.g., water + gas, water + gas + oil) through porous media, such as soil or rocks. In general, this section is focused on the permeability of rocks. Total fluid flow through rock mass is the accumulated effect of flow through the porous rock matrix and the flow through the joint network. In hard rocks (low porosity), the fluid flow is dictated through the passage of discontinuities. In this volume, the permeability of discontinuities is referred to as the fracture permeability or the intrinsic permeability, whereas intact permeability is referred to as the matrix permeability. Therefore, the combined permeability of rock mass is given by

$$k = k_f + k_m \qquad (3.1)$$

where k is the combined permeability of rock mass, and k_f and k_m are the individual fracture permeability and matrix permeability, respectively.

For crystalline rocks, fluid flow through rock matrix is much less than that through fractures, because the extent of interconnected pores and the pore sizes in hard rocks are generally small. The stress dependency of fluid flow through fractured rocks has been discussed earlier (Gale, 1975; Iwai, 1976; Raven & Gale, 1982). Permeability can greatly influence the mechanical behaviour of rocks, thereby increasing or decreasing the stability (failure mode) of rock structures. As shown in Fig. 3.3, fluid flow within a rock specimen can take place through either the rock matrix or interconnected discontinuities or combination of both. Depending on this nature of flow, the corresponding permeability term must be used in flow analysis. For example, if the fluid flow is governed by the rock matrix (e.g. intact coarse sandstone), the matrix permeability needs to be identified, whereas in the presence of joints particularly in low permeability rocks (e.g., granite, slate), the fracture permeability governs the fluid flow behaviour within the rock mass. Under single-phase fluid flow, permeability can be divided into three main categories:
(1) Matrix permeability;
(2) Fracture permeability; and
(3) Dual permeability.

In hydromechanics, the coefficient of permeability k has the dimensions of L^2 (units in m^2 or cm^2). In conventional soil mechanics, the permeability coefficient has different units, i.e., m/s. In order to avoid any confusion, it is important to

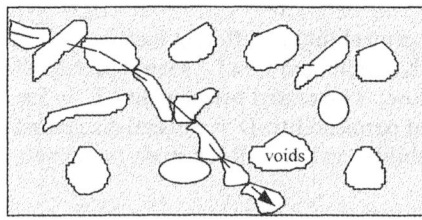
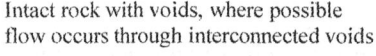
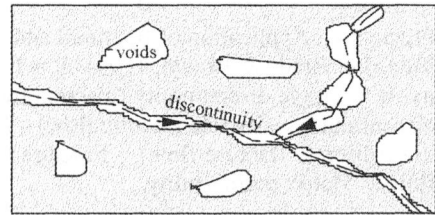

Intact rock with voids, where possible   Specimen with a major discontinuity, where
flow occurs through interconnected voids   flow occurs through discontinuity
                                              and any interconnected voids

Figure 3.3. Fluid flow paths in intact and fractured rock specimens.

Table 3.1. Common units used in permeability* of rocks.

Description	Units	Equivalent units
Coefficient of permeability of rock (k)	m^2	10^4 cm^2 10^{12} darcy 10^{15} millidarcy 10^{18} microdarcy 10^{21} nanodarcy
Conductivity of rock (K)	m/s $1K$ (m/s) $\cong 10^7$ m^2 k (for water) $1K$ (m/s) $\cong 10^6$ m^2 k (for air)	

*Term 'conductivity' used in rock mechanics is the same as 'permeability' in soil mechanics. Therefore conductivity of rock (K) (a) for water – $1K$ (m/s) $\cong 10^7$ m^2 k and (b) for air – $1K$ (m/s) $\cong 10^6$ m^2 k.

express the conductivity (K) in terms of the coefficient of permeability, where K has dimensions of LT^{-1}.

$$K = \frac{\rho g k}{\mu} \tag{3.2}$$

It is clear that the term 'conductivity' used in rock mechanics is the same as permeability used in soil mechanics. The common units used to express permeability are also listed in Table 3.1.

The permeability of rocks is measured using small-scale standard core specimens or it can be carried out at in the field using Packer tests in boreholes. Although the matrix permeability is relatively easy to determine from laboratory tests or in-situ tests, the fractured permeability measured from in-situ tests can vary depending on the test zone. This is because, it entirely depends on the number of fractures, which intersect the borehole. Also, the existing fractures may deform due to stress release or new fractures may open due to excessive hydraulic pressure or due to vibration of drilling tools. In spite of the effect of sample size, laboratory test results on fractured rock specimens can provide individual permeability of each fracture system for a given stress condition. The measured values may then be incorporated in numerical models to predict the permeability of the overall rock mass. In Sections 3.2 and 3.3, the factors which control matrix permeability and fracture permeability are discussed in detail.

Darcy's law is often employed for fluid flow calculation in soil and rock mechanics when the media is fully saturated with a single fluid phase. However, this law may even be extended to study the unsaturated flow through a media, and this is extensively discussed in Ch. 4. In a simplified Darcy's law, the hydraulic gradient is linear along the fluid flow path and given by the total energy difference between two points. In conventional fluid mechanics, the hydraulic head is defined as the energy on a stream line of fluid and the total energy at a point is the summation of the elevation head, pressure head and

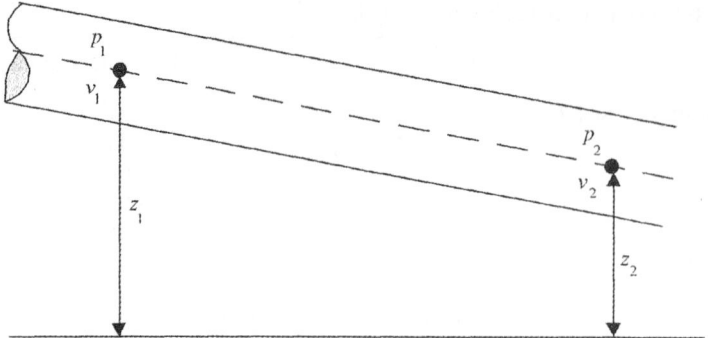

Figure 3.4. Flow along a pipe, showing total head at two points.

velocity head. As shown in Fig. 3.4, the total head at point 1 (h_1) and 2 (h_2) are written as follows:

$$h_1 = z_1 + \frac{p_1}{\rho g} + \frac{v_1^2}{2g} \tag{3.3a}$$

$$h_2 = z_2 + \frac{p_2}{\rho g} + \frac{v_2^2}{2g} - \text{losses} \tag{3.3b}$$

The hydraulic gradient along dx length can be expressed as

$$\frac{dh}{dx} = \frac{h_1 - h_2}{dx} = (z_1 - z_2) + \left(\frac{p_1 - p_2}{\rho g}\right) + \left(\frac{v_1^2 - v_2^2}{2g}\right) + \text{losses} \tag{3.3c}$$

where z is elevation head, p is fluid pressure and v is velocity.

3.2 FLOW THROUGH INTACT ROCK MATERIAL

'Rock material' is the term used to describe the intact rock between discontinuities; it might be represented by a hand specimen or a piece or drill core examined in the laboratory (Brady & Brown, 1994). In other words, there are no fractures existing within the rock matrix. However, in reality, microfractures can still exist in rock material, although they may not be important in a practical point of view. In this section, the authors attempt to discuss the important role of rock material on fluid flow behavior. For certain types of rock (e.g., coarse sandstone and limestone), it is found that significant flow takes place though the rock material (matrix), whereas negligible flow is expected to take place through low permeability rocks, such as granite and slate. Typical permeability values for various rocks are tabulated in Table 3.2 (Brace et al., 1968). Permeability measurements are based on either (a) steady-state flow measurements or (b) using a transient method, in which decay of pressure is observed. In the past, the transient method has been widely used by several researchers including Brace et al. (1968) and Kranz et al.

Table 3.2. Typical values of porous permeability (Brace et al., 1968).

Rock	Permeability, nanodarcy	Reference
Fine-grained dolomite, Tennessee	80	Ohle, 1951
Fine-grained limestone, Tennessee	30	Ohle, 1951
Coarse-grained dolomite, Tennessee	6000	Ohle, 1951
Granite, Barriefield, Ontario	50	Ohle, 1951
Granite, Quincy, Massachusetts	4600	Ohle, 1951
Diabase, Hudson, New York	0.8	Ohle, 1951

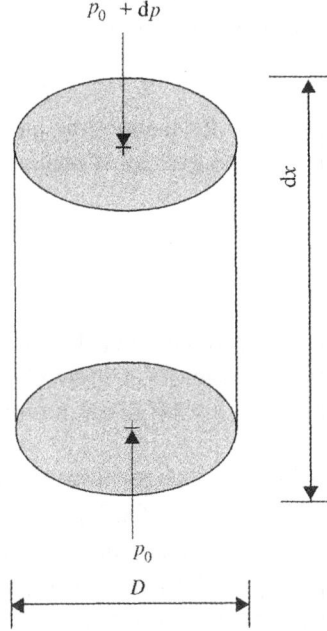

Figure 3.5. Pressure gradient along a rock specimen.

(1979). Currently, the steady-state flow rate method is the most popular among researchers because of the commercial availability of precise flow measuring equipment. Under steady-state flow rate approach, for a given cylindrical rock specimen, the coefficient of matrix/intact rock permeability (k_m) can be written using Darcy's law as given below:

$$k_m = \frac{4q\mu}{\pi D^2 (dp/dx)} \tag{3.4}$$

where q is fluid flow rate through the specimen, dp/dx is the pressure gradient along the length (dx) of the specimen (Fig. 3.5), μ is the dynamic viscosity of the fluid and D is the diameter of the specimen. Apart from the hydraulic gradient and surrounding stresses applied on the specimen, the matrix permeability depends on the properties the matrix, characterized by the pore size (voids), shapes and the interconnectivity of voids. If the fluid travelling through the porous rock is gas, then the component of the matrix coefficient of air permeability (k_m) is estimated according to the following equation:

$$k_m = \frac{2qp_e\mu L}{(p_i^2 - p_e^2)A} \tag{3.5}$$

where q = gas flow rate, μ = dynamic viscosity of gas, L = length of the specimen, A = cross-section area of specimen, p_i = inlet pressure of gas and p_e = exit pressure of gas.

If the permeability measurement is based on the transient method, the following expressions are used to calculate the matrix permeability coefficient. The pressure pulse decays with time according to the expression given below:

$$p_t = p_0 e^{-\alpha t} \tag{3.6}$$

where p_t and p_0 are final and initial pressure, respectively, t is the decay time and α is an empirical constant. The matrix permeability coefficient for transient method is given by Kranz et al. (1979):

$$k_m = \frac{\alpha\beta\mu L\, V_1 V_2}{A(V_1 + V_2)} \tag{3.7}$$

where β = isothermal compressibility of fluid, A = cross-sectional area, V_1 and V_2 = volume of the pore fluid at the top and the bottom of the sample, respectively, L = length of the specimen and α is calculated using Eqn. 3.6 for given initial pressure, and for the time period, t.

A number of experimental studies on the variation of matrix rock permeability with stress can be found in the literature (Gangi, 1978; Brace et al., 1968; Kranz et al., 1979; Indraratna et al., 1999; Ranjith, 2000). In order to investigate the effects of confining pressure on the matrix rock permeability, Gangi (1978) attempted to develop a theoretical model, in which he assumed that the porous rock was made up of packed, uniform spherical grains, and that the deformation of pores was based on the Hertz theory. The matrix permeability in this approach is given by

$$k_m = k_0[1 - C_0((\sigma_c + \sigma_i)/p_0)^{2/3}]^4 \tag{3.8}$$

where k_0 = initial permeability, σ_c = confining pressure, σ_i = the equivalent pressure due to the cementation and permanent deformation of the grains and P_0 = the effective elastic modulus of the grains.

From past work, it can be seen that the matrix permeability of hard rocks is negligible, and it is in the order of 10^{-21} m^2 (Brace et al., 1968; Indraratna et al., 1999). Based on Eqn. 3.7, Brace et al. (1968) have illustrated the variation of permeability of intact Westerly granite rocks with respect to the confining pressure. As expected, the matrix permeability decreases significantly with the increase in confining pressure (Fig. 3.6). This is because, the continuous deformation of pores results in reduced interconnectivity of the flow paths. Similar test results on granite specimens were obtained by Indraratna et al. (1999), whose experiments

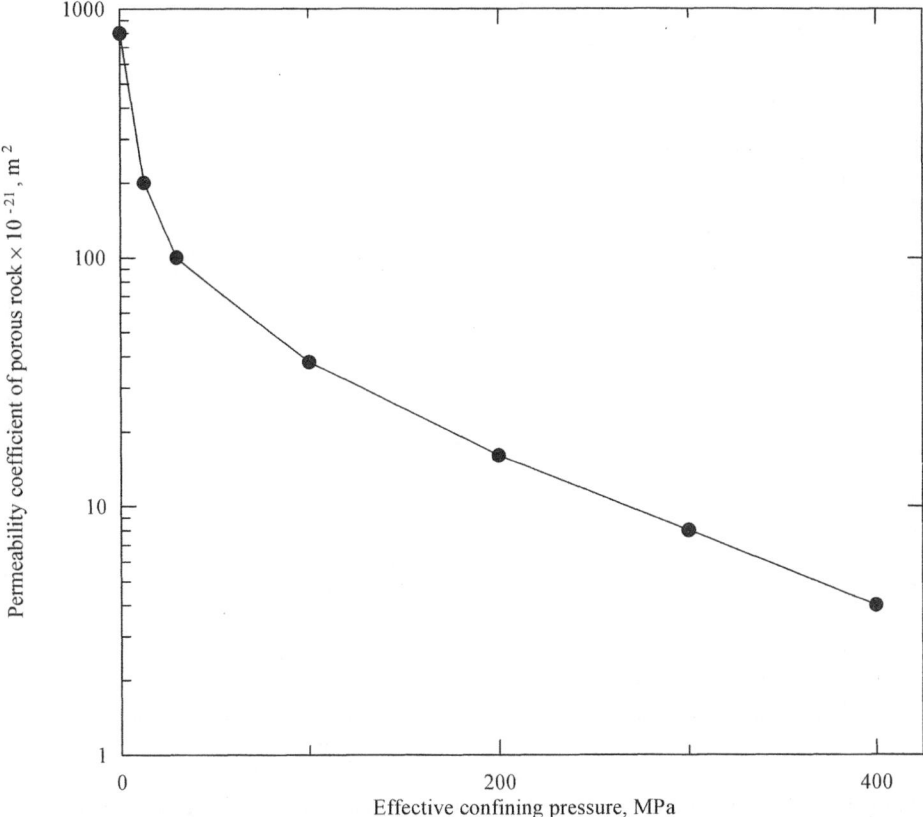

Figure 3.6. Effect of confining pressure on Westerly granite rock (data from Brace et al., 1968).

were based on individual air and water flow through specimens. As seen in Fig. 3.7, when the effective confining pressure exceeds 10 MPa, the variation of permeability becomes negligible. The permeability is calculated using Eqns. 3.4 and 3.5 for water and air flow, respectively. In contrast, increased matrix permeability of rock is achieved with an increase in axial stress (Fig. 3.8). This is attributed to the fact that new cracks are formed within the sample, resulting greater connectivity of fluid flow paths.

3.3   FLUID FLOW THROUGH A SINGLE JOINT

According to ISRM (1978), a joint is defined as a break of geological origin in the continuity of a body of rock along which there has been no visible displacement. A group of parallel joints is called a set, and different sets intersect to form a joint system, which may include joints that are open, filled or healed. Joints frequently form parallel planes, e.g., bedding planes, foliation and cleavage. The mechanical

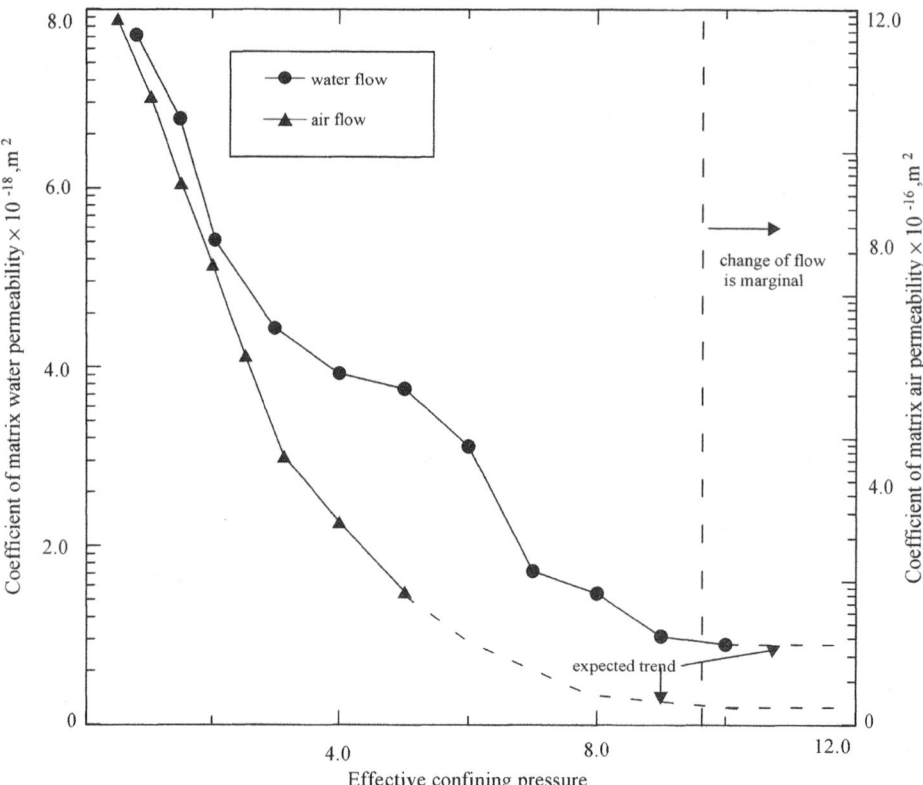

Figure 3.7. Effect of confining pressure on matrix permeability of intact granite specimens.

and geometrical characterisation of a single rock joint provides the basis to understand the fluid flow deformation behaviour in a fractured rock mass. It is difficult
to give a comprehensive description of flow behaviour even in a single joint,
because of the number of variables involved in three-dimensions. Therefore,
much analysis is based on plane strain (two-dimensional). Apart from external
boundary conditions, the variable void geometry in shape and size determines the
fluid flow through the fractures.

The main factors controlling fluid flow through a single rock joint are shown in
Fig. 3.9. Out of these controlling factors, the magnitude of the joint aperture is the
major parameter, which is a function of external stress, fluid pressure and geometrical properties of the joint. In many early studies, flow through a single joint was simulated as flow through a channel or pipe, in which no deformation was considered
due to external stress (Lomize, 1951). However, in reality, the deformation of fractures associated with external stress changes the flow rate of fluid, and the resulting
pore pressures affect the subsequent deformation of the discontinuities. The historical work on flow characteristics through a single joint was conducted experimentally
by Lomize (1951) and Louis (1969). In a single discontinuity, fluid flow is a function
of surface roughness, variable aperture, the magnitude of external loads and their

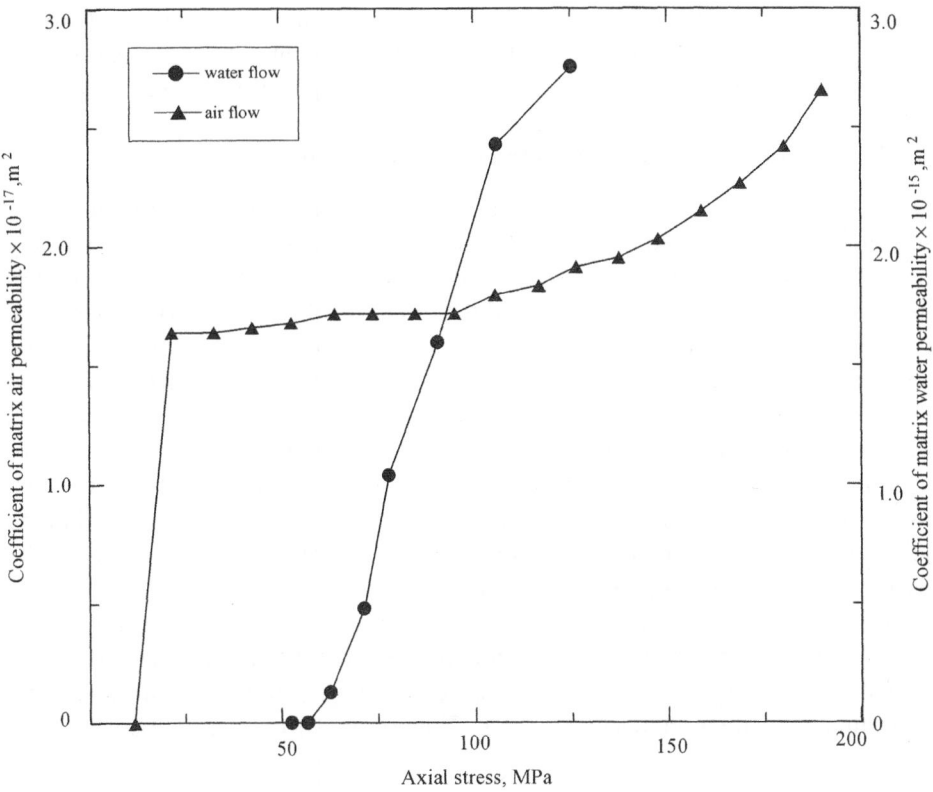

Figure 3.8. Effect of axial stress on matrix permeability (tests on intact granite).

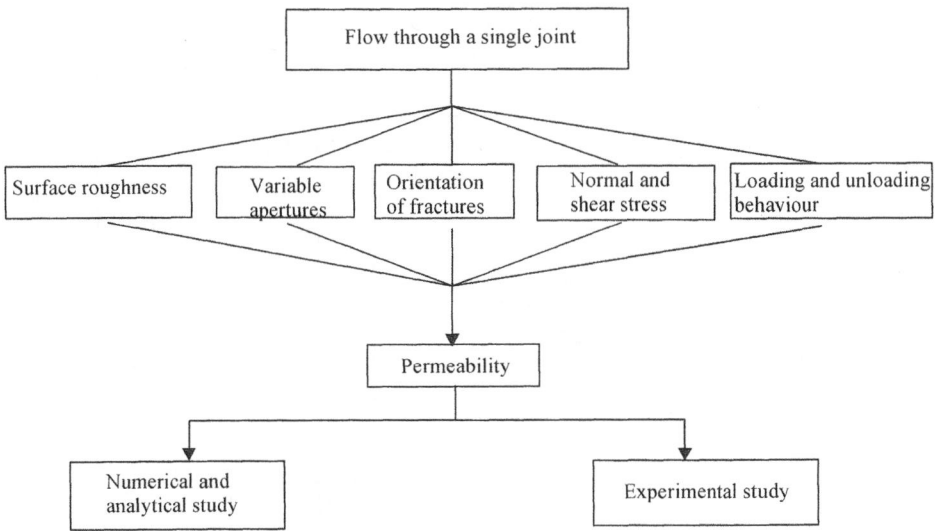

Figure 3.9. Factors which control permeability of a single joint.

direction relative to the orientation of joint, as well as the infill materials (Brown, 1987; Neuzil & Tracy, 1981; Tsang, 1984). Usually, the joint surface roughness plays a major role when the joint apertures are small or if the joints are sealed.

3.3.1   Effect of stress on permeability of single discontinuity

Discontinuities in a rock mass are usually subjected to surrounding in-situ stress, seismic loading and fluid pressure. In general, the resulting effective stress acting on a discontinuity consists of normal stress, shear stress and fluid pressure components. Depending on the magnitude and the direction of stress, a varying degree of mechanical behaviour of joint can be expected. As shown in Fig. 3.10, the applied stress will influence the joint to dilate or close, to create new contacts points, and even to crush the rock material depending on the surface geometry of the joint, magnitude of normal and shear strength and deformability of rock material. For example, normal stress closes the fracture, whereas the shear stress causes mismatch between the joint surfaces, resulting in the change of void space within the joint. The deformability of discontinuities is of paramount importance in the design of large structures, such as tunnels, underground nuclear storage plants, dams and bridges. The deformability of a discontinuity is usually expressed in terms of stiffness, which is defined as the ratio of stress to displacement. The stiffness has two components, based on normal stress (σ_n) and shear stress (τ_s) and their directional displacements, δ_n and δ_s. The normal stiffness is defined as

$$K_n = \frac{\sigma_n}{\delta_n} \tag{3.9}$$

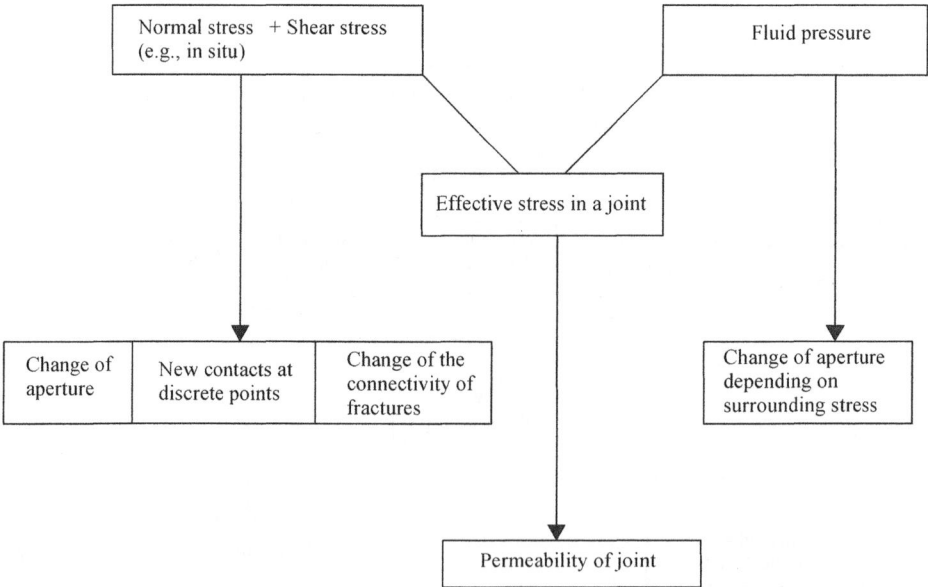

Figure 3.10. Effect of stress on a single rock.

The shear stiffness is defined as

$$K_s = \frac{\tau_s}{\delta_s} \tag{3.10}$$

The study of the deformation of discontinuities and intact material is a vast subject area, which has been discussed extensively by various researchers over the past five decades. The stress dependency of fluid flow through fractured rocks has been investigated in the past (Gale, 1975; Iwai, 1976; Raven & Gale, 1982). Therefore, the aim of this section is confined to evaluate the role of deformability on the permeability characteristics of a single joint.

A single rock fracture subjected to normal and shear stress components is shown in Fig. 3.11. At given stress conditions, if the geometry of fracture is defined by $F_T(x, y)$ and $F_B(x, y)$, then the fracture permeability for laminar fluid

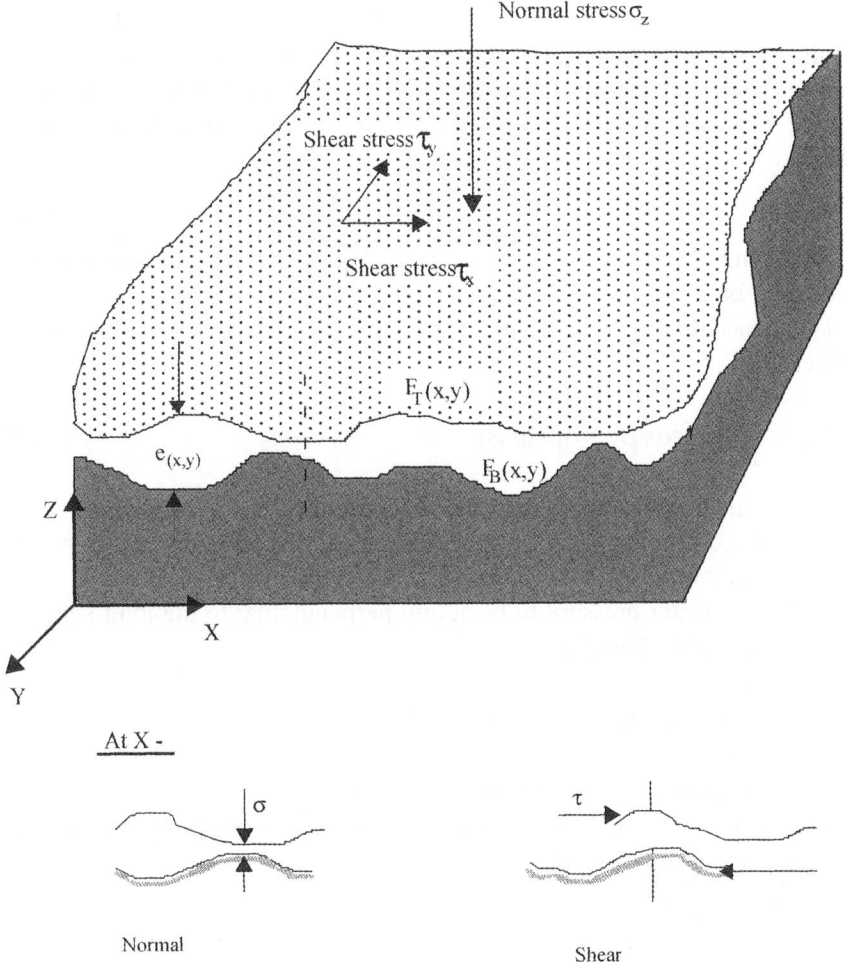

Figure 3.11. Effect of normal and shear stress on a typical joint.

flow is given by

$$k = \frac{(F_T(x, y) - F_B(x, y))^2}{12} \tag{3.11}$$

where $F_T(x, y)$ and $F_B(x, y)$ are the surface profiles of the discontinuity relative to the cartesian coordinate axis system, and k is called coefficient of fracture permeability, having units of m^2 or cm^2. It is usual to employ Eqn. 3.12 for fluid-flow analysis when flow takes place within smooth planar walls (i.e., $F_T(x, y) - F_B(x, y)$ = constant). The gap between the two surfaces is called the joint aperture, and the estimation of joint aperture is discussed in detail in Section 3.3.2. For a smooth, planar joint having an aperture of magnitude e, the fracture permeability for laminar flow is given by

$$k = \frac{e^2}{12} \tag{3.12}$$

The joint aperture e is mainly dependent on the normal and shear stress acting on the joint. Assuming the rock matrix to be isotropic and linear elastic, obeying Hooke's law, the following aperture–stress relationship can be formulated:

$$e = e_0 \pm \delta e \tag{3.13}$$

where e_0 is the initial joint aperture and δe is the change of the joint aperture due to stresses (i.e., both normal and shear components) acting on the joint. In conventional rock mechanics, the normal deformation component is given by Jaeger and Cook (1979):

$$\delta e_n = \frac{1}{K_n} [\sigma_z \cos \beta + \sigma_h \sin \beta] \tag{3.14}$$

where σ_z = vertical stress applied to the discontinuity, σ_h = horizontal stress applied to the discontinuity, K_n = normal stiffness of discontinuity and β = orientation of discontinuity.

Considering the water pressure to be acting perpendicular to the joint surface, Eqn. 3.14 can be modified to give

$$\delta e_n = \frac{1}{K_n} [\sigma_1 \cos \beta + \sigma_3 \sin \beta - p_w] \tag{3.15}$$

where p_w = water pressure within the discontinuity.

Combining Eqns. 3.12, 3.13 and 3.14 for planar and smooth joints, the permeability of single fracture is given by

$$k = \frac{(e_0 + \delta e_n)^2}{12} \tag{3.16}$$

Based on the initial hydraulic aperture and the closure of joint, Detoumay (1980) suggested the following relationship to determine the fracture permeability:

$$k = \frac{e_0^2 \left(1 - \dfrac{v}{v_0}\right)^2}{12} \tag{3.17}$$

where e_0 = hydraulic aperture at zero stress, v_0 = closure of the joint when the hydraulic aperture becomes zero and v = normal deformation of the joint, which may be computed using Eqn. 3.15. Snow (1968a) observed an empirical model to describe the fracture fluid flow variation against the normal stress, as described by

$$k = k_0 + K_n \frac{e^2}{s} (\sigma - \sigma_0) \tag{3.18}$$

where k_0 = initial fracture permeability at initial normal stress (σ_0), K_n = normal stiffness, s = fracture spacing and e = hydraulic aperture. From the test results obtained for carbonate rocks, Jones (1975) suggested the following empirical relation between the fracture permeability and the normal stress:

$$k = c_0 (\log(\sigma_{ch}/\sigma_c))^3 \tag{3.19}$$

where σ_{ch} = confining healing pressure in which the permeability is zero and σ_c = effective confining stress. The constant (c_0) depends on the fracture surface and the initial joint aperture. Nelson (1975) proposed the following empirical relationship for the permeability of fractured sandstone:

$$k = A + B\sigma_c^{-m} \tag{3.20}$$

where A, B and m are constants which are determined by regression analysis. These constants may vary from one rock to another, and even for the same rock type, depending on the topography of the fracture surface. Some values determined by Nelson (1975) are tabulated in Table 3.3. From this data, it can be clearly seen that there is no consistency of any of these parameters based on the above equation. By simulating a rock surface as a bed of nails, Gangi (1978) reported a theoretical model for fracture permeability as a function of the confining pressure, as represented by

$$k = k_0 (1 - (\sigma_c / P_1)^m)^3 \tag{3.21}$$

Table 3.3. Constants determined by regression analysis (Nelson, 1975).

Sample no.	Constant A	Constant B	Constant m
9-13	1494.0	4311.0	0.1
11-10	101.07	35800.0	0.7
16-17	−434.4	3410.0	0.2
19-15	−1600.0	3780.0	0.1

where P_1 = effective modulus of the asperities and m = constant which describes the distribution function of the asperity length. This expression gives a better prediction if the effect of surface roughness on flow is negligible, which of course is not reasonable in practice. Although Gangi (1978) could obtain a good fit for the flow data (Nelson, 1975), Tsang and Witherspoon (1981) encountered difficulties in fitting data from Iwai (1976) into Gangi's model. Having identified the demerits in the existing models, Tsang and Witherspoon (1981) developed a physical model, incorporating the joint roughness to see the effect of normal stress on fracture permeability. Considering the effects of surface roughness, Walsh (1981) derived an analytical expression for permeability against confining pressure as given by

$$k = k_0 \left[1 - \left(2\frac{h}{a_{02}} \ln \frac{\sigma_c}{\sigma_p} \right)^{0.5} \right]^3 \left[\frac{1 - b(\sigma_c - u_w)}{1 + b(\sigma_0 - u_w)} \right] \tag{3.22}$$

where $b = [3f/E(1 - v^2)h]^{0.5}$, f = autocorrelation distance, E = elastic modulus, v = Poisson's ratio, h = root mean square value of the height distribution of the fracture surface, k_0 = permeability at reference confining pressure (σ_p), σ_c = confining pressure, and a_{02} = half aperture at the reference pressure. Any of the permeability–stress relationships discussed above can be used for predicting fracture permeability depending on the required accuracy and availability of data. However, Eqn. 3.16 has been widely used in analytical and numerical models because of its simplicity and general reliability.

Laboratory results on normal stress on permeability
Using the triaxial test, effect of stress on fracture permeability can be investigated experimentally. A description of various kinds of triaxial apparatus was given earlier in Ch. 2. Some triaxial units are capable of applying three different stress levels (i.e., $\sigma_1 \neq \sigma_2 \neq \sigma_3$) in perpendicular directions, while the others are suitable for applying only two stress levels (i.e., σ_1 and $\sigma_2 = \sigma_3$). In the past, several researchers including Walsh (1981), Singh (1997) and Ranjith (2000) have investigated the effects of axial stress and confining pressure on permeability. All laboratory results show that the permeability is greatly affected by stress. Depending on the magnitude of stress levels and the relative orientation of fractures, increased or decreased permeability can be expected.

Singh (1997) found that the permeability decreases markedly at the beginning of the cycle due to crack closure, and followed by increasing permeability (Fig. 3.12). The increased permeability is due to the formation of new cracks. From the authors' experience in laboratory modelling (Ranjith, 2000), it is observed that there are three possible stages of permeability variation caused by an increase in axial compression (Fig. 3.13), which are as follows:
(1) Constant permeability;
(2) Decreasing permeability; and
(3) An increasing permeability.

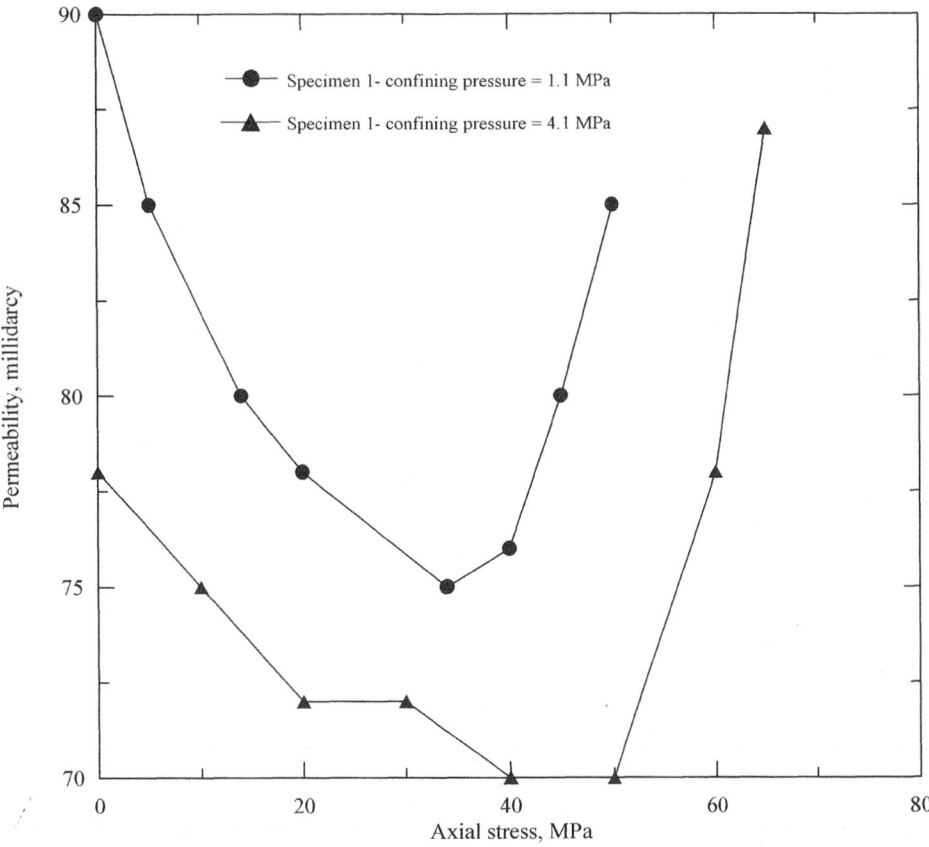

Figure 3.12. Effect of axial stress on permeability (data from Singh, 1997).

In Fig. 3.13, the permeability is calculated using Eqn. 3.12. Some specimens undergo all three stages while the others undergo combinations of either (1) and (2) or (1) and (3) or (2) and (3). Some rock samples show an almost constant permeability at the beginning of the application of axial stress. This behaviour is expected if the fractures are oriented in the same vertical direction as that of the axial load. The constant permeability is dependent on the magnitude of the effective confining stress applied to the rock specimen. The decrease in permeability associated with the increase in axial stress is attributed to possible crack closure. With further increase in axial load, micro cracks may start to develop and existing fractures may begin to dilate, thereby forming a new interconnected fracture network within the sample. As a result, a higher permeability is expected to occur until the sample fails.

As shown in Figs. 3.14 and 3.15, past and current laboratory test results have clearly indicated that the permeability through jointed rock specimens decreases as a function of increased effective confining stress, irrespective of the joint orientation (Kranz et al., 1979; Walsh, 1981; Raven & Gale, 1985; Ranjith, 2000).

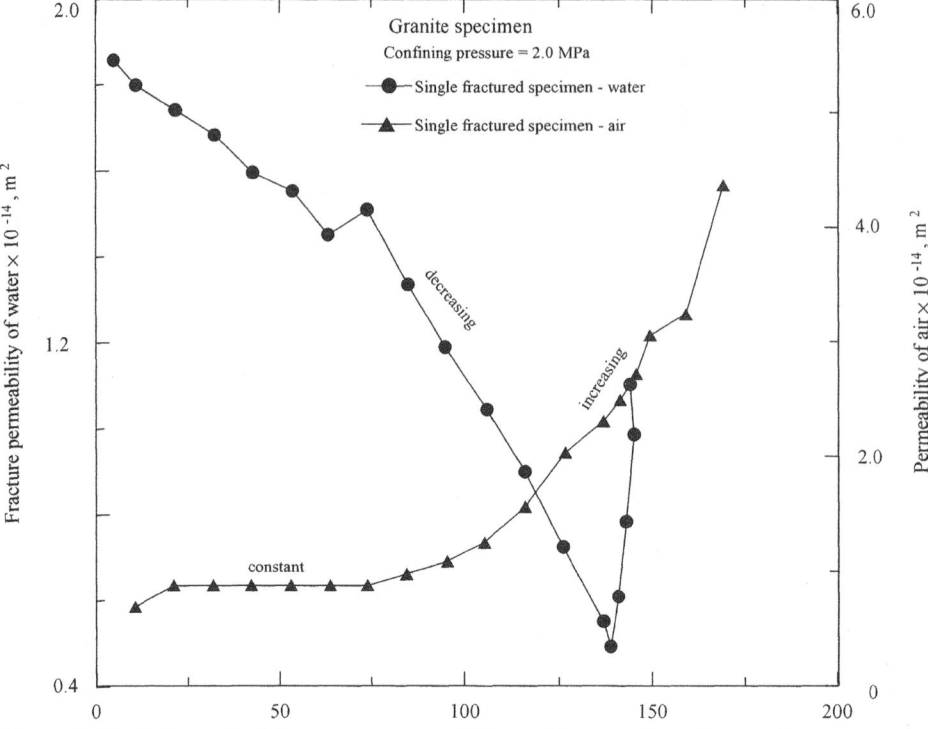

Figure 3.13. Effect of axial stress on permeability of fractured granite specimens.

According to Fig. 3.15, when the confining stress increases from 0 to 8 MPa, the average permeability decreases by more than 90%. Beyond an effective confining pressure of 8 MPa or so, further reduction in permeability is marginal for both air and water. This is attributed to joint apertures attaining their residual values, which are unaffected by further increase in confining stress. The residual aperture is a function of external stress conditions, the initial rock surface profile and the material and geometrical properties. It is also observed that greater the roughness, the lower the rate of permeability reduction under a given confining pressure. In other words, for smooth planar joints, the permeability decreases more significantly with the increase in confining pressure in comparison to rough joints. Tests conducted on Barre granite specimens by Kranz et al. (1979) also show that the permeability decreases significantly with the increase in confining pressure.

Effect of shear deformation on permeability
Although much focus has been given to the effect of normal stress on fracture permeability, the role of shear stress on permeability has not been considered to the same extent. Effect of shear deformation on permeability does not produce a simple relationship, as in the case of increased normal deformation where the permeability tends to decrease in most cases. The change of fracture permeability against shear stress entirely depends on the magnitude of shear displacement, the

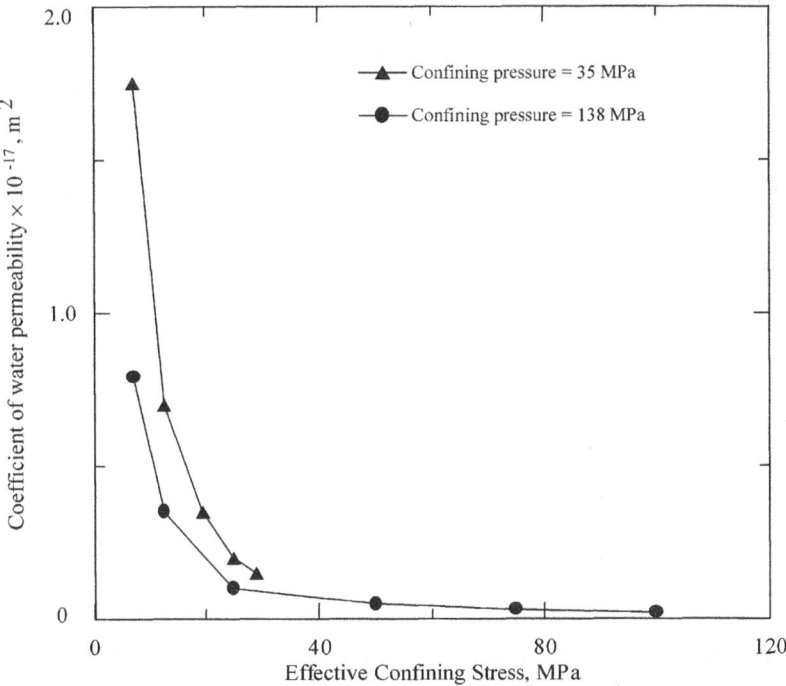

Figure 3.14. Effect of confining pressure on fracture permeability (data from Kranz et al., 1979).

surface topography of joints and the shear failure of asperities. In reality, a significant level of normal stress is generally applied on a rock joint due to gravity load (overburden). As a result, it is often difficult to isolate the role of shear stress on permeability, given the usual external boundary load conditions. Experimental work carried out by Makurat et al. (1990) reveals that the permeability of natural fractures can decrease or increase during the course of shear displacement as shown in Fig. 3.16. It seems that joints having a small *JRC* are characterized by a relatively unchanging permeability with increasing shear displacement.

3.3.2   Measurement techniques of joint apertures

Background
The study of fluid flow through fractured rocks has been the focus of various by researchers in different fields including, petroleum engineering, mining, hydrogeology and nuclear-waste engineering. Fluid flow through low permeability rocks is mainly dominated by interconnected fractures. Apart from the fluid properties and applied hydraulic head, the critical factor which controls the flow quantity is the joint aperture. The size and interconnectivity of fractures influence the volume and rate of flow through jointed rock. Accurate estimation of joint apertures is very difficult because of their irregular surfaces, that are characterized by many contact

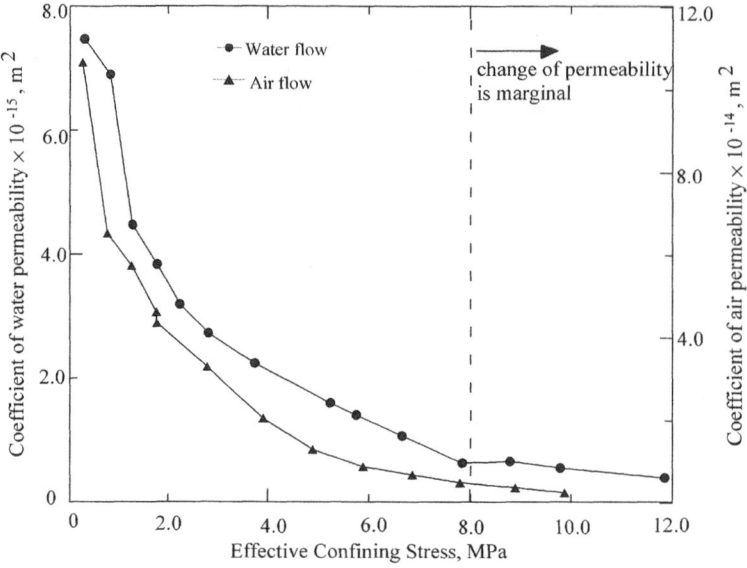

Figure 3.15. Effect of confining stress on permeabilty.

points between joint surfaces. For estimating joint apertures, several approaches can be adopted. They can be mainly categorized as direct and indirect measurements (Fig. 3.17). Direct measurements can provide the local mechanical aperture at a particular location. Mechanical apertures can develop due to shear displacement along highly irregular joint surfaces. Joint apertures can extend from say 1 μm to 1 m in a fractured rock mass. The dramatic opening of fractures is often due to the development of local tensile stresses within the rock mass. Once a discontinuity is created, its aperture can increase or decrease depending on the stress history, such as deposition of infill, washing out of infill, erosion or change of adjacent stresses due to blasting, or increased fluid pressures surrounding the rock mass. Figure 3.18 shows various forms of discontinuities in rock mass, such as planar uniform aperture, variable/rough aperture with or without contacts points between joint walls, and joints filled with foreign materials. In practice, although joints with uniform apertures rarely exit, for simplicity of flow analysis through a single joint or network of joints, smooth open joints are usually considered as shown in Fig. 3.18a. According to the method of Barton (1973), classification of apertures by size has been recorded by Lee and Farmer (1993), as given in Table 3.4. From the in-situ permeability tests conducted in Colorado, Snow (1968a, b) found that the fracture aperture could range from 50 to 350 μm at a depth of 10 m. At greater depths exceeding 30 m, the apertures decrease to 40 to 100 μm (Fig. 3.19).

Direct measurement of joint apertures
The physical measurements of joint apertures, which are exposed to the surface, give a rough but quick indication of the separation between joint walls at local points. The measured opening is termed as the mechanical aperture of the discontinuity at a

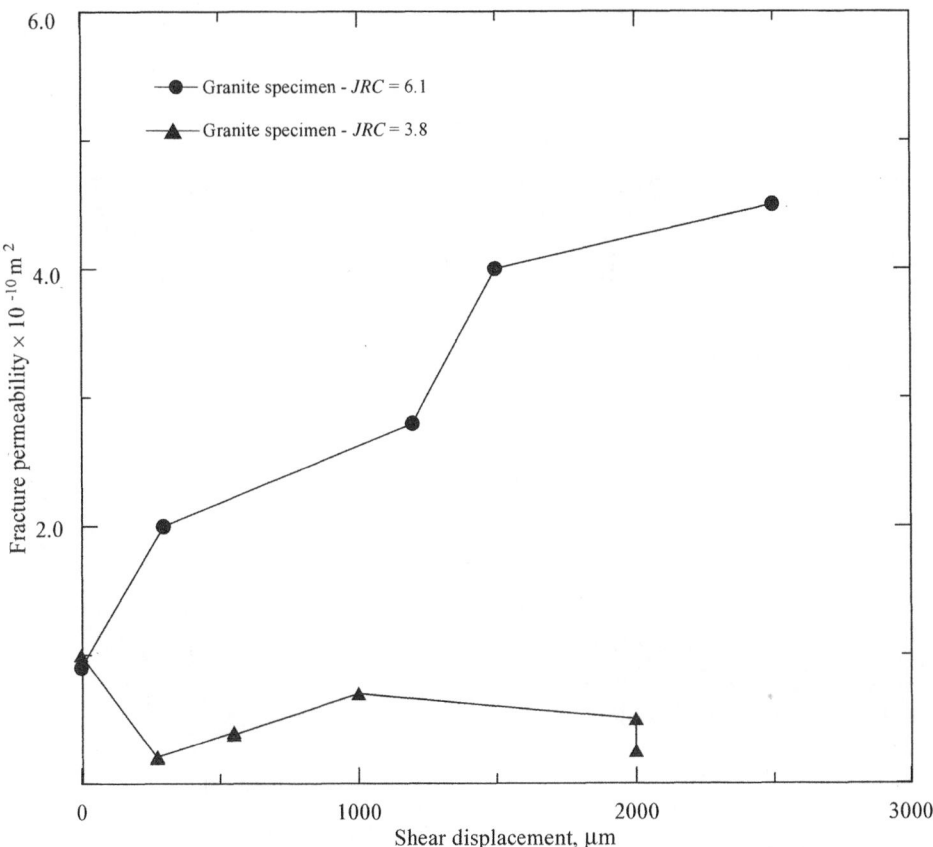

Figure 3.16. Effect of shear stress on fracture permeability (data from Makurat et al., 1990).

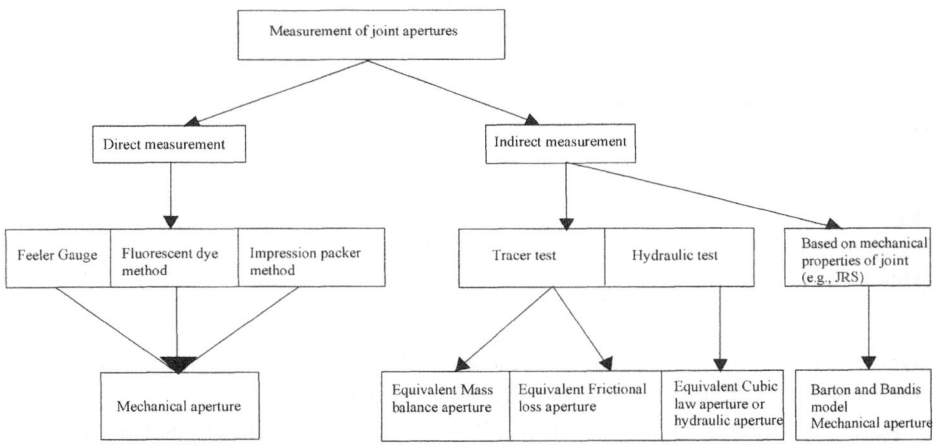

Figure 3.17. Techniques for measuring joint aperture.

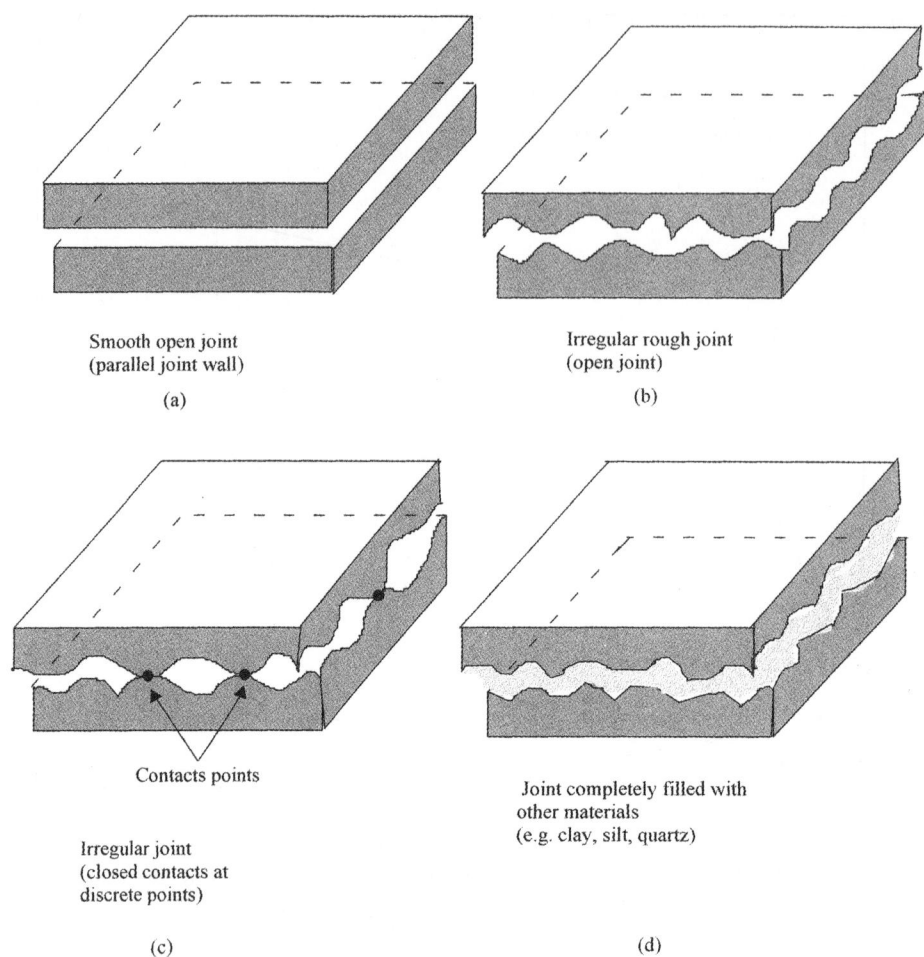

Figure 3.18.　Various forms of discontinuities within a rock mass.

Table 3.4.　Classification of joint apertures by its magnitude (after Barton, 1973).

Class	Aperture (mm)
Very tight	<0.1
Tight	0.1–0.25
Partly open	0.25–0.50
Open	0.50–2.5
Moderately wide	2.5–10.0
Wide	10

given stress level. In fluid flow calculations, the mechanical aperture is useful to determine the hydraulic aperture for a given joint profile.

Based on fluorescent dye, Snow (1970) described a direct measurement of mechanical apertures of discontinuities. The dye is first sprayed on to the specimen

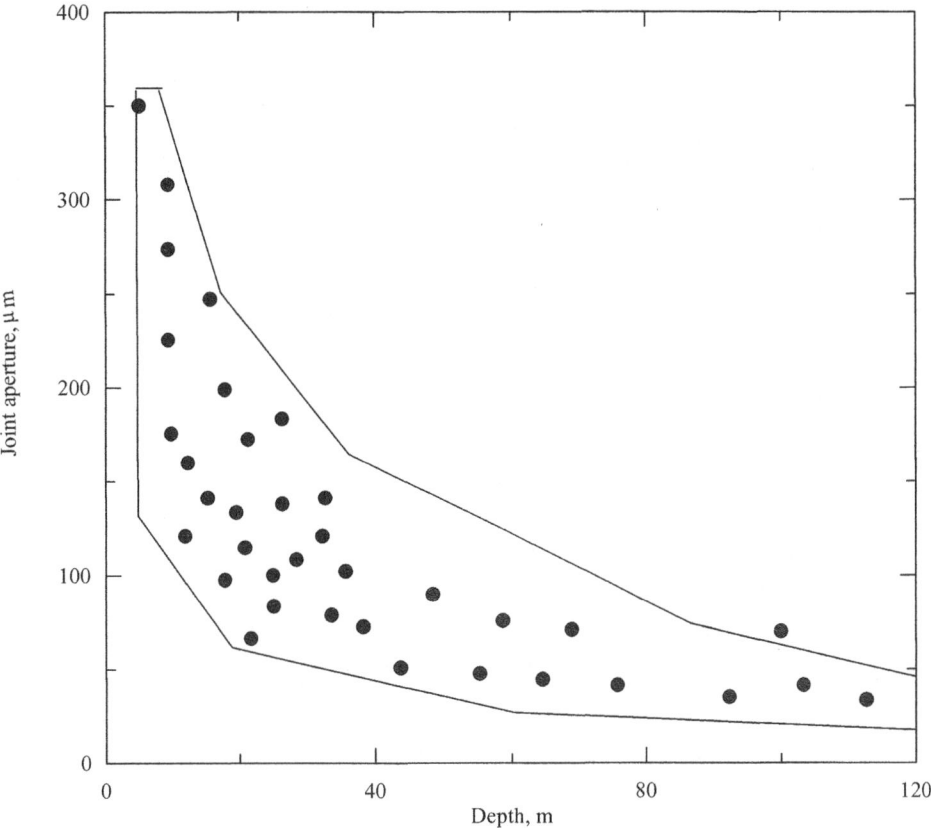

Figure 3.19. Effect of depth on the change of joint aperture (data from Snow, 1968a,b).

until it flows out from the edges of the discontinuity. Once the excess dye is removed from the surface of the specimen, a thin coating of developer is sprayed on the rock specimen in order to get a clear photo image of the discontinuity, from which the aperture is estimated. However, the accuracy of the aperture estimated in this manner greatly depends on the quality (sharpness) of the photo images.

Barr and Hocking (1976) have used an alternative approach to measure the mechanical aperture in boreholes using the impression packer test. In this method, a plastic wax is pressed on the surface of the borehole, and subsequently the borehole surface is photographed. This method provides not only the mechanical aperture of several joints that are connected to borehole surface, but also the discontinuity spacing and orientations. Bandis (1980) used a direct approach with a tapered feeler gauge to measure the joint aperture. Although this direct measurement is simple, the tapered feeler gauge must be small enough to be inserted into the very small joint which is in the order of $10^{-5} - 10^{-9}$ m. Around 65 jointed specimens were measured using the feeler gauge under zero applied stress. This technique can be easily employed in the laboratory as well as in the field to measure the apertures of exposed discontinuities in a rock mass. Particularly, for the joint aperture measurement of

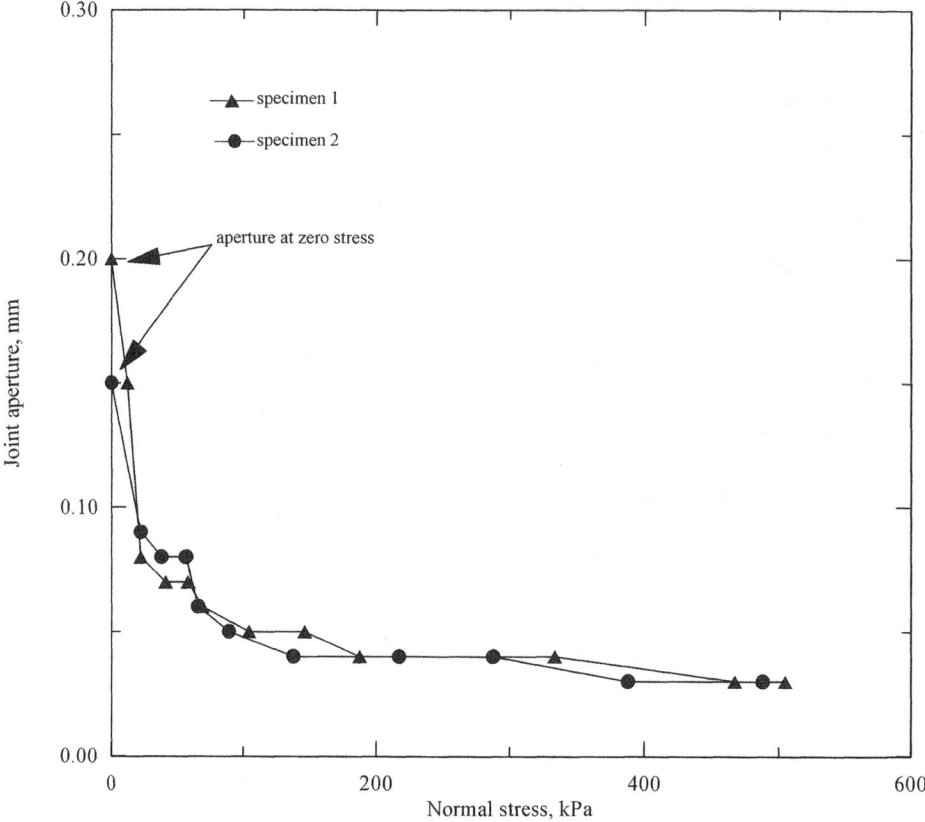

Figure 3.20. Variation of mechanical aperture with different normal stress.

roof tunnels or walls in the field, the feeler gauge method can easily be implemented with little cost. Authors have also used this method to estimate the initial hydraulic aperture calibrated with the Joint Roughness Coefficient. In addition, typical test results on mechanical apertures of naturally fractured specimens under a range of normal stress are shown in Fig. 3.20.

As an alternative approach to the above procedures, Hakami and Barton (1990) measured joint apertures at local points of artificially simulated joint replicas. The joint surfaces are first replicated using silicon rubber, and subsequently, the silicon rubber joint mould is used to copy the profiles on to the transparent epoxy. A measured quantity of water is placed on one surface of the joint surface, and the other surface is firmly fitted together. The water will cover a certain surface area (*a*) of the aperture and this area is then measured (Fig. 3.21). If the quantity of water placed at a point 'A' on the joint surface is *Q*, the aperture at point 'A' is then given by

$$e_A = \frac{Q}{a} \tag{3.23}$$

where *a* = area of the water.

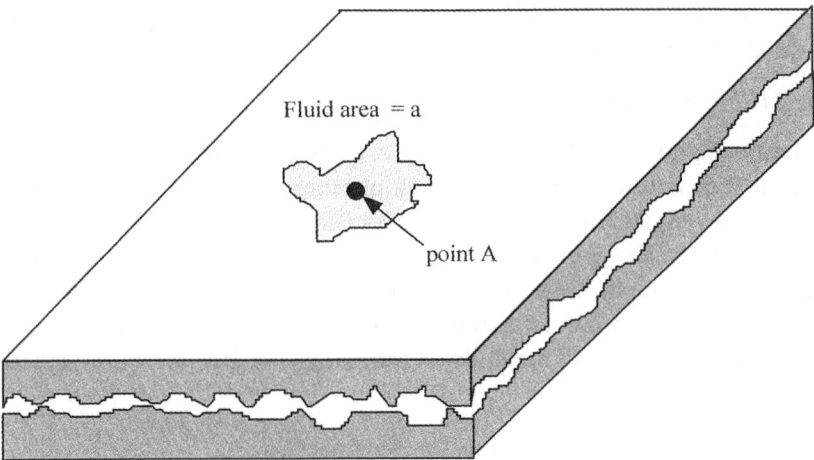

Figure 3.21. Estimation of fracture aperture at a given point (after Hakami & Barton, 1990).

Apart from the practical difficulties involved in this exercise, the evaluation of aperture at two different stresses (two directions) and for different fluid pressures is not feasible. It should also be noted that one cannot guarantee that the transparency epoxy joint profile would be the same as the natural profile, because this method operation involves copying of the profile twice.

Indirect measurement of joint apertures
Under indirect approach, the joint apertures may be estimated either by mechanical properties of the discontinuities or by the fluid flow measurements through discontinuities. Depending on the techniques employed, the mechanical aperture or the hydraulic aperture of the joint is computed, where the mechanical aperture is not the same as the hydraulic aperture. Usually the mechanical aperture is larger than the hydraulic aperture, except when the joint has reached its residual aperture at very high stress levels. The relationship between the mechanical aperture and the hydraulic aperture is discussed at the end of this section.

Measurement of Mechanical aperture based on JRC
Based on Joint Compressive Strength (*JCS*) and Joint Roughness Coefficient (*JRC*) values, Barton and Bakhtar (1983), and Bandis et al. (1983, 1985) proposed an empirical approach to perform indirect measurements of the mechanical aperture, as discussed below:

$$e_{\mathrm{m}} = \left[\frac{JRC}{5} \right][0.2\sigma_{\mathrm{c}}/JCS - 0.1] \tag{3.24}$$

where e_{m} = mechanical aperture (mm) and σ_{c} = unconfined compression strength.

Based on Schmidt hammer rebound number, the following empirical relationship is used to determine the *JCS*.

$$\log_{10}(JCS) = 0.00088 \gamma R + 1.01 \tag{3.25}$$

where γ = unit weight of rock (kN/m³) and R = rebound number.

It is of interest to note that the mechanical aperture depends only on surface roughness when σ_c equals *JRC* (i.e. applied load is not enough to change the deformation of joint). The joint roughness coefficient ranges from 0 (smooth joint) to 20 (extreme rough). Several researchers including Barton (1973) have discussed analytical and empirical techniques of predicting *JRC* for in-situ and laboratory conditions. Available methods of estimating of *JRC* are presented in detail in Section 3.3.3.

Measurement of hydraulic aperture based on the fluid flow

There are two other common indirect techniques to estimate the hydraulic aperture based on fluid flow through a rock mass. They are (a) based on laboratory steady-state flow measurement under triaxial test conditions and (b) using in-situ tests (e.g. borehole pumping test/tracer tests). Under laboratory conditions, the rock specimens with a single fracture or multiple fractures are tested using a triaxial apparatus for a given confining pressure, inlet fluid pressure and axial stress. Steady-state flow rates are used to calculate the hydraulic aperture using the Darcy's (cubic) law. It is important to note that the hydraulic aperture is not the same as the mechanical aperture, because the natural fractures are dissimilar to ideal parallel plates. However for plane, smooth joint surfaces with *JRC* = 0, the magnitude of mechanical aperture tends to approach the hydraulic aperture. When interpreting flow data for estimating joint aperture, various kinds of fluid flow theories may yield different values of hydraulic apertures for the same joint. These are termed as equivalent mass balance aperture, equivalent frictional loss aperture, equivalent cubic law aperture (i.e. hydraulic aperture), tracer joint aperture, etc.

'Equivalent' joint aperture is generally used for rough fractures with irregular surfaces. At any point (x, y) along the joint (Fig. 3.22), the joint aperture at that point is given by

$$e_{(x, y)} = F_{T(x, y)} - F_{B(x, y)} \tag{3.26}$$

where $F_{T(x, y)}$ and, $F_{B(x, y)}$ are the surfaces of top an bottom joint walls relatively to a given coordinate axis system. This expression yields joint aperture at any local point, which does not represent the mean aperture of the entire joint. Therefore, this is called the local joint aperture. However, in terms of practical applications, the equivalent fracture aperture is of greater importance than the local joint aperture, particularly in fluid flow calculations. Tsang (1992) has extensively discussed the equivalent fracture aperture based on hydraulic and tracer tests. The equivalent aperture may be calculated using the mass balance concept, friction loss due to the roughness and surface irregularities, and the cubic law theory. Therefore, in order to avoid any confusion, the equivalent aperture has to be

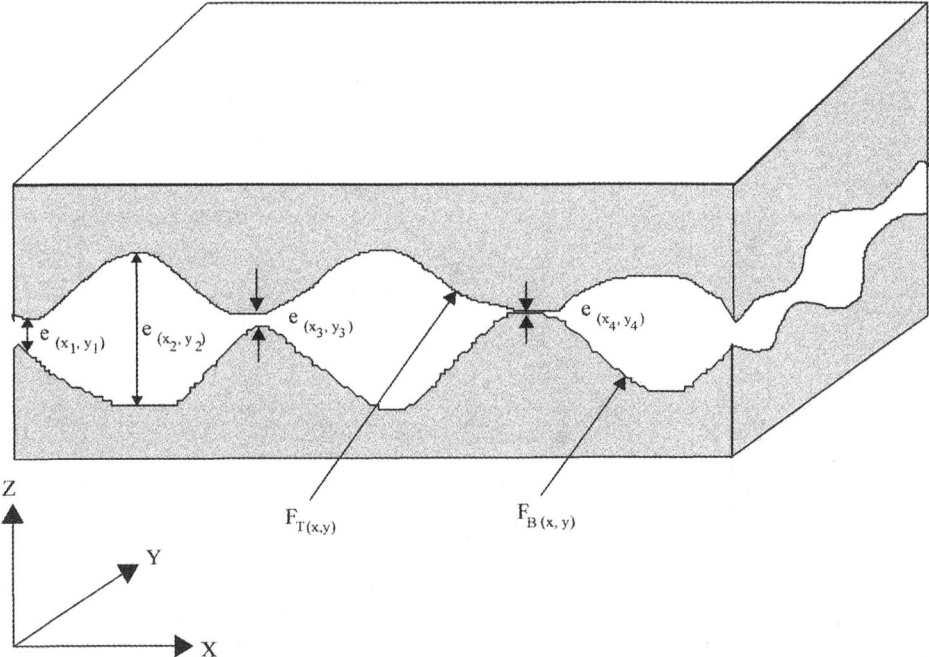

Figure 3.22. Typical natural rock fracture showing highly irregular joint walls.

referred to as either the mass balance equivalent aperture, or the friction loss equivalent aperture or the cubic law equivalent aperture. Mass balance and friction loss based apertures are usually associated with the tracer test, whereas the cubic law equivalent aperture is always predicted using the hydraulic test data.

It is found from the literature that many researchers seem to be more comfortable with the equivalent cubic law aperture (Iwai, 1976; Witherspoon et al., 1980; Ranjith, 2000) based on hydraulic tests. Frequently, and for simplicity, the equivalent cubic law aperture is referred to as the hydraulic aperture or mean aperture. For laminar fluid flow through parallel joint walls, the equivalent cubic law aperture (e_c) is defined by

$$e_c = \left[\frac{12q\mu}{b(dp/dx)} \right]^{1/3} \tag{3.27}$$

where q = steady-state flow rate, b = width of the fracture, dp/dx = pressure gradient between two ends of the specimen and μ = dynamic viscosity of fluid.

According to Tsang (1992), for radial flow, the equivalent cubic law aperture (e_c) is given by

$$e_c = \left[\frac{6\mu q}{\pi dp} \ln\left(\frac{r}{r_0} \right) \right]^{1/3} \tag{3.28}$$

where r_0 = well radius and r = distance to the tracer collection point.

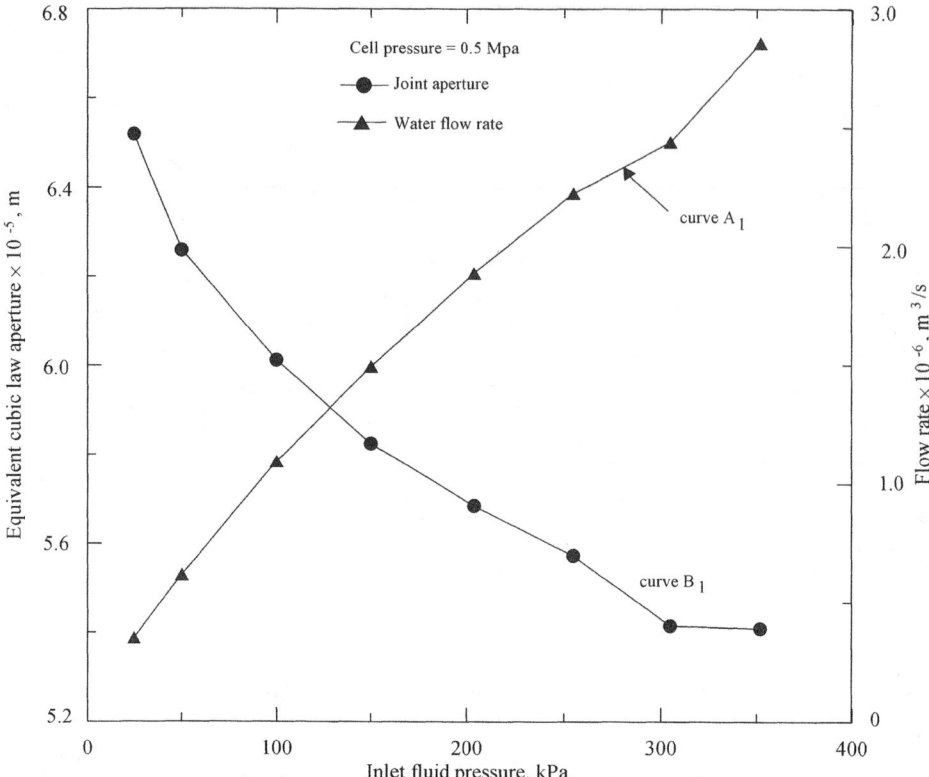

Figure 3.23. Equivalent hydraulic aperture estimated from cubic law at confining pressure of 0.5 MPa.

For given boundary conditions and for measured steady-state flow rate following cubic theory, Eqns. 3.27 and 3.28 are used to calculate the equivalent hydraulic aperture. In the text below, the simple terminology, hydraulic aperture is used in lieu of equivalent cubic law hydraulic aperture.

Apart from the material and geometrical properties of the joints, the magnitude of e depends on external stress, such as the confining pressure, axial stress and fluid pressure within the joint. Typical test results on naturally fractured granite specimens subjected to high pressure triaxial testing are shown in Figs. 3.23–3.26. For a given axial stress and confining pressure, a steady-state flow rate is observed at different inlet water pressures, as presented in Figs. 3.23 and 3.24. The curves A_1 and A_2 in Figs. 3.23 and 3.24 show that the flow rate linearly varies with the inlet fluid pressures, confirming the applicability of Darcy's law for fluid-flow analysis through naturally fractured rock samples. The validity of cubic law was investigated by Witherspoon et al. (1980), and they suggested that the cubic law was valid for joints with apertures varying from 250 μm to as small as 4 μm. The data represented by curves B_1 and B_2 in Figs. 3.23 and 3.24 are plotted using Eqn. 3.27. It is noted that the flow rate increases with the increasing inlet fluid

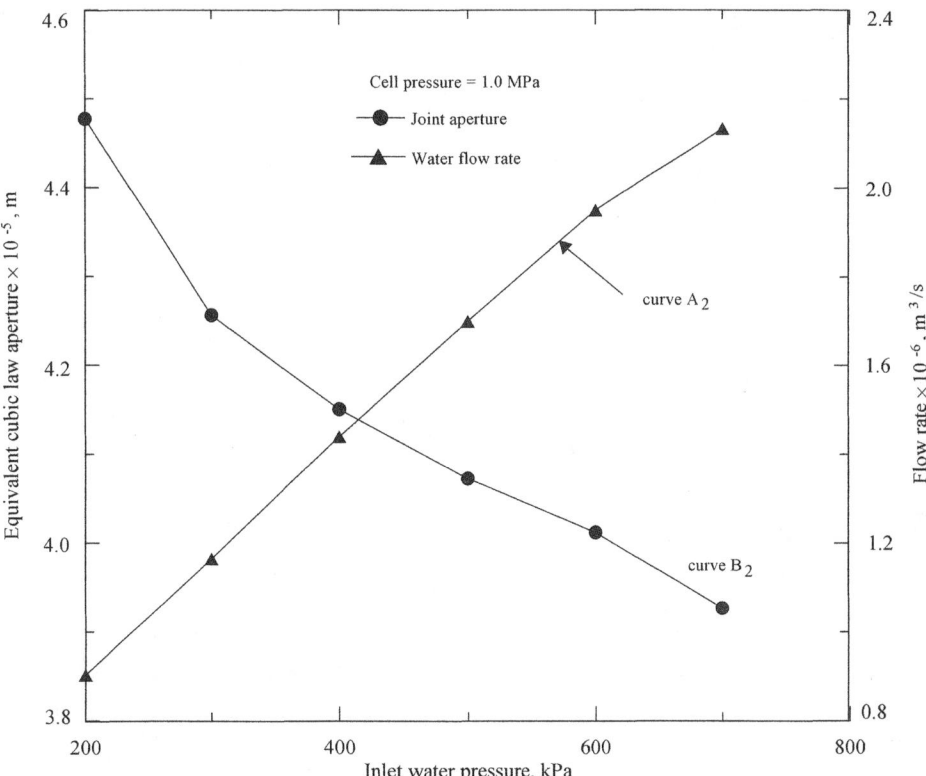

Figure 3.24. Equivalent hydraulic aperture estimated from cubic law at confining pressure of 1.0 MPa.

pressure, but the estimated joint aperture decreases (non-linear) against the inlet fluid pressure. This is because the fluid flow is directly proportional to the inlet fluid pressure, whereas the joint aperture is proportional to the cube root of the flow rate and inversely proportional to the inlet fluid pressure.

The confining pressure can significantly influence the magnitude of the joint aperture (Fig. 3.25). As expected, steady-state flow rate and the estimated joint apertures decrease with the increase in confining stress. When the confining pressure exceeds 6 MPa, the change of aperture becomes very small. This relatively constant aperture is called the residual aperture of the joint. Therefore, even at high stress levels, a minimum flow corresponding to the residual aperture is expected, provided the fluid pressure is large enough to cause the flow through the fracture. It is relevant to note that the residual aperture exists only when the fractures are irregular, but not in the case of planar and smooth joints.

Effect of axial stress on joint aperture is different to what one may observe against confining pressure. According to Fig. 3.26, the deviator stress may close or dilate or create new fractures depending on the magnitude of stress levels and the orientation of fractures in a rock mass. When the deviator stress level exceeds

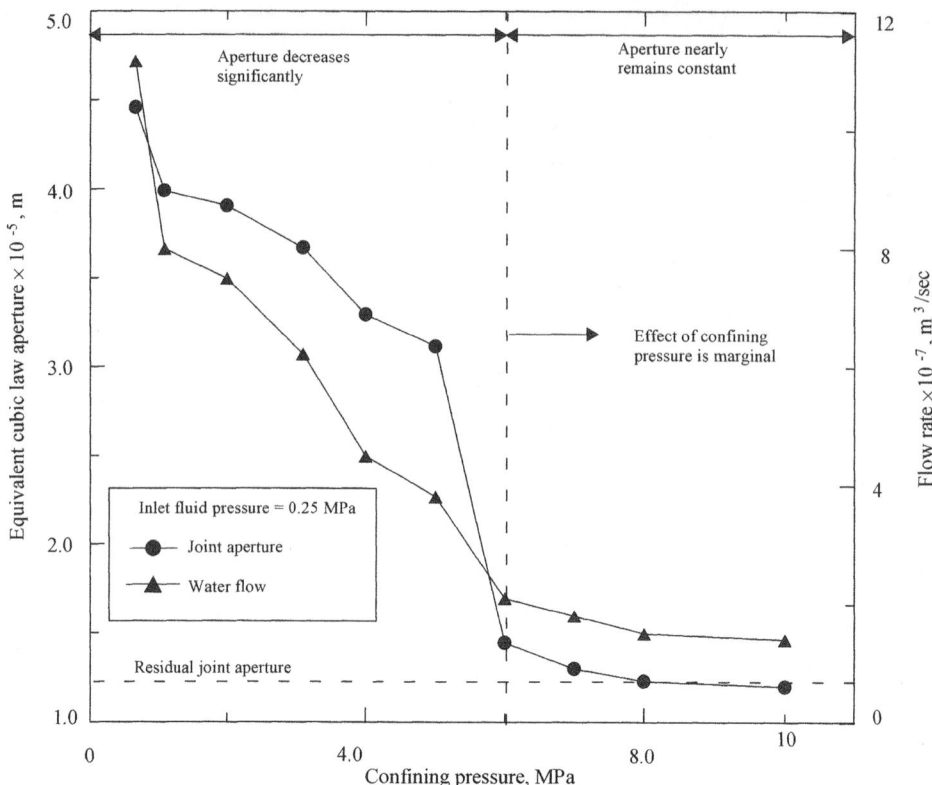

Figure 3.25. Equivalent hydraulic aperture estimated from cubic flow theory for various confining pressure.

130 MPa, the increased aperture size or the created new fractures dramatically increase the flow rate (Fig. 3.26). Generally, three distinct zones of aperture change with deviator stress are observed, as follows:

(1) Constant joint aperture – deviator stress is not sufficient to affect the deformation of fracture;
(2) Decreasing joint aperture – due to closure of existing fractures; and
(3) An increasing joint aperture – formation of new fractures or dilation of existing fractures or both.

The hydraulic aperture computed from Eqn. 3.27 does not account for the effects of surface roughness of the joint, therefore resulting in an over-estimated aperture. If the surface roughness effect is considered, the cubic law can be modified by introducing a correction factor as follows:

$$e_c = \left[\frac{12q\mu f}{b(\mathrm{d}p/\mathrm{d}x)} \right]^{1/3} \qquad (3.29a)$$

The parameter f has been evaluated by several researches in the past, including Lomize (1951), Louis (1969) and Quadros (1982). Their findings are summarized

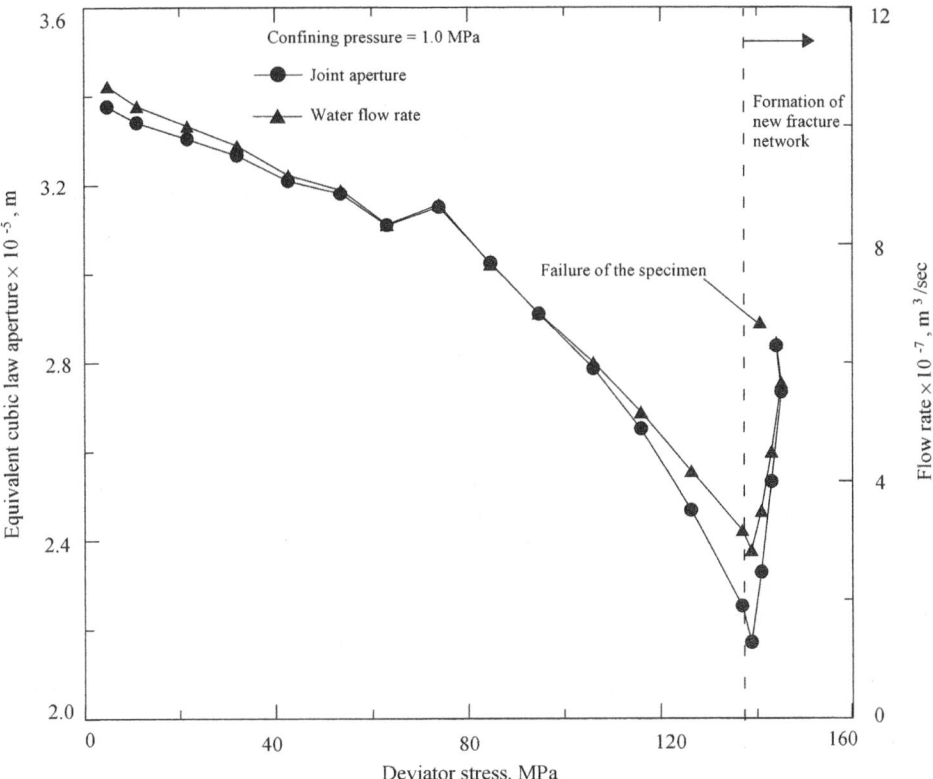

Figure 3.26. Equivalent hydraulic aperture estimated from cubic flow theory at various deviator stress.

by the following algebraic expression:

$$f = \left[1 + m\left(\frac{r_a}{2e}\right)^{1.5} \right]$$ (3.29b)

where $m = 17$ from Lomize (1951), $m = 8.8$ from Louis (1969), and $m = 20.5$ from Quadros (1982).

According to Louis (1969), r_a is the difference between the highest 'peak' and the lowest 'valley' of the physical wall roughness, and e is the joint hydraulic aperture. For a smooth parallel joint, r_a becomes zero, thus f becomes unity. By combining Eqn. 3.29a with Eqn. 3.29b, the following quadratic equation can be derived for estimating the joint aperture, taking into account of joint roughness:

$$e_c^{4.5} - Ae_c^{1.5} - Am\left(\frac{r_a}{2}\right)^{1.5} = 0$$ (3.30)

where

$$A = \left(\frac{12q\mu}{b(\mathrm{d}p/\mathrm{d}x)}\right)$$

and *m* is defined in Eqn. 3.29b.

For a given single fracture, having a length *l* and width *b*, if the flow rate through the fracture is *q* and the mean tracer transport time is *t*, then the mass balance equivalent aperture (e_m) may be written as follows:

$$e_m = \left(\frac{qt}{lb}\right) \tag{3.31}$$

The mass balance aperture is sometime referred to as the tracer aperture, because the data used to calculate the aperture are taken from tracer tests (Novakowski et al., 1985; Moreno et al., 1988). Smith et al. (1987) studied the measurements of joint aperture using hydraulic and tracer tests for a given network of fractures. For a given well with radius r_0 and for the tracer collection point at a distance of radius *r* (Fig. 3.27), Smith et al. (1987) refer to the equivalent mass balance aperture as 'volume balance aperture', given by

$$e_m = \left[\frac{qt}{\pi(r^2 - r_0^2)}\right] \tag{3.32}$$

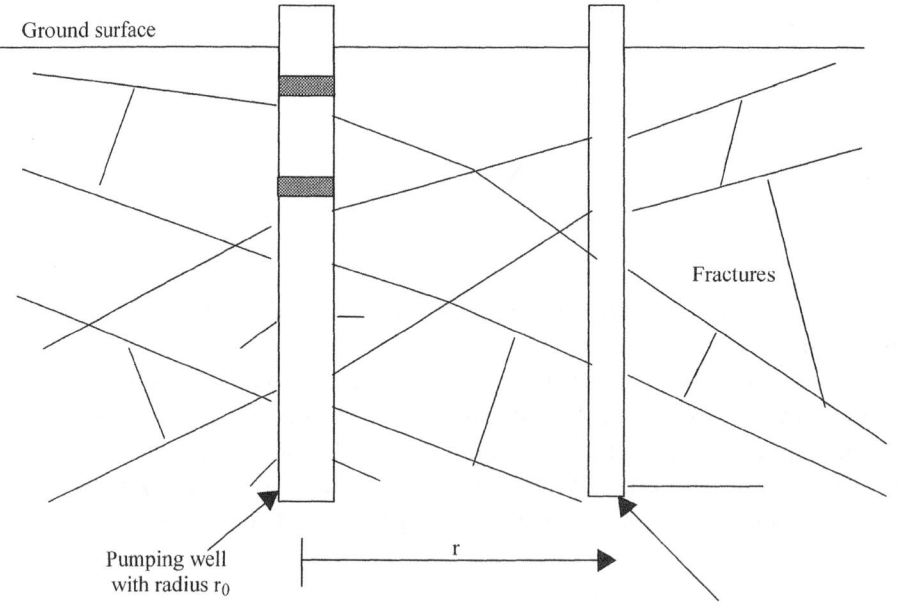

Figure 3.27. Pumping well and tracer collection point in tracer test.

where the symbols are the same as in Eqn. 3.31. Tsang (1992) concluded that the mass balance aperture is greater than or equal to the cubic law aperture, where the cubic law aperture is greater than or equal to the frictional loss aperture.

In order to incorporate the surface roughness and irregularities of joints, the joint aperture can be estimated based on the pressure drop in the tracer test; thus, the calculated joint aperture can be termed as equivalent frictional loss aperture. According to Abelin et al. (1985), the equivalent frictional loss aperture is expressed in terms of mean time (t) for tracer transport from injection l_1 to the collection l_2, and the velocity of flow. If the velocity through parallel joint walls is expressed using Darcy's law for laminar flow, then the equivalent frictional loss aperture is given by

$$e_{\mathrm{f}} = L \left[\frac{12\mu}{(\mathrm{d}p)t} \right]^{1/2}$$
(3.33)

where μ = dynamic viscosity of fluid, L = length of the flow path (i.e., $l_1 - l_2$), $\mathrm{d}p$ = pressure difference over the length of L, and t = mean tracer time over a length of L and given by $\int_{l_1}^{l_2} \frac{1}{v} \mathrm{d}l$, where v is Darcy's velocity. Unlike in the case of equivalent mass balance aperture, the frictional loss aperture depends on tracer time, fluid flow path and the pressure drop. With regard to measurable quantities, the main difference between the methods of mass balance aperture and the frictional loss aperture is that the pressure drop is measured in the latter, whereas the flow rate is measured in the former. From in-situ test results, Smith et al. (1987) suggested that the equivalent aperture for radial flow could be estimated from

$$e_{\mathrm{f}} = \left[\frac{6\mu \ln(r_0/r_1)}{t(\mathrm{d}p)} (r_0^2 - r_1^2) \right]^{1/2}$$
(3.34)

where t = time taken by a tracer to move from the injection point to the withdrawal point (r_0 to r_1), which is given by $t = \int_{r_0}^{r_1} \frac{1}{v} \mathrm{d}l$. The calculated aperture using Eqn. 3.33 or 3.34 is referred to as the tracer aperture by various researches including Smith et al. (1987). However, Tsang (1992) used the term equivalent frictional aperture, because the effects of surface roughness were also taken into account via the measurement of pressure drop ($\mathrm{d}p$) and the tracer time (t).

From borehole flow tests, the equivalent smooth joint wall aperture can be estimated using the equation given below (Barton et al., 1985):

$$e = (12k)^{0.5}$$
(3.35)

where k = the conductivity.

Mechanical aperture versus hydraulic aperture
As mentioned earlier, the hydraulic aperture is different to the mechanical aperture of a joint, because natural discontinuities are dissimilar to ideal parallel plates. Several investigators including Barton (1982), Witherspoon et al. (1979) and Elliot et al. (1985) studied the relationship between hydraulic aperture and mechanical aperture of natural rock joints. Witherspoon et al. (1979) and Elliot

et al. (1985) proposed a simple linear relationship between the two types of aperture based on initial hydraulic aperture and the mechanical deformation of the joint due to stresses, as reported by Nguyen and Selvadurai (1998):

$$e_h = e_{ho} + f\Delta e_m \tag{3.36}$$

where e_h = hydraulic aperture at time t, e_{ho} = initial hydraulic aperture, Δe_m = change of mechanical aperture and f = proportionality factor which varies between 0.5 and 1 (Benjelloun, 1993). The factor f accounts for the surface irregularities of the joint, and for smooth joint parallel walls, the value of f tends to become unity.

Detoumay (1980) and Elliott (1985) developed another linear relationship between the hydraulic and mechanical apertures, as also reported by Wei and Hudson (1988):

$$e_h = \left(\frac{e_{ho}}{\delta}\right)[e_m - (e_{max} - \delta)] \tag{3.37}$$

where δ = closure of the joint when the hydraulic aperture is zero, e_{max} = maximum mechanical aperture at zero stress and $e_{max} - \delta$ = the residual mechanical aperture when the hydraulic aperture is zero (i.e. no fluid flow takes place through the rock joint).

Although Eqn. 3.37 is a simple linear relation, the practical difficulty is the prior prediction of the maximum mechanical aperture for a given joint. The maximum mechanical aperture at zero stress may be measured using the feeler gauge, whereas the closure of the joint when the hydraulic aperture is zero, may be determined using the triaxial apparatus.

Based on Joint Roughness Coefficient (*JRC*), Barton et al. (1985) proposed following relationship between the mechanical and hydraulic apertures:

$$e_h = \frac{e_m^2}{JRC^{2.5}} \tag{3.38}$$

where e_h = hydraulic aperture and e_m = mechanical aperture.

Compared to Eqns. 3.36 and 3.37, Eqn. 3.38 is easier to employ in the field, as well as in the laboratory. For example, in the field, the joint is first mapped using the string line method to determine the *JRC*. Subsequently, the feeler gauge is used to measure the mechanical aperture. This will facilitate the calculation of the hydraulic aperture using Eqn. 3.38. By considering the normal deformation of the joint, Eqn. 3.38 can be modified as follows:

$$e_h = \frac{(e_m - \delta)^2}{JRC^{2.5}} \tag{3.39}$$

where δ = normal deformation of the joint which is given by the following expression $\sigma_n/(a + b\sigma_n)$, where σ_n = normal stress. The parameters a and b are constants which are determined using the uniaxial compression test. For a given roughness coefficient (e.g., *JRC* = 11), a typical graph of hydraulic aperture against the mechanical aperture is plotted in Fig. 3.28, based on Eqn. 3.38.

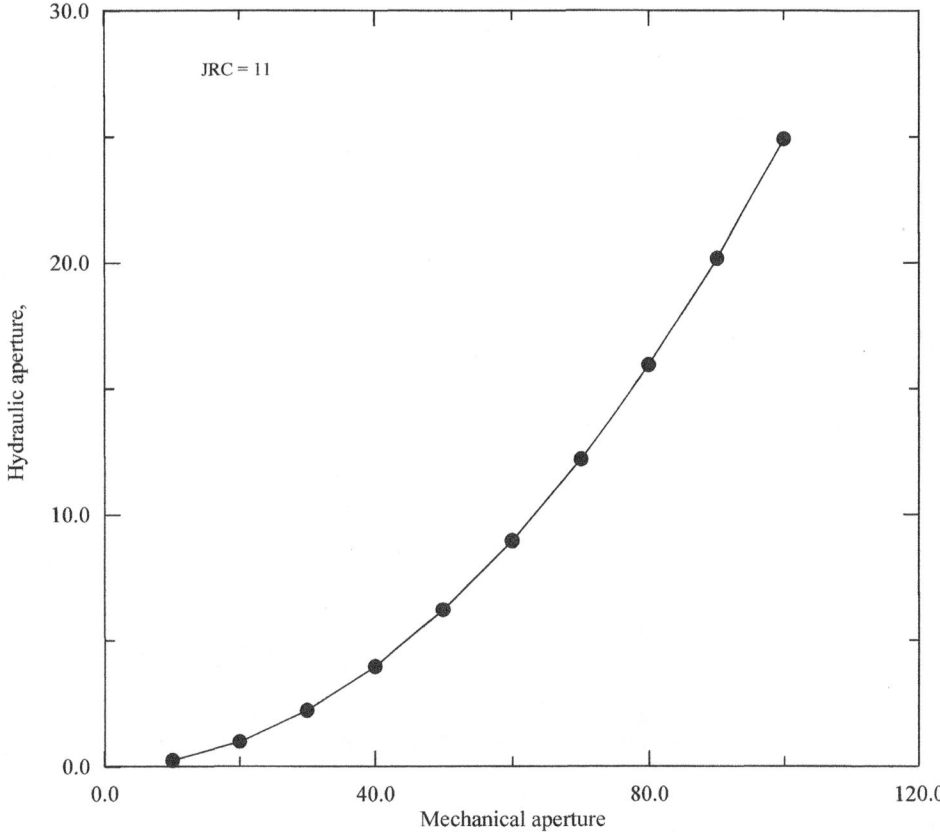

Figure 3.28. Relationship between hydraulic aperture and mechanical aperture.

Barton et al. (1985) suggest that for a range of rock types, the mechanical aperture tends to become equal to the hydraulic aperture for smooth joint walls having relatively large apertures. In contrast, when the roughness increases, the mechanical aperture is as high as seven times the hydraulic aperture. In the case of $e_h = e_m$, $e_h = 0.5e_m$ and $e_h = 0.1e_m$, Fig. 3.29 shows the variation of hydraulic aperture against *JRC*. It is clear that at low levels of *JRC*, the increase in hydraulic aperture is near-linear, whereas at increased values of *JRC*, the hydraulic aperture–*JRC* relationship becomes increasingly non-linear.

3.3.3   Effect of surface roughness on fluid flow

Discontinuity properties such as joint openings, roughness, dip direction and infill material influence the behaviour of a rock mass. During the past two to three decades, more focus has been given to surface roughness estimation by various researches (Patton, 1966; Barton, 1973; Barton & Choubey 1977; Brown & Scholz, 1985; Xie & Pariseau, 1995; Kwasniewski & Wang, 1997; Barton & Quadros, 1997). This is because, the joint roughness directly affects the fluid flow

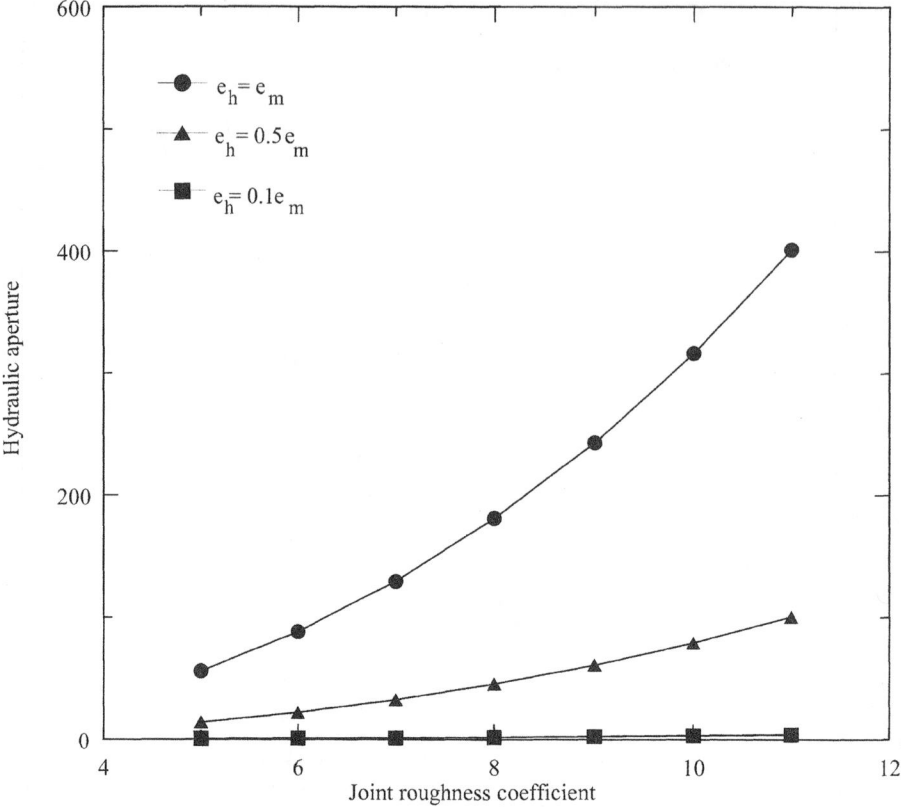

Figure 3.29. Change of hydraulic aperture against *JRC*.

characteristics as well as the shear strength of rock mass. Roughness is defined as the surface waviness (i.e., large scale undulations) and unevenness (i.e., small scale undulations) of a given discontinuity relative to its mean plane. Measurement of geometrical roughness of joint is often represented by the joint roughness coefficient, which is purely a numerical index, but not the effective frictional angle of the joint roughness. Factors such as mineral properties of rocks and mode of fracture inception (e.g. attributed to tensile stresses) govern the magnitude of surface roughness of rock joints. The effect of roughness varies with the change of joint aperture, thickness of infill material and the relative displacement of joint planes. Also, in the presence of water, the shear strength of rough joints diminishes, which results in a reduced joint roughness. ISRM commission (1978) has quantified the degree of roughness, based on unevenness and waviness of joints, as shown in Fig. 3.30. Rough joints are normally classified as planes having surface irregularities with a wavelength of less than 100 mm (Priest, 1993).

Barton (1973) introduced the term 'Joint Roughness Coefficient' (*JRC*) to express the roughness of a joint, which usually varies between 0 and 20. The

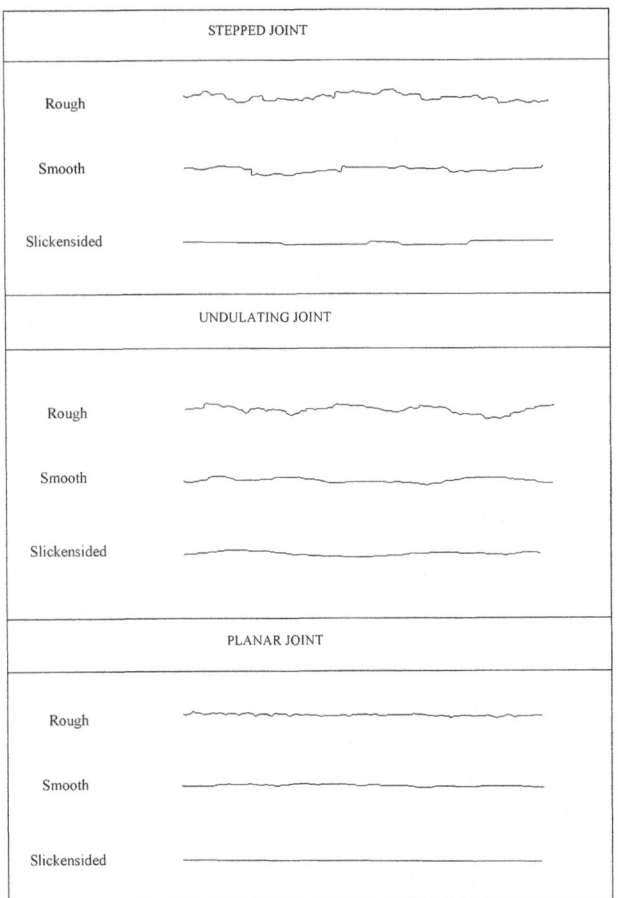

Figure 3.30. Varying degree of roughness (suggested by ISRM, 1978).

higher the *JRC* value, greater the surface irregularities. For an ideally planar (smooth) joint surface, the *JRC* value can be assumed to be zero. Because of irregularities (asperities), a rough joint shows a higher friction angle than a smooth joint for the same rock type. The increased friction angle associated with the surface roughness is called the 'effective roughness angle' whose magnitude depends on the surface profile (e.g. geometry of asperities), contact area and the degree of indentation (Schneider, 1974). Different methods of measuring surface roughness have been proposed by various researches, including Fecker and Rengers (1971), Barton (1973) and Weissbach (1978). In order to study the effect of roughness on fluid flow through jointed rocks and mechanical behaviour of rock mass, the first step is to map the correct surface texture of discontinuities. For simplicity, some researches have idealised the rock joint surfaces profiles as saw tooth profiles or sinusoidal profiles. In reality, one may question whether such idealised profiles behave as natural joints or not, and if not, how accurate the subsequent analysis will be. The following section describes different techniques of mapping surfaces.

Mapping surface profiles

Surface texture is perhaps the most important variable to governing the magnitude of friction, which controls the relative motion between surfaces and the fluid flow. Although the investigation of surface texture is of paramount importance in a variety of engineering applications, the discussion presented here is confined to the surface of jointed rocks only. Particularly in mining and rock engineering, the surface texture of rocks significantly influences the stability of mine roofs, jointed slopes and fluid flow through discontinuities. Techniques of modelling joint surfaces generally adopt one or more of the following alternatives:

(1) Graphical representation of the surface profile;

(2) Mathematical representation of the mapped surface profile; and

(3) Statistical features.

Various techniques have been used in the past to obtain the graphical representation of surface profiles in macroscopic and microscopic levels. On the basis of measurement options, such methods can be divided into four categories: (a) mechanical, (b) hydraulic, (c) optical and (d) laser techniques. Figure 3.31 describes different types of surface profilometers, on the basis of principal modes of measurement. In the past, mechanical methods such as brush gauges (Barton, 1973) were used; however, the laser and optical techniques have currently become the most popular, because of both convenience and accuracy. Some digitised rock surfaces mapped using the digital co-ordinates equipment are presented here.

All granite rocks samples tested by the authors were natural, and they were obtained at different depths from the Appin Mine, NSW, Australia. All 55 mm diameter cores used for testing contained a single fracture. The measurements were made using the mechanical profilometer, which could digitise the joint surface in three dimensions by taking a series of parallel lines along the joint, where the X–Y–Z system of axes is illustrated in Fig. 3.32. A typical mechanical profilometer

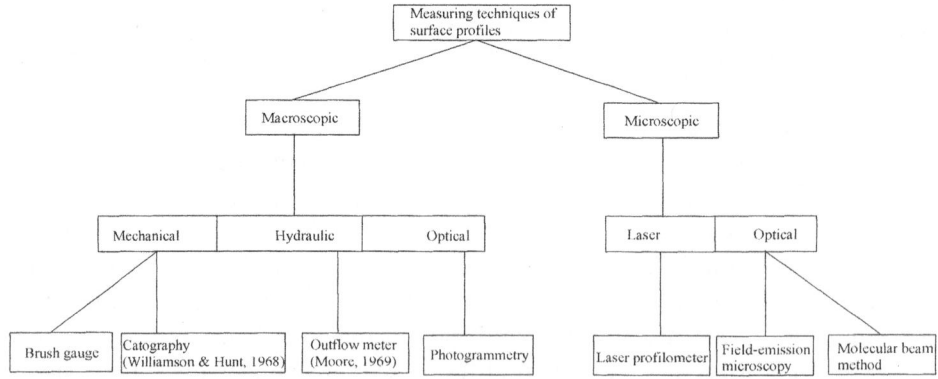

Figure 3.31. Available profilometers based on modes of measurement.

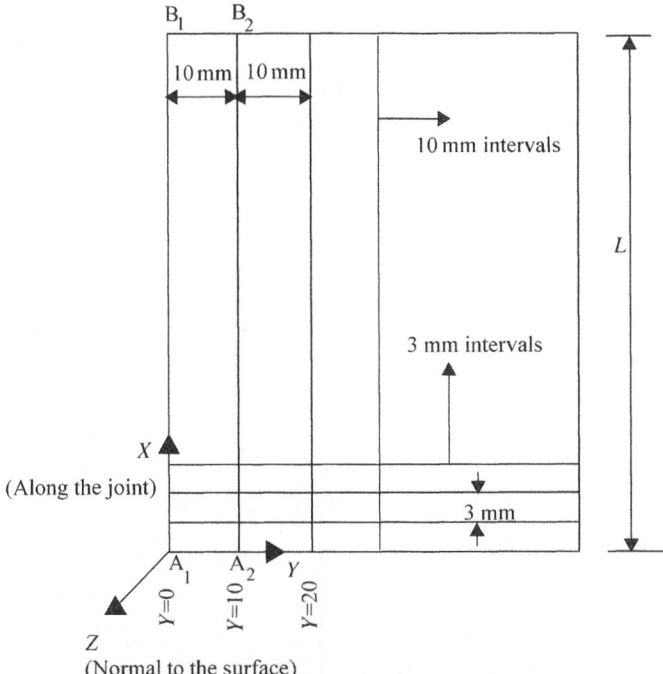

Figure 3.32. Plan view of selected grids for joint surface.

is shown in Fig. 3.33. In the course of measurements, the digital pointer is lowered to the surface to obtain *x*, *y* and *z* coordinates, and by moving the pointer systematically, many points of the surface are mapped. The readings of each point are automatically analysed by the computer to map the joint surface. Typical surface profiles of jointed cores that are mapped using this method are shown in Fig. 3.34.

The individual joint profiles along each line segment (e.g., A_1B_1 and A_2B_2 in Fig. 3.32) are also plotted in order to estimate the *JRC* values for each profile (Fig. 3.35a–c). This enables a better representation of the joint geometry. However, the procedures discussed above are more suitable for mapping profiles in laboratory conditions. For in-situ measurement, the simple string-line method is used to map the joint profiles as shown in Fig. 3.36. In this method a string is placed approximately parallel to the general surface of joint, and at certain intervals (*x*), the perpendicular distance from the line to the fracture surface is recorded. Depending upon the degree of irregularities, the number of measurements can be increased or decreased.

Measurement of surface roughness coefficient
There are several techniques that can be used to measure the surface roughness of a joint to quantify the Joint Roughness Coefficient, which include:
(1) Visual comparison of joint profiles with the standard profiles (Fig. 3.37) proposed by Barton (1973);

(2) Tilt tests to determine the basic friction angle;
(3) Use of maximum roughness amplitude, r_a;
(4) Calibration with the mechanical and hydraulic joint apertures;
(5) Analytical approach based on mathematical functions; and
(6) Fractal models.

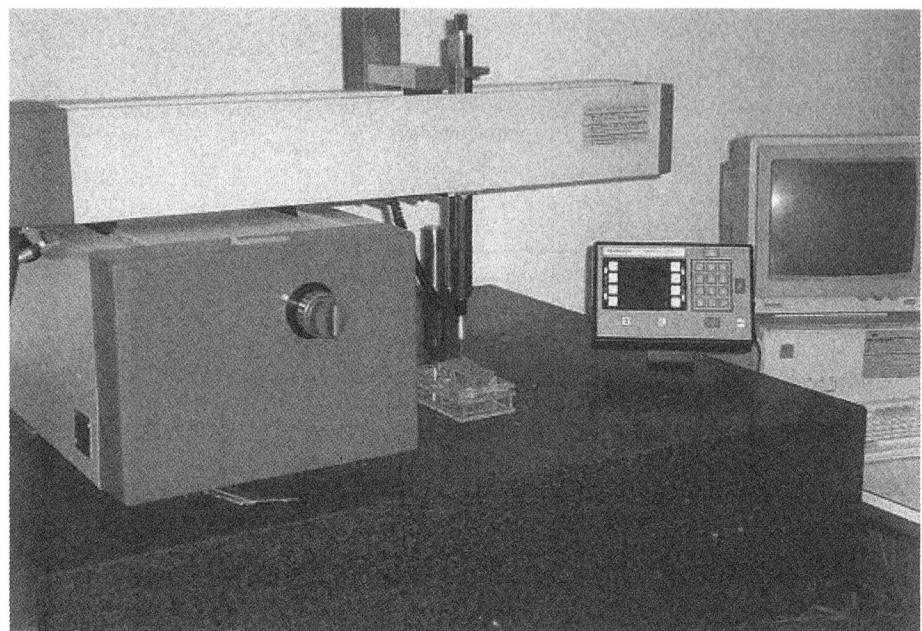

Figure 3.33. Sample placed in the profilometer (University of Wollongong)

(a)

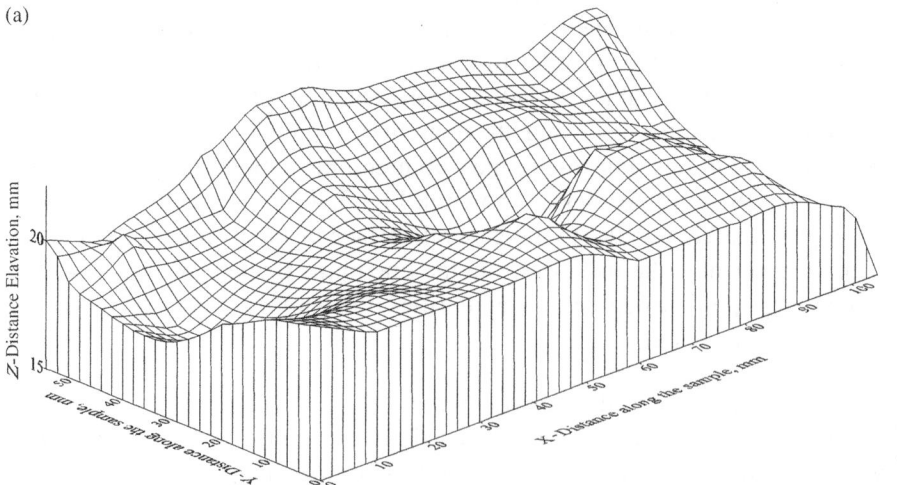

Figure 3.34. Joint profile of (a) specimen IF 16, (b) specimen IF 17 and (c) specimen IF 20.

(b)

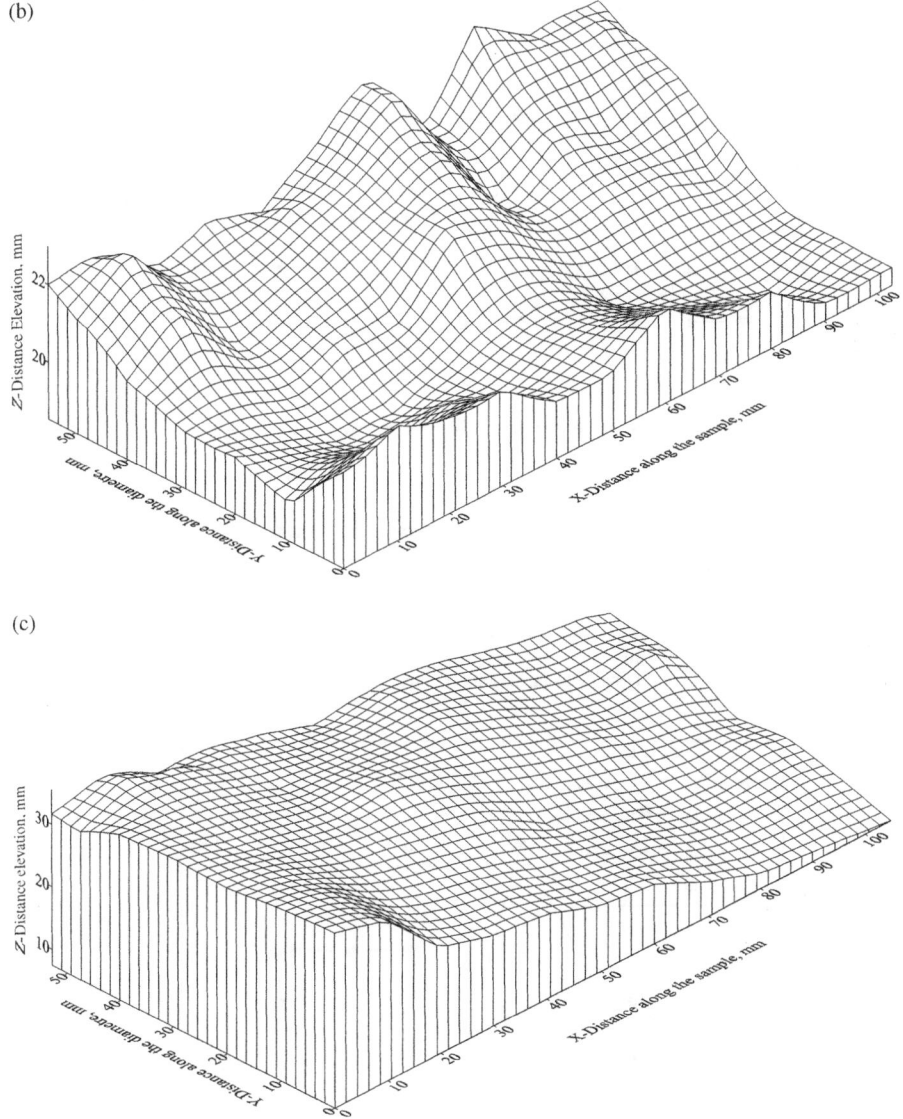

(c)

Figure 3.34. (continued)

Matching of joint profiles with standard profiles (Barton, 1973). For preliminary design, a quick estimation of *JRC* can be carried out by matching the joint surfaces with the standard profiles as proposed by Barton (1973). The standard profiles
and the *JRC* values are given in Fig. 3.37. The magnitude of *JRC* varies usually
from 0 to 20; however in fractal dimension analysis, some joint profiles are characterised by *JRC* values greater than 20 (Carr, 1989). For a smooth planar joint, *JRC*
tends to become zero, whereas for highly irregular surface, *JRC* values tend to
approach 16–20. This method of roughness determination has been supported by

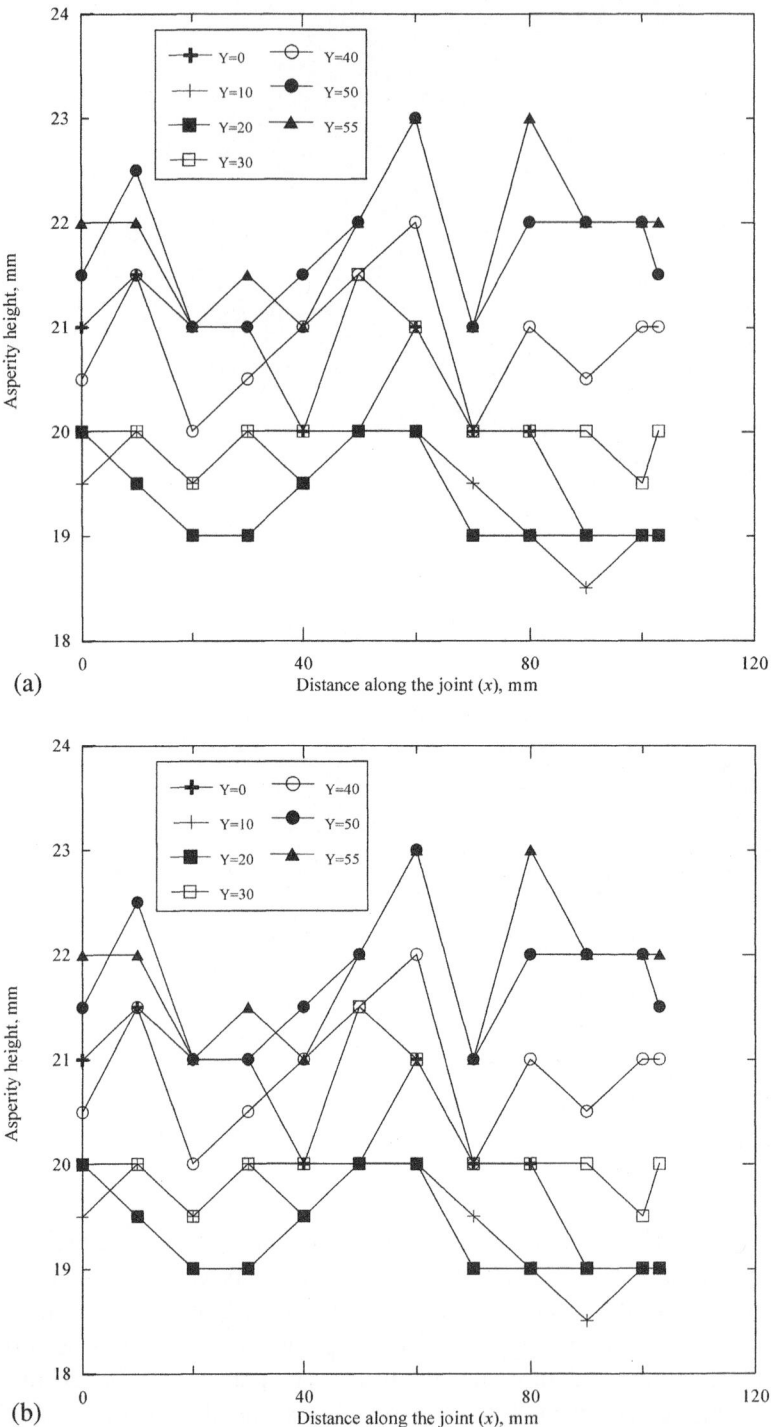

Figure 3.35. Individual profiles for (a) specimen IF 16, (b) specimen IF 17 and (c) specimen IF 20.

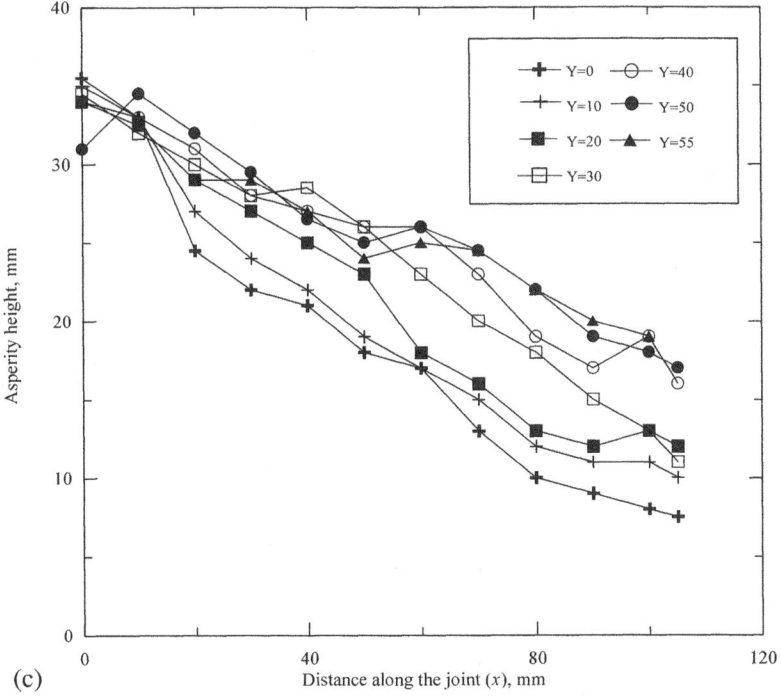

(c)

Figure 3.35. (continued)

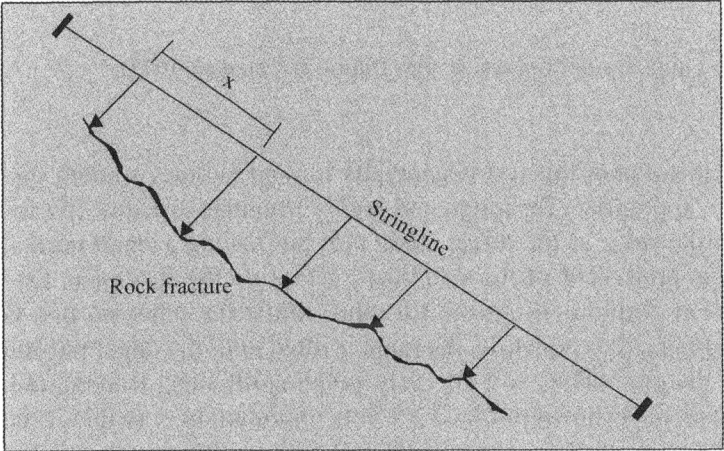

Figure 3.36. Mapping of fracture profile in the field (e.g. tunnel roof).

ISRM (1978), and due to its practicality, Barton's method is probably one of the most widely used in preliminary design in jointed rock strata. In order to use this method, the measured profiles (Fig. 3.35a–c) need to be drawn to the standard scale, and subsequently, these scaled profiles are compared with the standard profiles. The predicted values are tabulated in Table 3.5, for two granite specimens.

Description of joint	Standard joint profiles	JRC range
Smooth, planar: cleavage joints		0–2
Smooth, planar: tectonic joints		2–4
Undulating, planar: foliation joints		4–6
Rough, planar: tectonic joints		6–8
Rough, planar: tectonic joints		8–10
Rough, undulating: bedding joints		10–12
Rough, undulating: tectonic joints		12–14
Rough, undulating: relief joints		14–16
Rough, irregular: bedding joints		16–18
Rough, orregular: artificial tension		18–20
	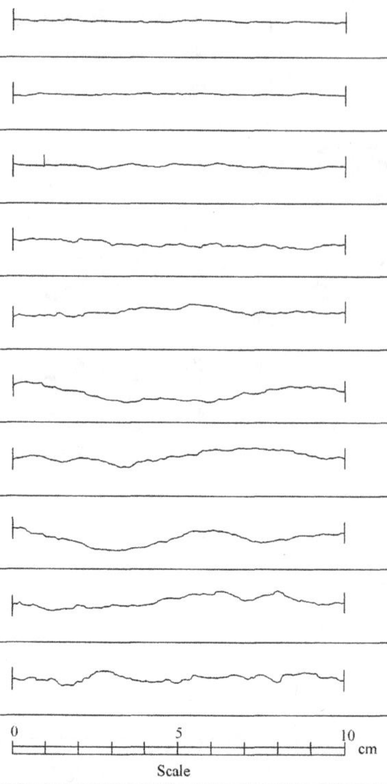 0 5 10 cm Scale	

Figure 3.37. Standard joint profiles (modified after Barton & Choubey, 1977).

Tilt tests. The application of this test is generally limited to fairly smooth surfaces, and often not appropriate for rough and highly irregular surfaces. In concept, the tilt test is the same as the direct shear test, but having a small normal stress (weight of the upper half of the specimen) acting on the specimen. One portion of the joint is firmly held on the tilt table, while the other portion is placed on the joint plane. Subsequently, the table is tilted until the upper portion starts to slide over the joint plane, and the corresponding tilt angel is measured. A typical test equipment is shown in Fig. 3.38. It is important to note that if the upper portion of the joint starts to overturn, then the tilt method is not suitable to measure *JRC*. This can occur when testing rough and highly irregular joint surfaces. Using the tilt test, the *JRC* value is estimated using the following equation:

$$JRC = \frac{\alpha - \phi}{\log\left(\dfrac{JCS}{\sigma_{n0}}\right)} \tag{3.40a}$$

Table 3.5. Estimation of *JRC* using standard profiles.

Profile location (mm)	*JRC* based on standard profiles	Scaled-joint profiles
Specimen 1 (IF16)		
Y = 0	1	
Y = 10	5	
Y = 20	7	
Y = 30	13	
Y = 40	7	
Y = 50	5	
Y = 55	5	
Mean *JRC*	6.2	
Specimen 2 (IF17)		
Y = 0	7	
Y = 10	6	
Y = 20	5	
Y = 30	8	
Y = 40	9	
Y = 50	10	
Y = 55	11	
Mean *JRC*	8	
Specimen 3 (IF20)		
Y = 0	11	
Y = 10	8	
Y = 20	9	
Y = 30	7	
Y = 40	8	
Y = 50	4	
Y = 55	8	
Mean *JRC*	7.8	

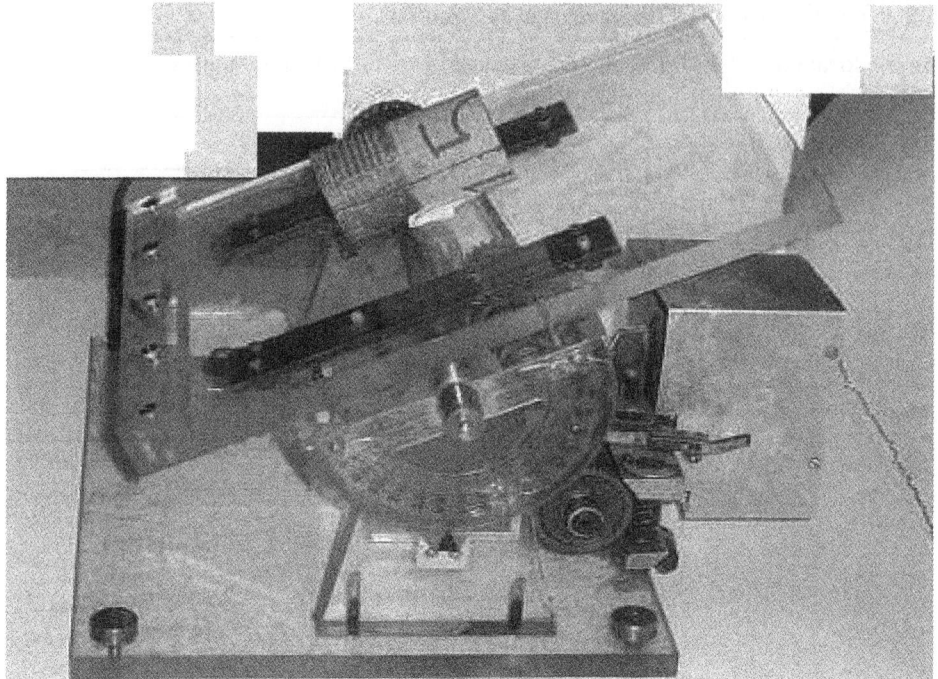

Figure 3.38. Photograph of the tilt test apparatus.

where α = tilt angle measured from the tilt test, ϕ_r = residual friction angle, σ_{n0} = effective normal stress on the joint and JCS = joint compressive strength which is usually measured by Schmidt Hammer Index Test. Miller (1965) discussed the suitability of Schmidt Hammer for recording the unconfined compressive strength of rock. Based on the following relationship between rebound number and the compressive strength, Miller produced a useful graph to estimate the compressive strength for rocks of different density (Fig. 3.39). The relationship between JCS and rebound number is given by

$$\log_{10}(JCS) = 0.00088\gamma R + 1.01 \tag{3.40b}$$

where γ = density of rock (kN/m^3) and R = rebound hammer number.

For a quick estimation of JCS, ISRM (1978) has provided a classification of soft rock and hard rock, based on JCS, as given in Table 3.6. Moreover, Barton and Choubey (1977) proposed a simple relationship to estimate the value of ϕ_r based on the basic friction angle, as given below:

$$\phi_r = (\phi_b - 20°) + 20(r/R) \tag{3.40c}$$

where ϕ_b = basic friction angle estimated from the residual tilt test, r = Schmidt rebound on wet joint and R = Schmidt rebound on dry joint surface.

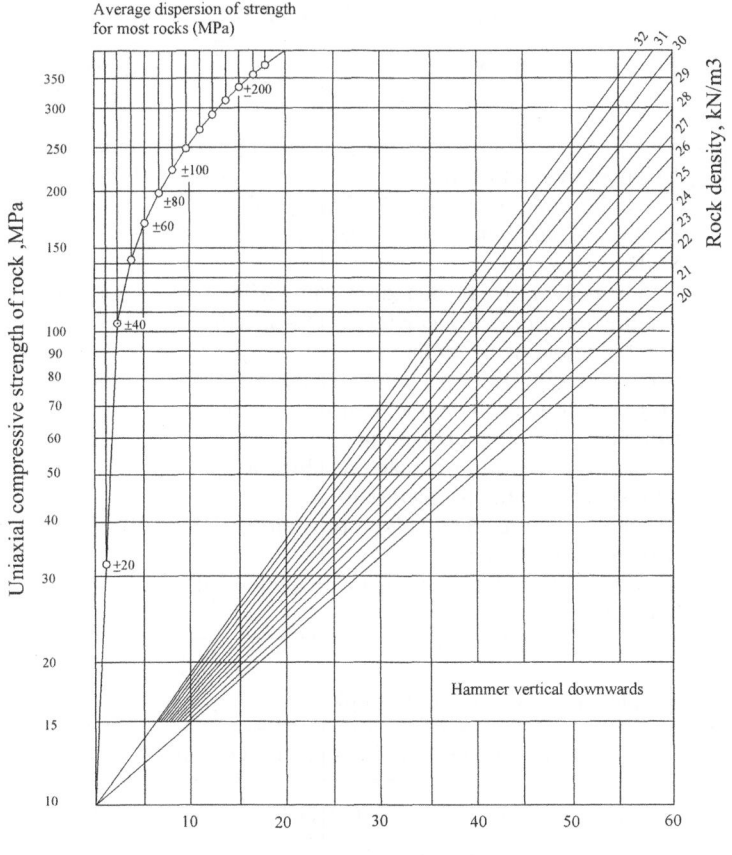

Figure 3.39. Estimation of *JCS* based on rebound hammer (Barton, 1973).

Maximum roughness amplitude (r_a).   Barton (1982) suggested an alternative empirical approach to estimate *JRC* based on the maximum amplitude (*a*) of a given profile length (*L*). On a 100 mm scale, an empirical relationship to estimate *JRC* is defined by

$$JRC = 400 \frac{a}{L} \tag{3.41}$$

For given three specimens (Fig. 3.34a–c), *JRC* values computed using Eqn. 3.41 are listed in Table 3.7. The computed *JRC* values vary from 7.0 to 9.3 for all these granite specimens. Barton and Choubey (1977) report that for granite specimens, in general, *JRC* varies from 6.7 to 9.5, which indicate that the authors' results are in agreement with this methodology.

Mechanical and hydraulic joint apertures.   Although this method is theoretically sound, in practice, its application is limited. For a given mechanical aperture

Table 3.6. Classification of rock based on Joint Compressive Strength (*JCS*) (ISRM, 1978).

Description	Field identification	Approximate values of *JCS* (MPa)
Extremely weak rock	Indented by thumb nail	0.25–1.0
Very weak rock	Crumbles under firm blows with point of geological hammer, can be peeled by a pocket knife	1.0–5.0
Weak rock	Can be peeled by a pocket knife with difficulty, shallow indentations made by firm blow with point of geological hammer	5.0–25
Medium strong rock	Cannot be scraped or peeled with a pocket knife, specimen can be fractured with single rim blow of geological hammer	25–50
Strong rock	Specimen requires more than one blow of geological hammer to fracture it	50–100
Very strong rock	Specimen requires many blows of geological hammer to fracture it	100–250
Extremely strong rock	Specimen can only be chipped with geological hammer	>250

Table 3.7. Joint roughness calculated for each profile lines.

Distance along the *Y*-axis, shown in Fig. 3.32	Joint roughness coefficient based on Eqn. 3.41		
	Specimen 1 (IF16)	Specimen 2 (IF17)	Specimen 3 (IF20)
0	0	9.7	10.3
10	8.8	5.8	9.5
20	8	3.8	8.3
30	17.78	7.7	8.9
40	12.8	8.54	6.87
50	6.6	7.7	6.67
55	11.1	7.6	3.4
Average	9.3	7.3	7.7
Standard deviation	5.5	1.9	2.3

(*E*) and hydraulic aperture (*e*) of joint, *JRC* can be predicted by the following equation:

$$JRC = \sqrt[2.5]{\frac{E^2}{e}} \qquad (3.42)$$

The units of *E* and *e* are given in microns.

Empirical approach. Based on the values of joint surface parameters, Tse and Cruden (1979) proposed two linear relationships to estimate *JRC* for relatively rough joints, subjected to low normal stress, as follows:

$$JRC = 32.2 + 32.47 \log\left[\frac{1}{L}\int_{x=0}^{x=L}(dy/dx)^2\right] \tag{3.43a}$$

and

$$JRC = 37.28 + 16.58 \log\left[\int_{x=0}^{x=L}(f(x) - f(x + \Delta x))^2\right] \tag{3.43b}$$

where y = the amplitude of the roughness about the centre line and $f(x)$ = amplitude of asperity height at distance x, along the length L.

The parameter within square brackets in Eqn. 3.43a is calculated as follows:

$$Z = \frac{1}{L}\int_{x=0}^{x=L}(dy/dx)^2 = \left[\left(\frac{1}{M(dx)^2}\right)\left(\sum_{i=1}^{M}(y_{i+1} - y_i)^2\right)\right]^{0.5} \tag{3.43c}$$

where M = number of intervals over the length of the joint and dx = interval length.

Based on Eqns. 3.43a and 3.43c, the calculated *JRC* values are listed in Table 3.8. When the logarithm of Z is higher than -1, the estimated value of *JRC* takes negative values. This occurs when the asperity height difference between the adjacent intervals is small. In other words, for relatively smooth joints, the validity of the equation is uncertain.

Fractal model for roughness estimation. The fractal dimension method was originally developed in areas of physics and mathematics, and is now applied to other fields including discontinuous rock. In recent years, several studies (Carr & Warriner, 1987; Carr, 1989; Xie & Pariseau, 1995; Kwasniewski & Wang, 1997) have attempted to describe joint roughness using fractal dimensions. Mandelbrot (1967) first applied the fractal dimension concept for measuring the length of British coastline. This method is applied to describe precisely, the geometry of an irregular object using fractional numbers. According to the definition given by Mandelbrot (1982), fractal is a 'shape made of parts similar to the whole shape in some way. It can also be described as an object, which is highly irregular or possesses roughness than the objects considered' (Xie & Pariseau, 1995). The fractal object can be a line, surface or volume. Basically, there are two methods of applying fractal dimensions for evaluating joint roughness, as follows:

(1) Spectral method, which was initially developed by Berry and Lewis (1980) and applied to rock surfaces by Brown and Scholz (1985);
(2) Divider method, based on the procedure suggested by Mandelbrot (1967).

Here, the fractal dimension is simply explained considering a line, in which the length of the line is estimated accurately using the divider method. As indicated in Fig. 3.40, the length of the irregular line (L) can be expressed in terms of step

Table 3.8. Estimated *JRC* based on Tse and Cruden's approach for specimens shown in Fig. 3.34a–c (Tse & Cruden, 1979).

Joint profile location (Fig. 3.32)	Log Z	Estimated *JRC*
Specimen 1 (IF 16)		
Y = 0		Not defined
Y = 10	−1.07	−2.78
Y = 20	−1.21	−7.1856
Y = 30	−0.9	2.7
Y = 40	−1.07	−2.8
Y = 50	−1.06	−2.5
Y = 55	−1.3	−10
Specimen 2 (IF 17)		
Y = 0	−1.18	−6.1
Y = 10	−1.26	−8.8
Y = 20	−1.28	−9.4
Y = 30	−1.18	−6.1
Y = 40	−1.02	−0.9
Y = 50	−1.01	−0.7
Y = 55	−0.97	0.6
Specimen 3 (IF 20)		
Y = 0	−1.39	−13
Y = 10	−1.39	−12.9
Y = 20	−1.35	−11.8
Y = 30	−1.39	−13
Y = 40	−1.27	−9.1
Y = 50	−1.25	−8.4
Y = 55	−1.81	−16.8

Figure 3.40. Stepwise measurement of joint profile.

length (x), whereby

$$L = \sum_{i=0}^{m} x_i \qquad (3.44a)$$

or

$$L = nx \qquad (3.44b)$$

where n number of steps.

The accuracy of the total length can be increased by selecting a smaller step length. According to Mandelbrot (1967), the fractal dimension (D) is introduced to Eqn. 3.44b such that

$$L = nx^D \qquad (3.44c)$$

For a example, D tends to become 1 for a perfect line, and 2 for a perfect plane. These integrals such as, 1, 2, 3 are called ideal dimensions, because in practice, no perfect line or plane or volume exits. Therefore, D lies intermediate between the ideal values. Because in rock mechanics, the joint roughness is usually expressed in terms of *JRC* (Barton, 1973), it is important to express the fractal dimensions using *JRC*. An empirical relationship between *JRC* and fractal dimension (D) has been proposed by Carr and Wariner (1987):

$$JRC = -1022.55 + 1023.92D \qquad (3.45)$$

Surface roughness on fluid flow
Analytical, experimental and numerical studies have shown that the joint roughness and aperture are the most important factors governing fluid flow through a simple joint or a network of joints. In this section, more attention is given to the effects of roughness on permeability characteristics of jointed rock mass. It is expected that a minimum fluid flow through a fracture occurs even at very high normal stress levels. Corresponding to this minimum flow, there is a nominal aperture characterising every rough joint. This nominal aperture is referred to as the residual aperture. The joint roughness can be incorporated into fluid flow equations via
(1) Mechanical joint roughness coefficient (*JRC*);
(2) Relative (hydraulic) joint roughness coefficient (*F*); and
(3) Mathematical function describing the joint surface.

The estimation of Joint Roughness Coefficient (*JRC*) was discussed in detail in the previous section. The relative joint roughness is a function of the hydraulic aperture and the asperity height of the joint surface. Lomize (1951), Louis (1969) and Quadros (1982) have described the effect of surface roughness on conductivity, in terms of relative joint roughness. The relative roughness (F) is defined as the ratio of the difference between maximum and minimum asperity height (k) divided by twice the hydraulic diameter, which is usually determined by hydraulic tests, and back-calculated using the cubic law.

$$F = \frac{k}{2e} \qquad (3.46)$$

Depending on the magnitude of F, Louis (1969) found experimentally that the flow within joints could either be parallel or non-parallel. When $F \leqslant 0.033$, the flow is assumed to be parallel, which is probably expected when $JRC < 8$. Depending on the viscosity of fluid, joint aperture and velocity of fluid within a joint, the flow may be either laminar or turbulent. The flow pattern can be

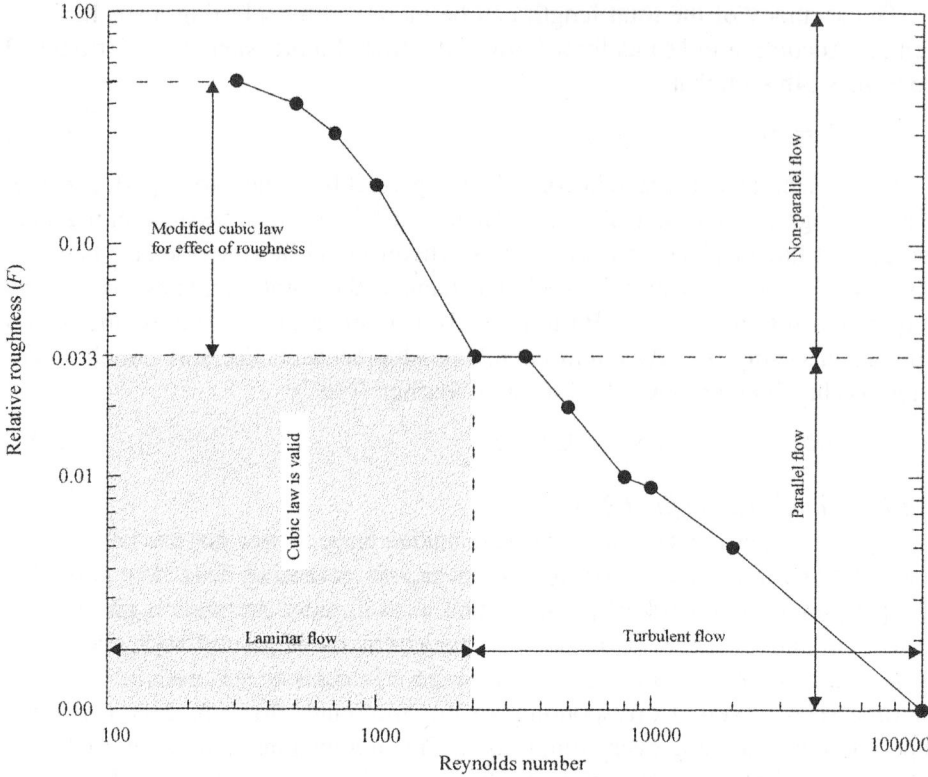

Figure 3.41. Flow pattern in a joint depending on relative roughness and Reynolds number (modified after Louis, 1976).

established considering the Reynolds number. If the Reynolds number is found to be more than 2000, the flow will most probably be turbulent. Based on the value of relative roughness and the Reynolds number, Fig. 3.41 shows possible types of flow within a single joint (Louis, 1976).

Role of relative roughness coefficient in flow equations
Effect of topology on fluid flow through a fracture was established through the pioneering efforts of Lomize (1951). For a smooth planar joint, the pressure drop coefficient (λ) depends on the Reynolds number (Re) only, such that

$$\lambda = \frac{96}{Re} \tag{3.47}$$

The well-known cubic law for planar smooth joint is based on the pressure drop coefficient (Eqn. 3.47), and it is given by

$$q = \frac{be^3}{12\mu}\left(\frac{\mathrm{d}p}{\mathrm{d}x}\right) \tag{3.48}$$

where b is the width of the joint, dp is the pressure variation along length of dx, e is the joint aperture and μ is the dynamic viscosity of fluid.

For an irregular joint, the pressure drop is expected to increase by a factor (C) due to roughness of the joint wall, as represented by

$$C = \left(1 + mF^{1.5}\right) \tag{3.49}$$

The magnitude of m has been evaluated by several researches (Lomize, 1951; Louis, 1969; Quadros, 1982) for various rough surfaces, where m varies from 8.8 to 20.5.

In general, for a rough joint, the pressure drop coefficient takes the following form:

$$\lambda = \frac{96}{Re}\left(1 + mF^{1.5}\right) \tag{3.50}$$

For a given joint aperture, the variation of C with the difference in asperity heights (k) between the maximum and minimum of joint surface is shown in Fig. 3.42.

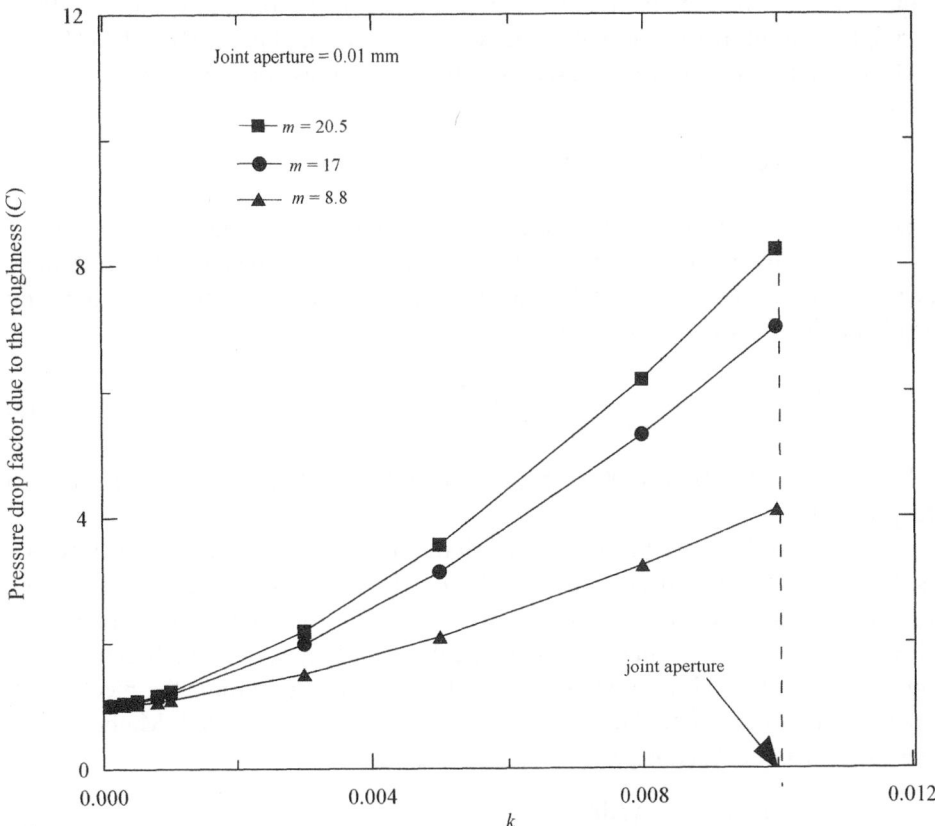

Figure 3.42. Effect of roughness on pressure drop coefficient.

When k approaches the magnitude of the joint aperture (e), the factor (C) associated with roughness becomes as high as 8. As seen in Fig. 3.42, a large pressure drop coefficient (λ) is expected when $m = 20.5$, whereas the least value of λ is given for $m = 8.8$.

By combining Eqn. 3.50 with Darcy's law, the effect of roughness is introduced into the flow equation, and represented by

$$q = \frac{be^3}{12\mu}\left(\frac{1}{1 + mF^{1.5}}\right)\left(\frac{dp}{dx}\right) \tag{3.51}$$

where b is the width of the joint, dp is the pressure variation along length of dx, μ is the dynamic viscosity of fluid, e is the joint aperture and F is relative joint roughness coefficient.

Role of JRC in flow equation

In the previous section, the introduction of relative roughness into fluid flow equations was discussed. The aim of this section is to look at how *JRC* can be coupled with the governing flow equations. Assuming the parameter a (see Eqn. 3.41) approximately equals k (see Eqn. 3.46), Barton and Quadros (1997) arrived at the following correlation between relative roughness (F) and *JRC*.

Combining Eqns. 3.41 and 3.42, the following expression is obtained:

$$F = \left(\frac{L}{800}\right)\left(\frac{JRC^{3.5}}{E^2}\right) \tag{3.52}$$

where E is the mechanical aperture of the joint and L is the length of the joint.

Instead of the relative roughness (F) in Eqn. 3.51, it is beneficial to introduce *JRC* because of the practical relevance of *JRC* as discussed earlier. Incorporating *JRC* in to the flow equation yields

$$q = \frac{be^3}{12\mu}\left(\frac{E^3}{E^3 + nL^{1.5}JRC^{5.25}}\right)\left(\frac{dp}{dx}\right) \tag{3.53}$$

where n takes 0.000751, 0.000375 and 0.000906 based on $m = 17$, 8.0 and 20.5 respectively.

Normalised flow rate versus *JRC* is plotted in Fig. 3.43. Normalised flow rate is defined as flow rate divided by $(be^3/12\mu)(dp/dx)$. Due to the increased *JRC*, the normalised flow rate is significantly decreased. Such dramatic reduction in flow rate is often questionable. In a practical point of view, one can directly incorporate *JRC* in the cubic formula using Eqn. 3.42, which requires the mechanical aperture (E) to compute the flow rate for a given hydraulic head. Mathematical rearrangement between Eqns. 3.42 and 3.48 gives

$$q = \frac{b}{12\mu}\left(\frac{E^6}{JRC^{7.5}}\right)\left(\frac{dp}{dx}\right) \tag{3.55}$$

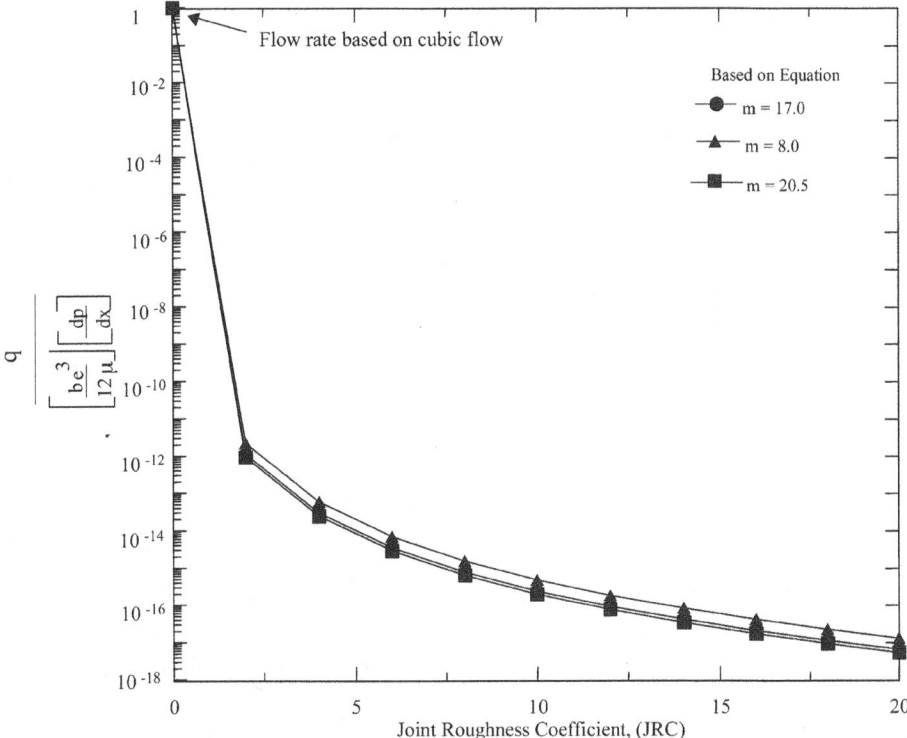

Figure 3.43. Effect of joint roughness on fluid flow rate.

The above formula is useful in a practical sense. For instance, consider the need for a quick estimate of flow rate through a fracture that is exposed at a tunnel roof. After mapping the fracture surface using the string line method, the fracture surface is matched with the standard profiles to obtain the appropriate *JRC* value. The next step involves measuring the mechanical aperture of the fracture, which can be carried out using a feeler gauge. The magnitude of the mechanical joint aperture depends on the stress redistribution due to excavation. For a given hydraulic head, it is now a matter of substituting these values into Eqn. 3.55 to compute the flow quantity.

Brown (1987) extended the results of Patir and Cheng (1978) who studied the effects of surface roughness on fluid flow, in their study dealing with hydrodynamic lubrication of rough bearings. The rock surfaces were simulated appropriately using fractal models, and the laminar flow between rough surfaces was simulated according to the Reynolds number. In this analysis, the lowest order of surface roughness is described as follows:

$$\nabla \cdot \left[\frac{d^3}{12} \nabla p \right] = 0 \qquad (3.56)$$

where d = average fluid film thickness and p = fluid pressure.

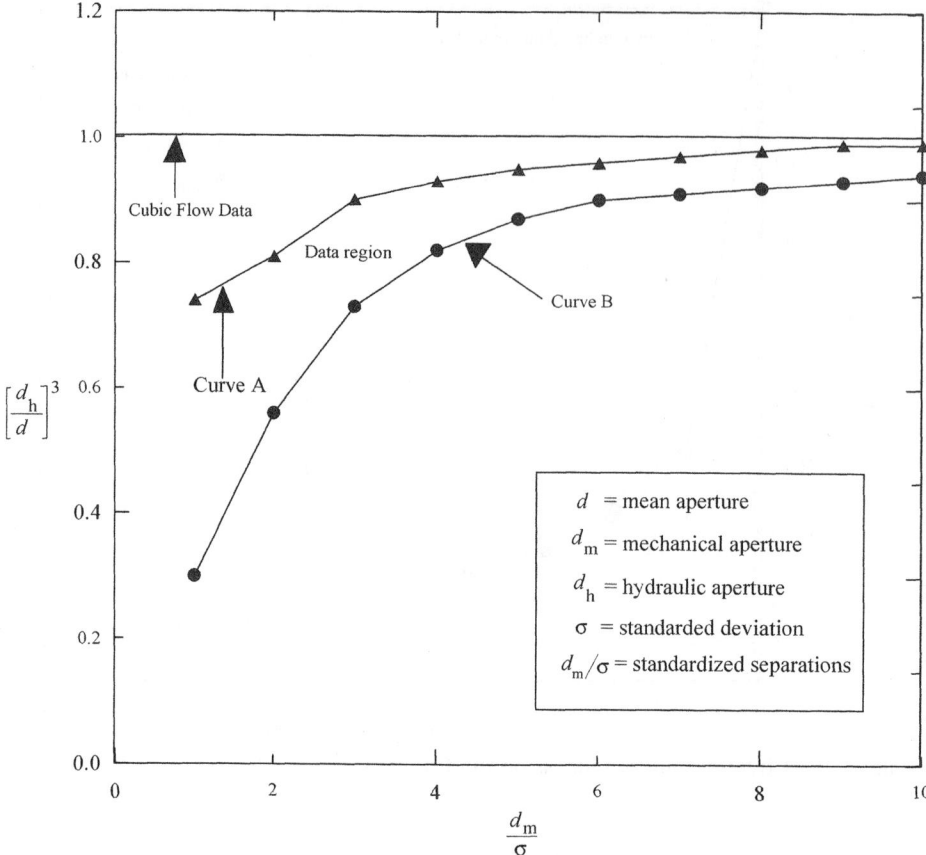

Figure 3.44. Comparison of flow rate from cubic law with flow rate computed for rough surface (data from Brown, 1987).

Assuming that the linear profiles of natural rock surfaces have power spectra of the form $G(k) \approx k^s$, in which $k = 2\,\pi/\lambda$ is the wave number, λ is the wave length and s is the slope of power spectrum, Brown (1987) generated numerical surfaces, in order to solve the Reynolds equation. Figure 3.44 shows the flow rate ratio (i.e. actual flow rate through a rough surface/cubic law flow rate) plotted against the mechanical aperture. Only the upper and lower bounds for the data provided by Brown (1987) are illustrated. Figure 3.44 reveals that for larger apertures, the deviation from the cubic theory is small. However, at much smaller apertures, the actual flow rate is 40–70% of the theoretical value predicted by the cubic law. The uncertainty of this model is attributed to the fact that Reynolds equation includes mainly the lowest order of roughness, whereas the actual joint roughness varies over a larger range. Instead of plotting the direct flow rate ratio, Brown (1987) plotted the ratio $(d_h/d)^3$ against different mechanical apertures, where d_h is the hydraulic aperture and d is the average aperture. When roughness is considered, one has to employ the correct value of λ which should be larger than that assumed in the cubic law.

3.3.4 Applicability of cubic law for natural rock joints

In the simplified form of Darcy's law, the hydraulic gradient is assumed to be linear along the fluid path, but this assumption is no longer valid when the fluid flow is non-linear. From the laboratory test results obtained for natural rock joints, it can be seen that the Darcy's law (i.e., linear relationship between the flow rate and the pressure gradient) does not hold at elevated confining pressures or at very high hydraulic gradients (Ranjith, 2000; Chitty and Blouin, 1995). Number of attempts have been made to find a relationship between the hydraulic gradient and the fluid velocity for non-linear flow (Elseworth and Doe, 1986; Witherspoon et al., 1980; Chitty and Blouin, 1995).

For linear flow,

$$v = k\left(\frac{dh}{dx}\right) \tag{3.57}$$

where v = average velocity of fluid in the discontinuity, k = permeability coefficient and dh/dx = hydraulic gradient which varies linearly along the joint length, dx. For non-linear laminar flow, Sharp and Maini (1972) described that the velocity depends not only on the hydraulic gradient, but also on the geometry of the fluid flowing surface, as represented by the following equation:

$$v = k\left[\frac{dp}{dx} - B\left(\frac{dp}{dx} - \frac{dp}{dx_{\text{lim}}}\right)^n\right] \tag{3.58}$$

where B and n = constants, which are determined empirically. The values of n and B depend on the properties of fluid and the geometry of the joint surface. Lee and Farmer (1993) have reported the following relationships for hydraulic gradient when the flow is non-linear. According to the Forchheimer law, the hydraulic gradient takes a quadratic form of velocity given by

$$\frac{dh}{dx} = (av + bv^2) \tag{3.59}$$

where v = average fluid velocity, a and b are constants. Missbach's law says that the hydraulic gradient is proportional to the power of the velocity, as follows:

$$\frac{dh}{dx} = Cv^m \tag{3.60}$$

where C = a proportionality constant and m = ranges between 1 and 2.

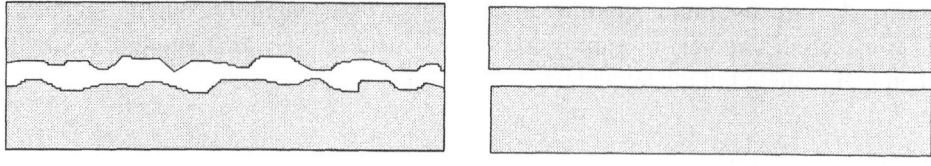

(a) Natural joint : fluid flow is not parallel
    due to rough surface

(b) Idealised joint : fluid flow is parallel
    to the smooth walls

Figure 3.45. Comparison of natural rock joint having a rough surface with an idealised joint with smooth parallel walls.

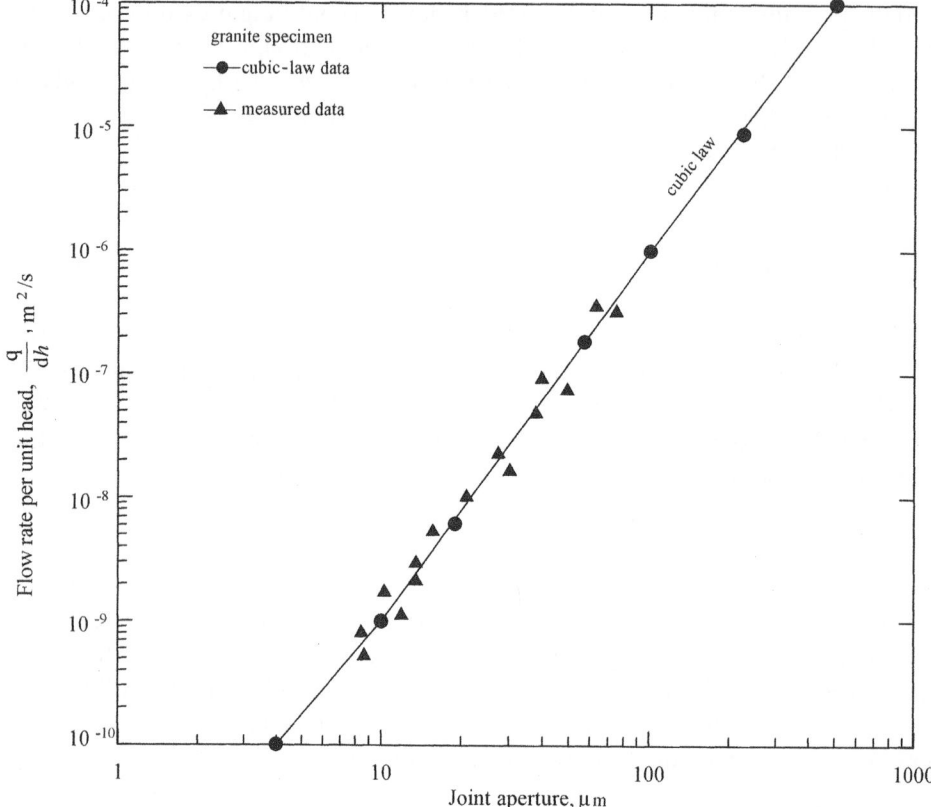

Figure 3.46. Deviation of cubic law from measured data for granite specimen (data from Witherspoon et al., 1980).

As discussed previously, natural fractures are generally rough and irregular (uneven walls) making contact at random (discrete) points. Over the past four to five decades, for simplicity, fractures have often been simulated as smooth and parallel joint walls in order to develop mathematical models and to analyse fluid flow data (Baker, 1955; Louis, 1969; Engelder & Scholz, 1981; Brown, 1987; ITASCA, 1996). Flow through a single discontinuity is often expressed in terms of the cubic law, which is based on smooth laminar flow between parallel plate

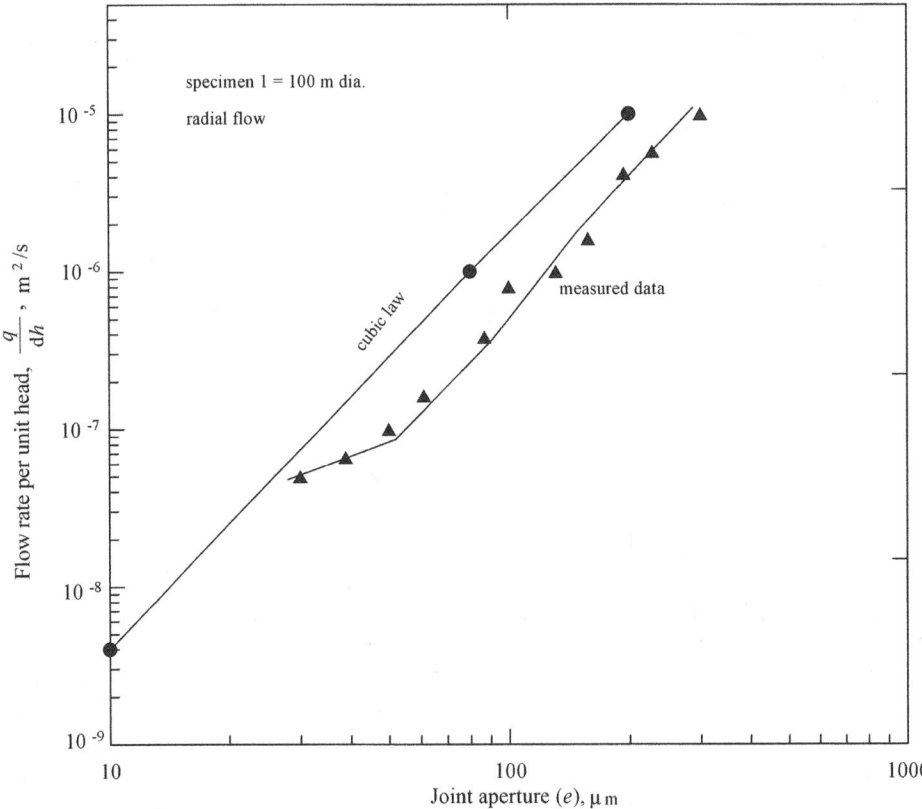

Figure 3.47. Deviation of cubic law from measured data for granite specimen 1 (data from Raven & Gale, 1985).

walls (Fig. 3.45). On the basis of cubic law, the velocity profile of fluid across the joint is simulated as a parabolic shape as elaborated later in this section. After extensive laboratory studies, Witherspoon et al. (1980) and Iwai (1976) suggested that the cubic law could still be used for rough, natural joints at low confining stress. However, at comparatively elevated stress, a significant deviation from the cubic law was noted, because of the increased contact area of the joint surface.

For steady-state flow through a parallel wall fracture with aperture (*e*), the flow rate (*q*) is usually expressed in terms of joint aperture and hydraulic head difference (d*h*) using Darcy's law for saturated, laminar and incompressible flow. In fluid mechanics, it is well known that the solution to Navier–Stokes equation for flow between parallel plates is referred to as the cubic law, which can be written as

$$\frac{q}{\mathrm{d}h} = Ce^3 \tag{3.61}$$

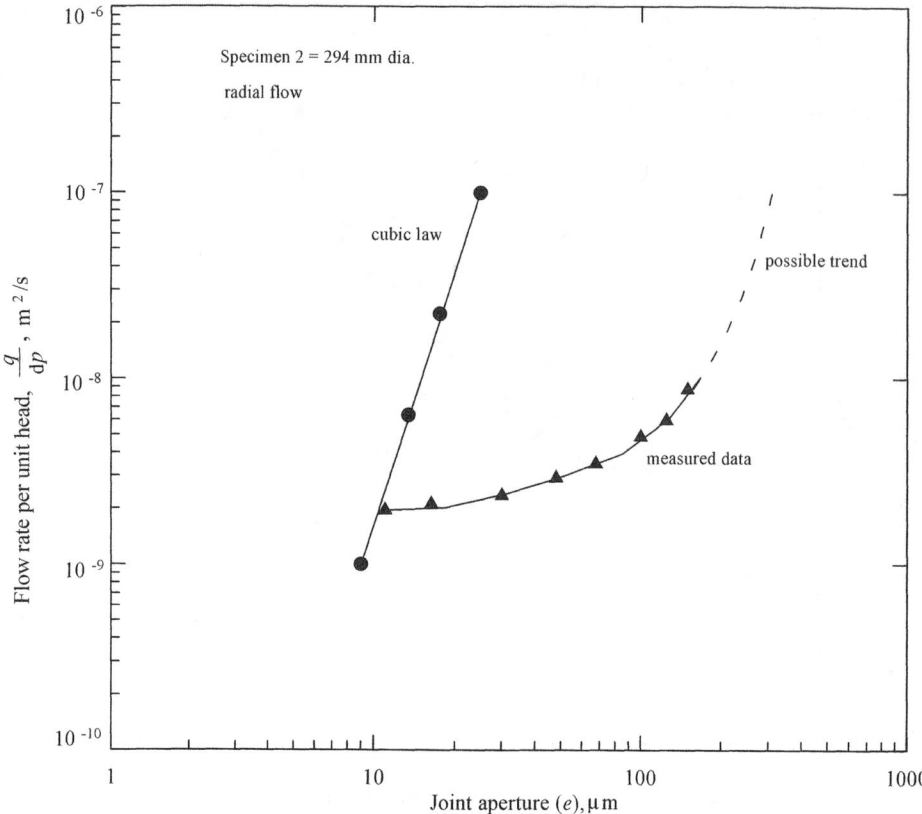

Figure 3.48. Deviation of cubic law from measured data for granite specimen 2 (data from Raven & Gale, 1985).

where C is a function of the fluid properties (e.g., Reynolds number) and the fracture length. The cubic law is sometimes referred to as Poiseuille's law or simply parallel plate theory.

For straight (planar) flow,

$$C = \frac{b}{12\mu L} \tag{3.62}$$

where b = width of the fracture, μ = dynamic viscosity of fluid and L = length of the fracture.

For radial flow, the above equation is represented by

$$C = \frac{1}{12\mu} \frac{2\pi}{\ln(r_o / r_w)} \tag{3.63}$$

where r_o is the outer radius and r_w is the well radius.

Taking the logarithm of Eqn. 3.61 yields the following:

$$\ln(q/dh) = \ln C + 3\ln e \tag{3.64}$$

The above relationship indicates that the plot of $\ln(q/\mathrm{d}h)$ against $\ln(e)$ should have a gradient of 3 with the intercept giving the value of C.

If joint aperture along the joint does not vary significantly, then the flow is well approximated by the cubic law. An attempt to investigate the validity of cubic law was carried out by Witherspoon et al. (1980) for artificial fractures created in granite, marble and basalt specimens. Under radial flow and planar flow conditions, these studies indicated that the cubic law was still reasonable for assessing flow through artificial fractures (Fig. 3.46). Regardless of the loading cycles and rock type, the cubic law could be employed to estimate flow rates. This is because artificial fractures have less surface irregularities than the natural rock fractures. The effect of roughness was found to reduce the flow rate by a factor of 1.04–1.65.

Raven and Gale (1985) studied water flow through naturally fractured granite specimens under different levels of normal stress (up to 30 MPa), and they found significant deviation from the cubic law as shown in Figs. 3.47 and 3.48.

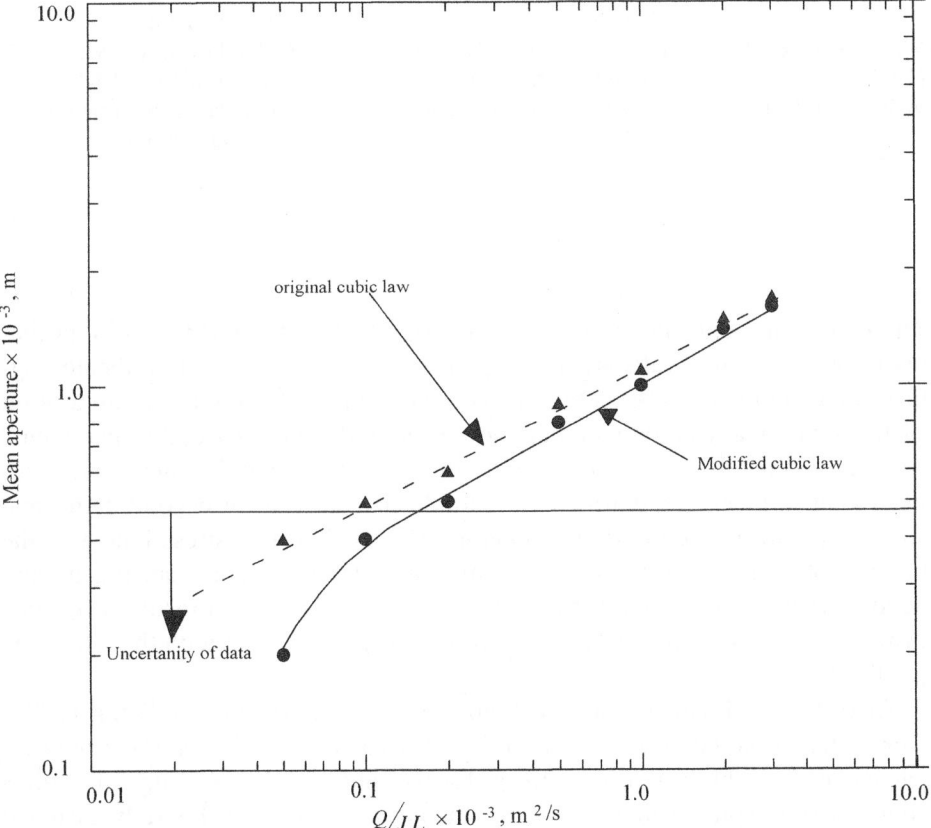

Figure 3.49. Experimental data from (Sharp, 1970) compared to modified Poiseuille equation (data from Neuzil & Tracy, 1981).

Table 3.9. Application of cubic law in research purposes and in numerical models.

Studies	Applications	Comments
Engelder and Scholz (1981)	Experimental investigation of flow through artificial fractures	Cubic law is found to be valid
Gale and Raven (1980)	Radial fluid flow through natural fractures – experimental	Depending on magnitude of roughness, sample size and repetition of load, the cubic law deviates significantly
Pyrak-Nolte et al. (1987)	Flow through natural low-permeability rocks – experimental	Cubic law is valid for effective stress below 20 Mpa
Brown (1987)	Effects of surface roughness on flow – numerical	Cubic law overestimates flow rate by 40–70%
Amadei and Illangasekare (1992)	Transient flow in single joint	Cubic law was modified to account for surface roughness
Iwai (1976)	Fundamental studies of fluid flow through a single joint	Cubic law is followed
ITASCA (1996)	Universal Distinct Element Code (UDEC)	Cubic law is used for flow-rate calculations
Witherspoon et al. (1980)	Experimental study of flow through artificial fractures	Cubic law may devite by a a factor of 1.04–1.64
Wilcock (1996)	NAPSAC fracture network code	Cubic law is assumed for flow calculations

The specimen 1 indicates less deviation from the cubic law (Fig. 3.47), probably because the joint walls are nearly planar. In contrast, Fig. 3.48 indicates significant deviation from the cubic theory, mainly due to the highly irregular joint surfaces with many random contact points along the joint. Also, the specimen 2 was approximately three times larger than the specimen 1. Therefore, apart from joint surfaces irregularities, the sample size and the corresponding normal stress could influence the flow behaviour. Elevated normal stress increases the number of contacts points between joint surfaces as well as causing greater deposition of gouge material by shearing the asperities. As a result, a reduced flow rate is expected for the same aperture, when comparing with the cubic law prediction.

Also, the work carried out by Neuzil and Tracy (1981) and Tsang (1984) shows that cubic law does not hold for flow calculations in natural fractures, particularly at high normal stress levels and for large rough fractures (Fig. 3.49). In spite of these demerits, the cubic model is still widely assumed in practical situations, because in the field, it is not feasible to incorporate the roughness of each joint in numerical models. Table 3.9 summarises some

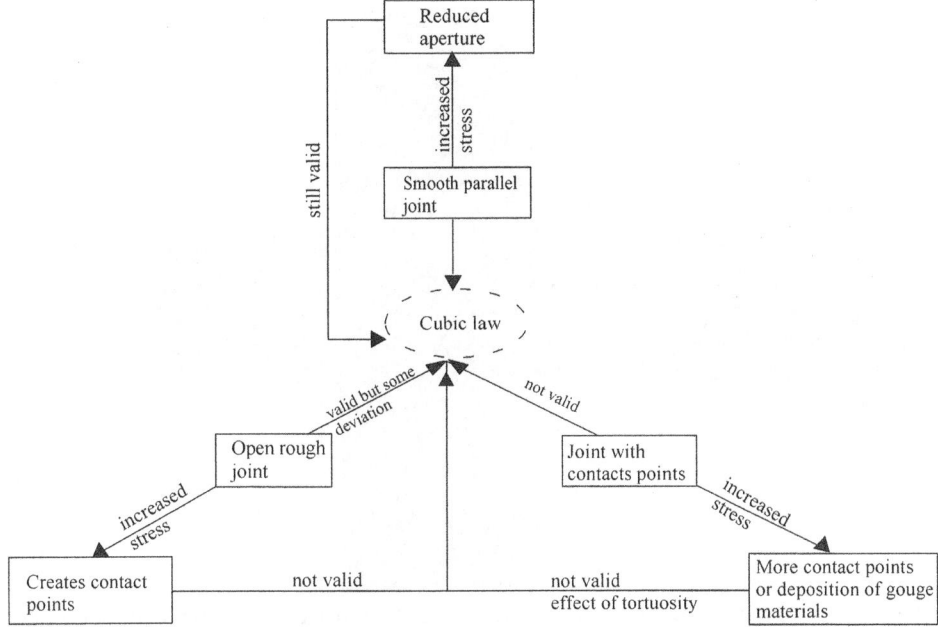

Figure 3.50. Validity of cubic law for different fractures.

applications of cubic law, including its use in commercially available numerical models.

Having identified the efforts made by past researchers, validity of the cubic law for joints with different geometry is depicted in Fig. 3.50. For instance, for open rough joints, cubic law may still hold with little deviation from the observed flow rates. However, with increase in stress, the turbulent flow may develop instead of laminar flow. Further increase in stress will result in greatly increased number of contact points, and ultimately crushing the asperities. Subsequently, the crushed material (gouge) will deposit within the joint, and under these circumstances, the cubic law cannot be assumed in flow calculations.

3.4   PERMEABILITY OF ROCKS WITH MULTIPLE FRACTURES: LABORATORY INVESTIGATION

Over the past four to five decades, considerable effort has been devoted to understand the complete flow deformation characteristics of a single fracture using experimental, analytical and numerical techniques. However, less effort has been made to study the effect of multiple fractures experimentally (Elsworth & Piggot, 1987; Shimo & Iihoshi, 1993; Ranjith, 2000). The

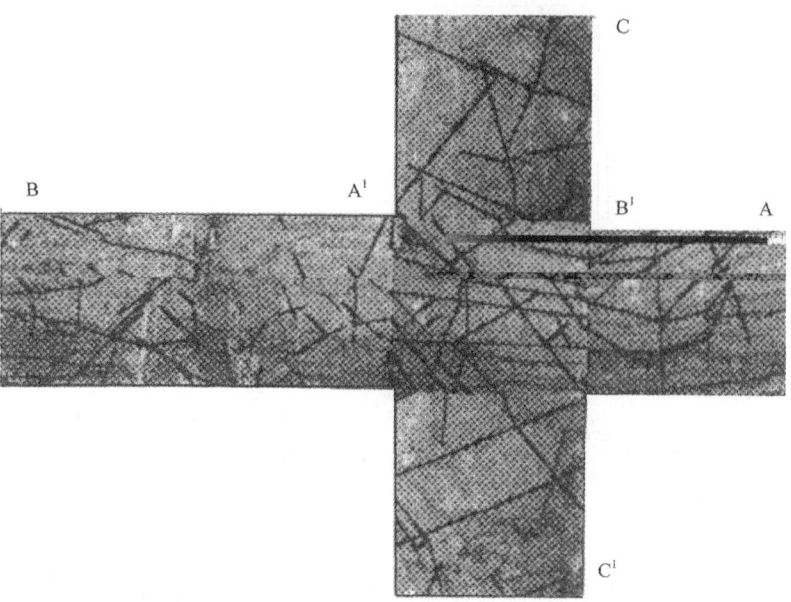

Figure 3.51. Mapped fractures on the surface of the chert rock cube (Shimo & Iihoshi, 1993).

Figure 3.52. Chert rock cube with 25 panels on each side (modified after Shimo & Iihoshi, 1993).

numerical study of flow through an interconnected fracture network is discussed in detail in Ch. 6. Shimo and Iihoshi (1993) used a large block of rock (30 cm cube) with natural fractures in their program. Figure 3.51 shows the mapped surface fractures on the chert rock cube used for testing. The hydraulic measurement of individual fluid flux or pressure can be carried out at several locations, once the fluid pressure is applied via small tubes to one side of the cube. Each side has 25 subpanels as shown in Fig. 3.52. One advantage of this test is that fluid boundaries can be effectively controlled than in a conventional triaxial testing method. However, the effects of confining pressure and axial stress, which govern fluid flow through the sample cannot be observed using this equipment. Typical values of the observed flow rate corresponding to the sub-panel number are plotted in Fig. 3.53. Different flow volumes indicate that the interconnectivity of flow paths is different even for a small block of 30 cm sides.

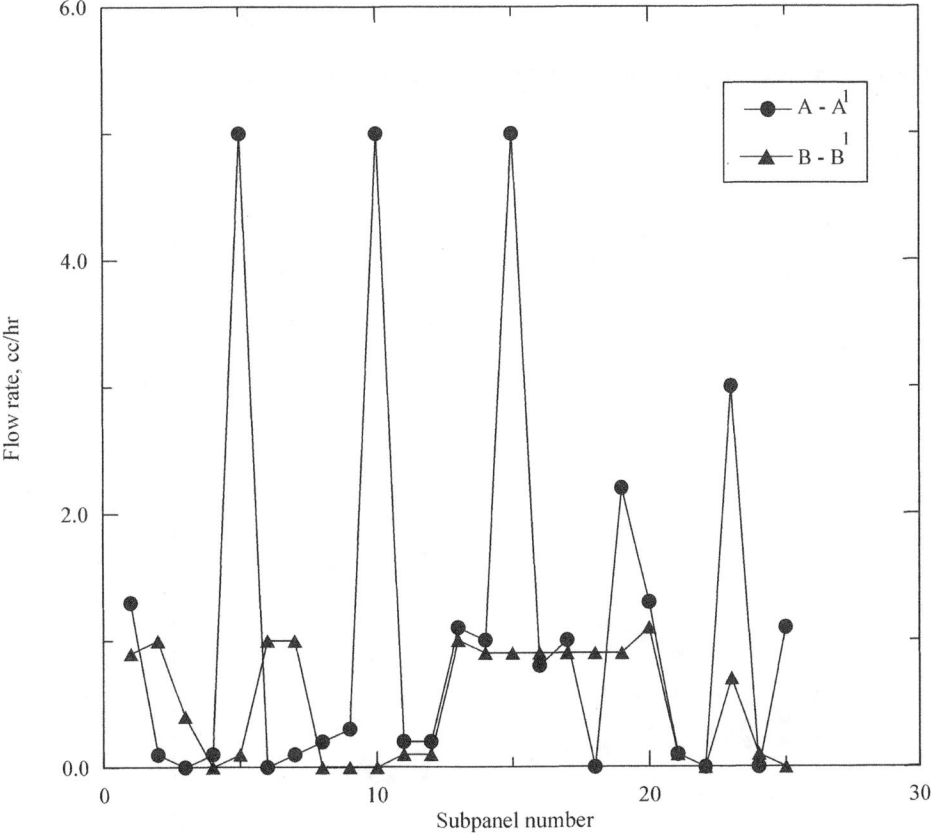

Figure 3.53. Flow rate distribution along sides of the sample (data from Shimo & Iihoshi, 1993).

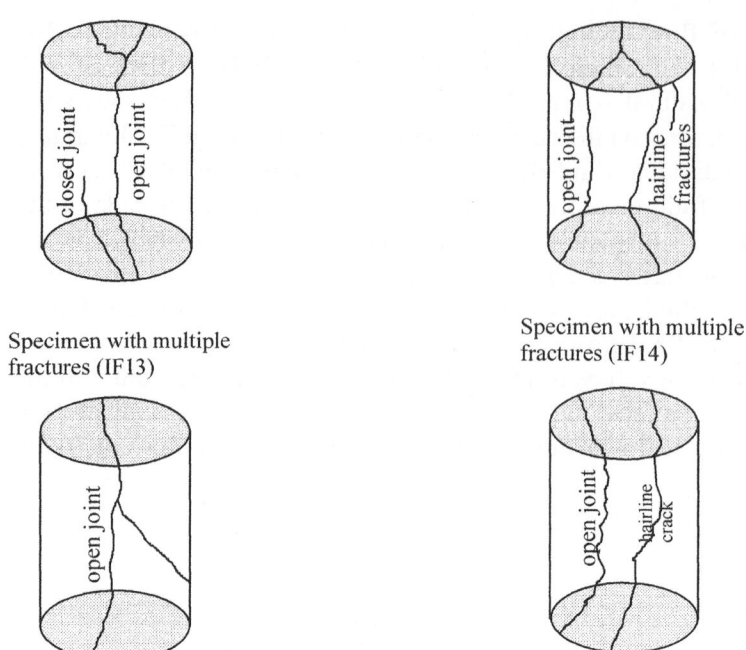

Figure 3.54. Naturally fractured rock specimens with multiple fractures.

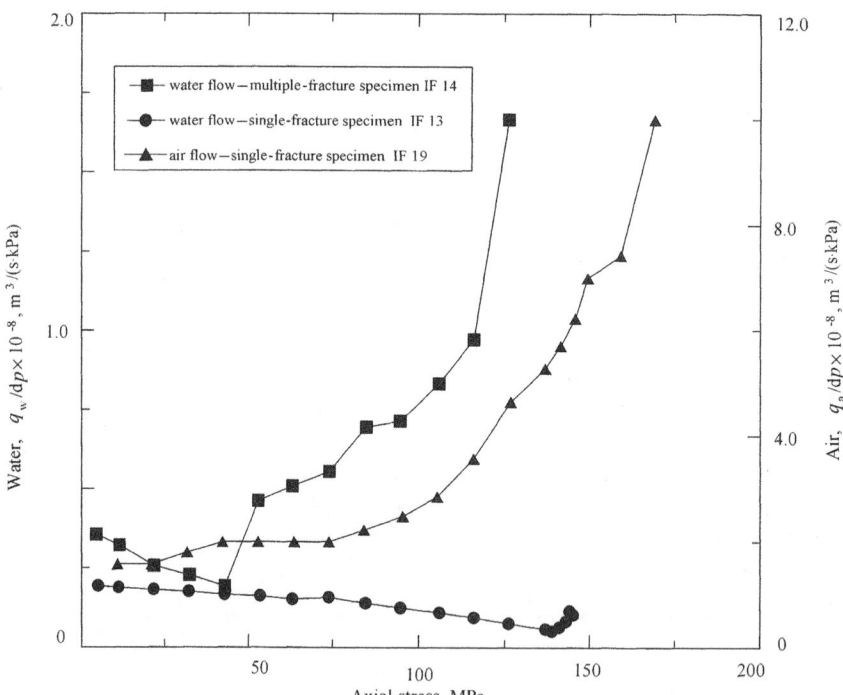

Figure 3.55. Comparison of fluid flow rate through jointed specimens.

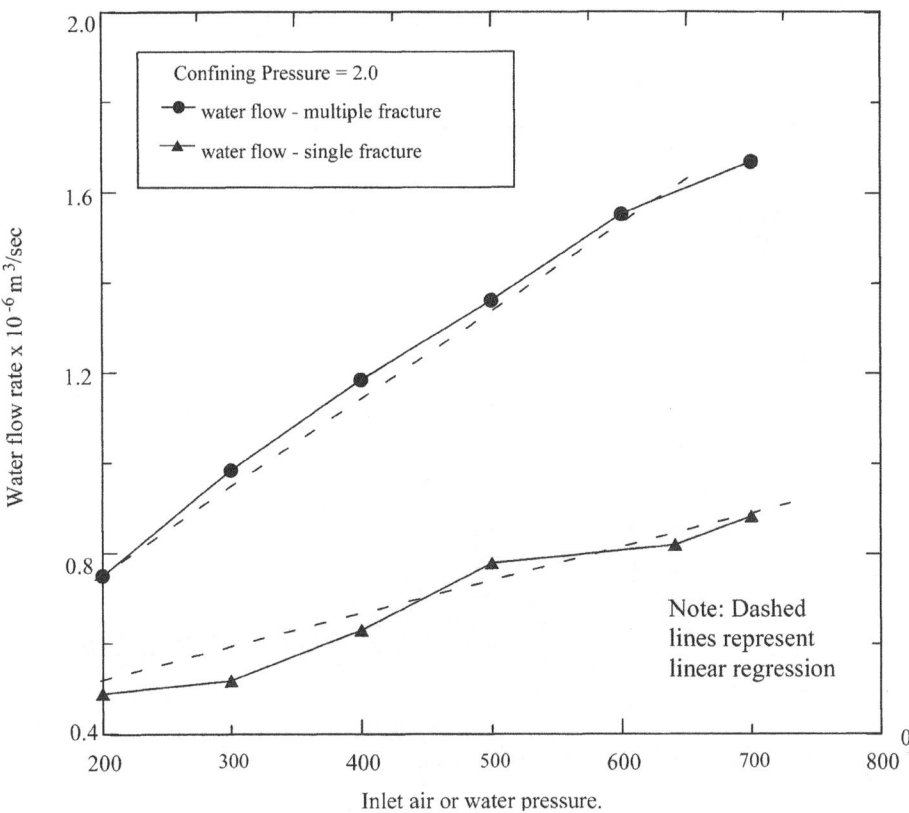

Figure 3.56. Effect of inlet pressure on flow rate through jointed specimens.

Authors have carried out air and water flow through standard size granite core specimens (45 mm dia. to 55 mm dia.) with multiple fractures, taken from Appin mine, Australia (Fig. 3.54). Some fractures within the specimens are found to be open, while the others are sealed. Closed joints may not contribute to any flow changes under low inlet fluid pressures depending on the relative orientation of joints to the principal stresses. As expected, specimens with multiple fractures carry a larger flow rate than a single fracture, because of the increased connectivity of fluid flow paths (Fig. 3.55). Irrespective of the number of flow paths, the well-known Darcy's law can be employed here, as the flow rate against the inlet fluid pressure shows an approximately linear relationship, for both single and multiple fracture specimens (Fig. 3.56). In Fig. 3.55, the specimen with multiple fractures (IF14) shows an initial decrease in flow rate, when the axial stress increases up to 40 MPa. This may be attributed to initial closure of joint apertures. Further increase in axial stress causes an increase in flow rate due to the formation of new cracks or dilation of existing cracks or both. According to Fig. 3.57, the fracture permeability coefficient of both single and multiple fracture specimens diminish exponentially with increased confining pressure. The fracture permeability (k) is calculated using the expression $k = e^2/12$, where e = hydraulic

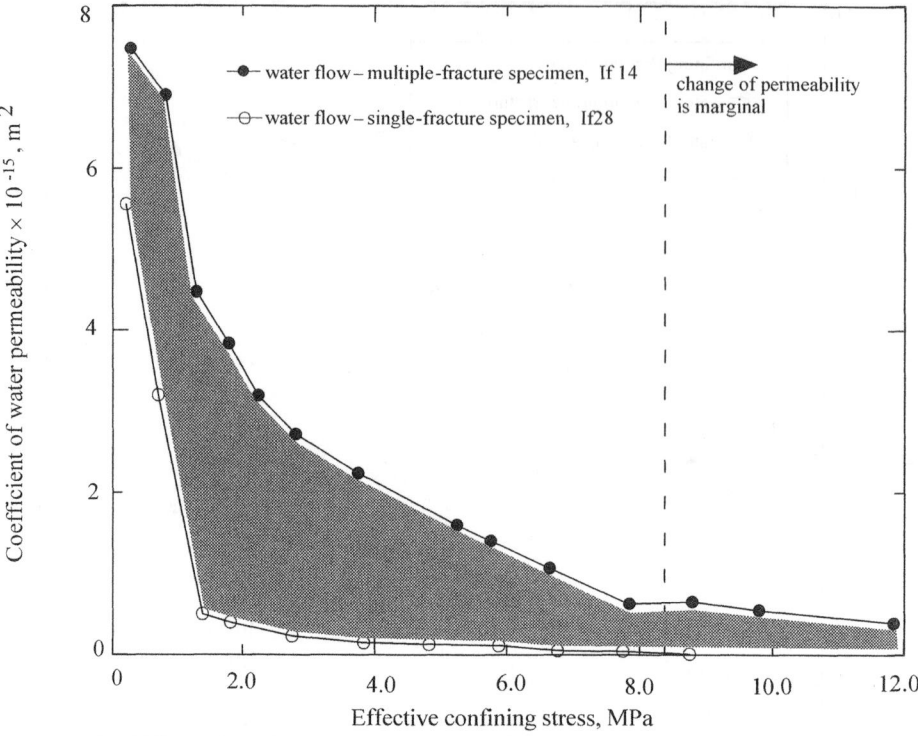

Figure 3.57. Effect of confining stress on permeability.

aperture based on the cubic law for single fracture. In the case of multiple fractures, the value of *e* can be evaluated as an average of all fractures carrying fluid. Apart from the external boundary conditions (e.g., applied stress), the permeability of the multiple fracture specimen is a function of the interconnectivity of joints, their individual apertures, joint orientations and the specimen size.

REFERENCES

Abelin, H., Neretnieks, I., Tunbrant, S. and Moreno, L. 1985. Migration in a single fracture: Experimental results and evaluation. Final Report, Stripa Project, Tech. Pub. 83-03, Nuclear Fuel and Waste Mangement Co. (SKB), Stockholm, Sweden.

Amadei, B. and Illangasekre, T. 1992. Analytical solutions for steady and transient flow in non-homogeneous and anistropic rock joints. *Int. J. Rock Mech. Min. Sci. Geomech. Abstr.,* 29: 561–572.

Baker, W.J. 1955. Flow in fissured formations. *Proc. 4th World Petroleum Congress,* Sect. II/E, Carlo Colombo, pp. 379–393.

Bandis, S. 1980. *Experimental Studies of Scale Effects on Shear Strength and Deformation of Rock Joints.* PhD thesis, Department of Earth Sciences, University of Leeds.

Bandis, S.C., Barton, N.R. and Christianson, M. 1985. Application of a new numerical model of joint behaviour to rock mechanics problems. *Proc. Int. Symp. on Fundamentals of Rock Joints,* Bjorkliden.

Bandis, S.C., Lumsden, A.C. and Barton, N.R. 1983. Fundamentals of rock joint deformation. *Int. J. Rock Mech. Min. Sci. Geomech. Abstr.,* 20: 249–268.

Barr, M.V. and Hocking, G. 1976. Borehole structural logging employing a pneumatically inflatable impression packer. *Proc. Symp. on Exploration for Rock Engineering,* Johannesburg, Balkema, Rotterdam, pp. 29–34.

Barton, N. 1973. Review of a new shear-strength criterion for rock joints. *Eng. Geol.,* 7(4): 287–332.

Barton, N. 1982. Modeling rock joint behavior from in situ block tests; implications for nuclear waste repository design. Office of Nuclear Waste Isolation Report ONWI-308, Columbus, Ohio.

Barton, N. and Bakhtar, K. 1983. Rock joint description and modelling for the hydro-thermomechanical design of nuclear waste repositories (Contract Report, submitted to CANMET). Mining Research Laboratory, Ottawa, Parts 1–4, 270 p.; Part 5, 108 p.

Barton, N., Bandis, S. and Bakhtar, K. 1985. Strength, deformation and conductivity coupling of rock joints. *Int. J. Rock Mech. Min. Sci. Geomech. Abstr.,* 22(3): 121–140.

Barton, N. and Choubey, V. 1977. The shear strength of rock joints in theory and practice. *Rock Mech.,* 10: 1–54.

Barton, N. and Quadros, Eda F. de 1997. Joint aperture and roughness in the prediction of flow and groutability of rock masses. *Int. J. Rock Mech. Min. Sci. Geomech. Abstr.,* 34(3–4): Paper No. 252.

Benjelloun, Z.H. 1993. *Etude Experimentale et Modèlisatioin du Comportement Hydromècanique des Joints Rocheux.* Thèse de doctorat, Université Joseph Fourier, Grenoble 1.

Berry, M.V. and Lewis, Z.V. 1980. On the Weierstrass-Mandelbrot fractal function. *Proc. R. Soc.* London, 459–484.

Brace, W.F., Walsh, J.B. and Frangos, W.T. 1968. Permeability of granite under high pressure. *J. Geophys. Res.,* 73(6): 2225–2236.

Brady, B.H.G. and Brown, E.T. 1994. *Rock Mechanics for Underground Mining.* 2nd ed., Chapman & Hall, London, 571 p.

Brown S.R. 1987. Fluid flow through rock joints: Effects of surface roughness. *J. Geophys. Res.,* 92(B2): 1337–1347.

Brown, S.R. and Scholz, C.H. 1985. Broad bandwidth study of the topography of natural rock surfaces. *J. Geophys. Res.,* 9: 12575–12582.

Carr, R.J. 1989. Fractal characterization of joint surface roughness in welded tuff at Yucca Mountain, Nevada. In *Rock Mechanics as a Guide for Efficient Utilization of Natural Resources,* Khai, ed., Balkema, Rotterdam, pp. 193–200.

Carr, R.J. and Warriner, J.B. 1987. Rock mass classification using fractal dimension. *28th US Symp. of Rock Mechanics,* Tucson, 29 June–1 July, Balkema, Rotterdam, pp. 73–80.

Chitty, D.E. and Blouin, S.E. 1995. Strength, deformation, and fluid flow measurement in porous limestone. In *Fractured and Jointed Rock Masses,* Myer, Cook, Goodman and Tsang, eds., Balkema, Rotterdam, pp. 375–383.

Detoumay, E. 1980. Hydraulic conductivity of closed rock fractures: An experimental and analytical study. *Proc. 13th Canadian Rock Mech. Symp.,* Toronto, pp. 168–173.

Elliot, G.M. 1985. An investigation of the mechanical and thermal properties of a Comish granite. A Report to the Department of Mineral Resources Engineering, Imperial College, London, Part B.

Elliot, G.M., Brown, E.T., Boodt, P.I. and Hudson, J.A. 1985. Hydromechanical behaviour of joints in the Carmenelis granite, SW England. *Proc. Int. Symp. on Fundamentals of Rock Joints,* Bjorkliden, Sweden, Balkema, Rotterdam, pp. 249–258.

Elsworth, D. and Doe, T.W. 1986. Application of non-linear flow laws in determining rock fissure geometry from single borehole pumping tests. *Int. J. Rock. Mech. Min. Sci. Geomech. Abstr.,* 23: 245–254.

Elsworth, D. and Piggot, A.R. 1987. Physical and numerical analogues to fractures media flow. *Proc. 6th Int. Congress on Rock Mechanics,* Vol. 1, pp. 93–97.

Engelder, T. and Scholz, C.H. 1981. Fluid flow along very smooth joints at effective pressures up to 200 MPa, in mechanical behaviour of crystal rocks. *Am. Geophys.,* 24: 147–152.

Fecker, E. and Rengers, N. 1971. Measurement of large-scale roughness of rock planes by means of profilograph and geological compass. *Rock Fracture, Proc. Int. Symp. on Rock Mechanics,* Nancy, France.

Gale, J.E. 1975. *A Numerical, Field and Laboratory Study of Fluid Flow in Rocks with Deformable Fractures.* PhD thesis, University of California, Berkeley, 225 p.

Gale, J.E. and Raven, K.G. 1980. Effect of sample size on stress-permeability relationship for natural fractures. Technical Information Report No. 48, LBL-11865, SAC-48, UC-70.

Gangi, A.F. 1978. Variation of whole and fractured porous rocks permeability with confining pressure. *Int. J. Rock. Mech. Min. Sci. Geomech. Abstr.,* 15: 249–257.

Hakami, E. and Barton, N. 1990. Aperture measurements and flow experiments using transparent replicas of rock joints. *Proc. Int. Symp. on Rock Joints,* Loen, Norway, pp. 383–390.

Indraratna, B., Ranjith, P.G. and Gale, W. 1999. Deformation and Permeability Characteristics of Rocks with Interconnected Fractures. *9th Int. Congress on Rock Mechanics,* Vol. 2, pp. 755–760.

ISRM 1978. International Society for Rock Mechanics Commission on Standardisation of Laboratory and Field Tests suggested methods for the quantitative description of discontinuities in rock masses. *Int. J. Rock Mech. Min. Sci. Geomech. Abstr.,* 15: 319–368.

ITASCA Consulting Group 1996. UDEC-Universal Distinct Element Code, User's Manual, Version 3.0, Vols. 1–3.

Iwai, K. 1976. *Fundamental Studies of Fluid Flow Though a Single Fracture.* PhD thesis, University of California, Berkeley, 208 p.

Jaeger, J.C. and Cook, N.G.W. 1979. *Fundamentals of Rock Mechanics.* 3rd ed., Chapman & Hall, London, 593 p.

Jones, F.O. 1975. A laboratory study of the effects of confining pressure on fracture flow and storage capacity in carbonate rocks. *J. Pet. Tech.,* 21: 21–27.

Kranz, R.L., Frankel, A.D., Engelder, T. and Scholz, C.H. 1979. The permeability of whole and jointed barre granite. *Int. J. Rock. Mech. Min. Sci. Geomech. Abstr.,* 16: 225–334.

Kwasniewski, M.A. and Wang, J.A. 1997. Surface roughness evolution and mechanical behaviour of rock joints under shear. *Int. J. Rock. Mech. Min. Sci. Geomech. Abstr.,* 34 (3–4): Paper No. 157 (CD-Rom).

Lee, C.H. and Farmer I. 1993. *Fluid Flow in Discontinuities Rocks.* Chapman & Hall, London, 169 p.

Lomize, G. 1951. *Fluid Flow in Fissured Formation.* Moskva, Leningrad (In Russian).

Louis, C. 1969. Etude des écolements d'éau dans les roches fissurées et de leur influence sur la stabilité des massifs rocheurs. *BRGM. Bulletin de la Direction des Études et Récherches, Serie A,* 3: 5–132. (These presentée a l'Univ. de Karlsruhe).

Louis, C. 1976. *Introduction de l'hydraulique des roches.* PhD thesis, Paris.

Makurat, A., Barton, N., Rad, N.S. and Bandis, S. 1990. Joint conductivity variation due to normal and shear deformation. In *Rock Joints,* Barton, N.S. and Stephansson, eds., Balkema, pp. 535–540.

Mandelbrot, B.B. 1982. *The Fractal Geometry of Nature.* Freeman, San Francisco, 468 p.

Mendelbrot, B.B. 1967. How long is the coast of Great Britain? Statistical self-similarity and the fractional dimension. *Science,* 156: 636–638.

Miller, R.P. 1965. *Engineering Classification and Index Properties for Intact Rock.* PhD thesis, University of Illinois, 332 p.

Moore, D. F. 1969. A history of research on surface texture effects. *Wear,* 13(June): 381–412.

Moreno, L., Tsang, C.F., Tsang, Y.W. and Neretnieks, I. 1988. Flow and tracer transport in a single fracture: A stochastic model and its relation to some field observations. *Water Resour. Res.,* 24(12): 2033–2048.

Nelson, R. 1975. *Fracture Permeability in Porous Reservoirs: Experimental and Field Approach.* PhD. dissertation, Department of Geology, Texas A&M University.

Neuzil and Tracy, J.V. 1981. Flow through fractures. *Water Resour. Res.,* 17: 191–199.

Nguyen, T.S. and Selvadurai, A.P.S. 1998. A model for coupled mechanical and hydraulic behaviour of a rock joint. *Int. J. Numer. Anal. Meth. Geomech.,* 22: 29–48.

Novakowski, K.S., Evans, G.V., Lever, D.A. and Raven, K.G. 1985. A field example of measuring hydrodynamic dispersion in a single fracture. *Water Resour. Res.,* 21(8): 1165–1174.

Ohle, E.L. 1951. The influence of permeability on ore distribution in limestone and dolomite. *Econ. Geol.,* 46: 667.

Patir, N. and Cheng H.S. 1978. An average flow model for determining effects of roughness on partial hydrodynamic lubrication. *J. Lubr. Technol.,* 100: 12–17.

Patton, F.D. 1966. Multiple modes of shear failure in rock. *Proc. 1st Int. Congress Soc. Rock Mechanics,* Lisbon, Vol. 1, pp. 509–513.

Priest, S.D. 1993. *Discontinuity Analysis of Rock Engineering.* Chapman & Hall, London, 473 p.

Pyrak-Nolte, L.J., Myer, L.R., Cook, N.G.W and Witherspoon, P.A. 1987. Hydraulic and mechancial properties of natural fractures in low permeability rock. *Int. Congress on Rock Mechanics (ISRM),* Montreal, Canada.

Quadros, E.F. 1982. *Determinação das características do fluxo de água em fraturas de rochas.* Dissert. de Mestrado, Department of Civil Eng., Polytech School, University of São Paulo.

Ranjith, P.G. 2000. *Analytical and Numerical Investigation of Water and Air Flow Through Rock Media.* PhD thesis, Department of Civil Engineering, University of Wollongong, Australia.

Raven, K.G. and Gale, J.E. 1985. Water flow in a natural rock fracture as a function of stress and sample size. *Int. J. Rock Mech. Min. Sci. Geomech. Abstr.,* 22(4): 251–261.

Schneider, H.J. 1974. Rock friction – a laboratory investigation. *Proc. 3rd Cong. Int. Soc. Rock Mechanics,* Denver, Vol. 2-A, pp. 311–315.

Sharp, J.S. and Maini, Y.N.T. 1972. Fundamental considerations on the hydraulic characteristics of joints in rock. In *Proc. Symp. on Percolation Through Fissured Rock,* Farmer, I., ed., Int. Soc. for Rock. Mech. and Int. Assoc. of Eng. Geology.

Shimo, M. and Iihoshi, S. 1993. Laboratory study of water flow though multiple fractures. *Int. J. Rock. Mech. Min. Sci. Geomech. Abstr.,* 30(7): 853–856.

Singh, A.B. 1997. Study of rock fracture by permeability method. *J. Geotech. Geoenviron. Eng.,* 123: 601–608.

Smith, L.C, Mase, C.W. and Schwartz, F.W. 1987. Estimation of fracture using hydraulic and tracer tests. *28th U.S. Symp. on Rock Mechanics,* University of Arizona, Tucson, June 29th to July 1st, pp. 453–463.

Snow, D.T., 1968a. Anisotropic permeability of fractured media. *Water Resour. Res.,* 5(6): 1273–1289.

Snow, D.T. 1968b. Rock fracture spacing, openings and porosity. *J. Soil Mech.* (Found. Div., ASCE), 94(SM 1): 73–91.

Snow, D.T. 1970. The frequency and apertures of fractures in rock. *Int. J. Rock Mech. Min. Sci. Geomech. Abstr.,* 7: 23–40.

Tsang, Y.W. 1992. Usage of equivalent apertures for rock fractures as derived form hydraulic and tracer tests. *Water Resour. Res.,* 28(5): 1451–1455.

Tsang, Y.W. 1984. The effect of tortuosity on fluid flow through a single fracture. *Water Resour. Res.,* 20: 1209–1215.

Tsang, Y.W. and Witherspoon, P.A. 1981. Hydro-mechanical behaviour of a deformable rock fracture subject to normal stress. *J. Geophys. Res.,* 86(B10): 9287–9298.

Tse, R. and Cruden, D.M. 1979. Estimating joint roughness coefficient. *Int. J. Rock Mech. Min. Sci. Geomech. Abstr.,* 16: 303–307.

Walsh, J.B. 1981. Effect of pore pressure and confining pressure on fracture permeability. *Int. J. Rock. Mech. Min. Sci. Geomech. Abstr.,* 18: 429–434.

Wei, Z.Q. and Hudson, J.A. 1988. Permeability of jointed rock masses. In *Rock Mechanics and Power Plants,* Romana, ed., Balkema, Rotterdam.

Weissbach, G.A. 1978. A new method for the determination of roughness of rock joint in the laboratory. *Int. J. Rock Mech. Min. Sci. Geomech. Abstr.,* 15: 131–133.

Williamson, J.B.P. and Hunt, R.T. 1968. Burndy Research Division. Research Report No. 59, U.S.A.

Wilock, P. 1996. The NAPSAC fracture network code. *Coupled Thermo-Hydro-Mechanical Process of Fractured Media,* Elsevier, New York, 575 p.

Witherspoon, P.A., Amick, C.H., Gale J.E. and Iwai, K. 1979. Observations of a potential size effect in experimental determination of the hydraulic properties of fractures. *Water Resour. Res.,* 15: 1142–1146.

Witherspoon, P.A., Wang, J.S.Y., Iwai, K. and Gale, J.E. 1980. Validity of cubic law for fluid flow in a deformable rock fracture. *Water Resour. Res.,* 16(6): 1016–1024.

Xie H. and Pariseau, W.G. 1995. Fractal estimation of joint roughness coefficients, In *Fractured and Jointed Rock Masses,* Myer, L.R., Cook, N.G.W., Goodman, R.E. and Tsang, C.F., eds., Balkema, Rotterdam, pp. 125–131.

CHAPTER 4

Unsaturated fluid flow through a jointed rock mass

4.1   INTRODUCTION

Usually, rock mass is in an unsaturated form, and it carries both water and air through interconnected fractures and pores. It is commonly observed that air can flow through very fine fractures, whereas the water phase generally prefers to flow through larger fractures which offer much less resistance. Generally speaking, multiphase flow in rock mass involves a set of complex phenomena, and to date, no comprehensive model has been developed to properly include factors such as the interaction between each phase, change of properties of fluid and associated joint deformation.

At present, satisfactory procedures are not available in the rock mechanics literature to estimate flow quantities occurring in multiphase flow in jointed rock media. The study described herein is an attempt to introduce a new direction in the analysis of multiphase flow, in the field of rock engineering. Although laboratory work has been carried out to understand the complete two-phase flow behaviour in the fields of chemical and mechanical engineering, the proper understanding of two-phase flow behaviour in jointed rock engineering still remains at infancy. During the past two decades, a number of numerical, analytical and experimental models have been developed for single-phase flow (water or gas) through a single joint or jointed rock mass. Very limited attempts to analyse coupled hydro-mechanical behaviour can be found in literature with respect to gas–water flow (Pruess & Tsang, 1990; Rasmussen, 1991; Fourar et al., 1993; Fourar & Bories, 1995). Rasmussen (1991) investigated the fracture flow under condition on partial fluid saturation. A fracture made of 2 glass panels was filled with water in one portion, and the remaining portion was filled with air. He estimated the air–water interface position in the fracture for different capillary pressures both analytically and experimentally. Fourar et al. (1993) observed different flow patterns via two parallel glass plates and two parallel brick layers which were not subjected to external loads. The aim of selecting different materials was only to simulate discrete fractures and porous media. The valuable experimental results ascertained by him may be directly applied for simplified two-phase flow analysis, but are not suited to characterize the coupled hydro-air-mechanical flow that usually takes place within rock joints under stress. Pruess and Tsang (1990) conducted a numerical analysis based on relative permeability of two-phase flow subject to pressure difference between air and water in real rock fractures. He observed that the nature and range of spatial correlation between apertures would have an influence on the relative permeability (i.e. permeability of air with respect to water and vice versa). It is assumed that single-phase flow phenomenon can be applied when discontinuities become fully saturated with a fluid (generally water or gas). However in practice, even under fully saturated conditions, some gas may still be dissolved in water and can come out of solution depending on the pressure changes, subsequently leading to complex flow patterns. Water–air or solid–water or water–air–solid mixtures are generally encountered in underground excavations, particularly in mining situations. Although in most cases, the transported slurries consist of small size particles, a coal–water mixture can contain much coarser particles. The above mixtures cannot be treated as single-phase flow, simply because they exhibit settling characteristics (e.g. clay, iron particles), thereby indicating time-dependent fluid densities.

To the knowledge of the authors, inter-relations of multiphase flows, relative permeability and the pressure variation through rock fractures have not been properly identified. Therefore, this chapter aims to shed light on the relatively complicated two-phase flow systems, and to provide a comprehensive mathematical model and experimental investigation to compute the quantities of each fluid travelling in a given joint domain. It is noted that the success of any numerical or analytical model depends on the basic experimental data (simulating the gas–water interface behavior) because of the necessity to furnish realistic flow pattern parameters to calibrate these models. There is no doubt that flow pattern inputs remain to be influenced by visual observations, hence, the governing equations that represent any mathematical model must be capable of predicting the correct flows associated with these observations.

4.2   NEED FOR UNSATURATED FLOW IN UNDERGROUND ACTIVITIES

Rock fractures are usually unsaturated and they often carry water, gas and sediments. On some occasions (e.g., petroleum recovery process), fractures may be partially saturated with three phases, such as oil–water–gas. Two-phase flow through jointed rock media has gained increasing interest in both industry and academia due to the important applications in two-phase flow transportation through joint networks. Importance of such flows extends from petroleum engineering, mining engineering and nuclear engineering to medical applications and food manufacturing. Two-phase flow through rock fractures can be in the form of water–gas, water–solid, solid–gas, water–oil and oil–gas combinations. This chapter is limited to the analysis of two-phase flow structures consisting of water and gas only.

A number of numerical, analytical and experimental models for single-phase flow (i.e. either water or gas) through a single joint can be found in the literature (Louis, 1976; Brown, 1987; Indraratna & Ranjith, 1999; Ranjith, 2000). These single-phase models have often been used to simulate the fluid flow (either two or three phase flow) in many underground applications because of the lack of knowledge on unsaturated flow through jointed rock. At present, it is difficult to predict the risk of groundwater inundation and outbursts in complex hydro-geological environments prevailing in underground mines. These potential outbursts become worst when fractures contain two-phase flows because of the decreasing desorption rate due to the presence of water. As an example, in coal mining, the existence of gas pockets (e.g., CO_2 and CH_4) within complex hydro-geological regimes present obvious safety hazards. Mining coal reserves under these circumstances always carries the risk of simultaneous groundwater inundation and gas outbursts, and sites affected by unexpected inflows have been reported in many parts of the world including Australia (e.g., Bulli and Westcliff Collieries mines, New South Wales). The hazardous nature of groundwater inundation, carbon dioxide intrusion, and the catastrophic consequences of methane explosions are well known. The polluted (acidic) mine water may either contaminate the surrounding groundwater or ultimately discharged to the nearby water sources (streams, etc.) causing a serious threat to the environment.

Under the increased environmental and regulatory controls, the evaluation of the quantity and quality of total inflow to excavations and the procedures for discharging polluted mine water are significant factors in the development and operational stages of underground mining. Moreover, in nuclear wastage storage plants, special attention should be given to prevent any radioactive contamination of groundwater. The accurate prediction of inflow volumes is useful to reduce the extent of environmental hazards and damage to mine equipment, as well as to reduce the time delay associated with dewatering.

Planning decisions concerning the water and gas flow control measures such as draining and removing gas, grouting and dewatering should be implemented in advance, so that the whole operation system would contribute more efficiently towards greater economies of scale within a safer work environment. For rational analysis and design, it is essential to understand the two-phase flow and mechanical behaviour of rock mass, the response of the natural joint system, and how the water and air phases response to changes induced by proposed excavation sequence. A proper understanding of such flow is of paramount importance in oil recovery process.

4.3 THEORETICAL ASPECTS OF TWO-PHASE FLOW

Similar to single-phase flows, two-phase flows also obey the basic laws of fluid mechanics. However, the mathematical formulation of two-phase flow is more difficult than the case of single-phase flow. This is because different flow patterns, such as stratified and mixed flows, may develop within a pipe or rock joint in different locations at different times, depending on the change of pressure or velocity of each phase, their interaction and the surface geometry of the fluid flow path. Under these circumstances, it is not easy to transfer the mathematical equations directly from one flow pattern to a newly formed flow pattern. Basically, there are two main approaches to analyse two-phase flow:

(1) Homogeneous flow model (i.e. considering only a single phase flow) and
(2) Multi-fluid flow models (i.e. treating the fluid phases separately).

4.3.1 Homogeneous steady-state flow model

From an analytical or numerical point of view, this is the most convenient way of modelling two-phase flows. However, the accuracy of results is subjected to variation, based on estimated properties of the homogeneous fluid. The crudest point in this method is the determination of average fluids properties. For some flow patterns (e.g. stratified flow), it may be easier to estimate the average properties of fluid such as density, velocity, temperature and viscosity. Nevertheless, more rigorous efforts supported by experimental work are certainly needed to determine properties of homogeneous flow models, particularly when the flow pattern takes a mixed or a complex form. Once the average properties are determined with a certain degree of confidence, the mixture is treated as an equivalent single-phase flow, in which all-basic fluid flow laws are considered to be applicable.

The density of the mixture can be expressed by either considering the volume fraction or the mass fraction. Based on the volume fraction, the mean density of the mixture is given by

$$\rho_m = \frac{\rho_\alpha q_\alpha + \rho_\beta q_\beta}{q_\alpha + q_\beta} \tag{4.1}$$

where ρ = density, q = flow rate and the subscripts m, α and β represent mean, phase α and phase β, respectively.

Dukler et al. (1964), Mcadams et al. (1942) and Cicchitti et al. (1960) suggested different expressions for the viscosity of the mixture, based on the volume fraction and the mass fraction. According to Mcadams et al. (1942), the viscosity of the mixture of water–gas flow depends on the individual component of the viscosity of each phase and the mass fraction as follows:

$$\frac{1}{\mu_m} = \frac{x}{\mu_\alpha} + \frac{1-x}{\mu_\beta} \tag{4.2a}$$

where μ = dynamic viscosity of fluid phase, x = mass fraction factor and given by $M_\alpha/(M_\alpha + M_\beta)$ and M = mass flow rate.

Based on the volume flow rate fraction, Dukler et al. (1964) proposed the mixture viscosity as given below:

$$\mu_m = \frac{\mu_\alpha q_\alpha + \mu_\beta q_\beta}{q_\alpha + q_\beta} \tag{4.2b}$$

Based on the average viscosity and density of the mixture, the Reynolds number (Re_m) of the mixture can now be formulated as given below:

$$Re_m = \frac{2hV_m\rho_m}{\mu_m} \tag{4.3}$$

where V_m = mixture velocity and given by $(q_\alpha + q_\beta)/A$. A is the cross-section area and h is the hydraulic diameter.

Using the momentum equation, the pressure gradient of the mixture $(dp/dx)_m$ for an inclined fluid flow surface is the summation of friction, acceleration and gravity terms. Therefore, the pressure gradient of the mixture $(dp/dx)_m$ is as follows:

$$\left(\frac{dp}{dx}\right)_m = -\left[\frac{dp_f}{dx} + \frac{dp_a}{dx} + \frac{dp_g}{dx}\right] \tag{4.4}$$

where p = pressure and the subscripts f, a and g represent friction, acceleration and gravity, respectively.

By substituting corresponding expressions in Eqn. 4.4, the following equation represents the pressure drop of the mixture in terms of fluid properties and geometrical properties of the fluid flow path.

$$\left(\frac{dp}{dx}\right)_m = -\left[\frac{S}{A}\tau_w + \left(\frac{\rho q}{A}\frac{dV}{dx}\right)_m + \rho_m g \sin\theta\right] \tag{4.5}$$

where S = perimeter of the mixture of flow path, A = cross-sectional area, θ = inclination of the fluid flow surface to the horizontal, and τ_w = wall shear stress.

The average wall shear stress may be expressed in terms of velocity, friction factor (C_f) and velocity of the mixture, as follows:

$$\tau_w = \frac{1}{2}C_f\rho_m V_m^2 \tag{4.6}$$

where $C_f = A/Re_m^\alpha$. In this equation A and α are variables. From experimental work carried out by Fourar et al. (1993) for gas and water flow through two horizontal glass plates, the following values for A and α were obtained for smooth and rough fractures.

For smooth fractures and $Re_m < 1000$, $A = 6806$ and $\alpha = 1.1$;
For smooth fractures and $Re_m > 1000$, $A = 0.75$ and $\alpha = 0.45$;
For rough fractures, $A = 6.46$ and $\alpha = 0.61$.

4.3.2 Multi-fluid steady-state flow models

In a comprehensive flow model, theories are first developed based on (a) individual equations for each phase and (b) equations at the boundary of the discontinuity. Subsequently, those two sets of equations are combined to develop the final expression. In the first and second categories, a set of equations are written for each phase and at the boundary of the discontinuity phases, considering the mass, momentum, enthalpy and energy theories. It is feasible to develop flow theories for flow patterns, if the given flow pattern remains constant along the flow path. In the following section, general equations for two-phase flow analysis based on mass, momentum, energy and enthalpy theories are presented.

Based on the mass conservation theory, for a control volume of two-phase flow, the net mass flux should be made equal to zero. The complete derivations of the following equations are found elsewhere (Hetsroni, 1982), and only a summary is given below.

$$\frac{d}{dt}\int_{V_\alpha} \rho_\alpha dV + \frac{d}{dt}\int_{V_\beta} \rho_\beta dV + \int_{A_\alpha(t)} \rho_\alpha v_\alpha \cdot n_\alpha dA + \int_{A_\beta(t)} \rho_\beta v_\beta \cdot n_\beta dA = 0 \qquad (4.7)$$

where V and A = volume and area, respectively, v = velocity of fluid, n = normal unit vector, t = time, and α and β = two phases in the control volume.

Due to the external forces, torques, and the velocity of the fluid flow, the control volume is subjected to linear and angular momentum. The linear momentum in the control volume equals the summation of the momentum of influx and the external forces, as given below:

$$\left[\frac{d}{dt}\int_{V_\alpha(t)} \rho_\alpha v_\alpha dV + \frac{d}{dt}\int_{V_\beta(t)} \rho_\beta v_\beta dV\right]$$

$$+ \left[\int_{A_\alpha(t)} \rho_\alpha v_\alpha(v_\alpha \cdot n_\alpha)dA + \int_{A_\beta(t)} \rho_\beta v_\beta(v_\beta \cdot n_\beta)dA\right]$$

$$- \left[\int_{V_\alpha} \rho_\alpha f dV + \int_{V_\beta} \rho_\beta f dV + \int_{V_\alpha} n_\alpha \cdot T_\alpha dA + \int_{V_\beta} n_\beta \cdot T_\beta dA\right] = 0 \quad (4.8)$$

where f = external forces and T = torque.

The kinematics and internal energy contribute to the total energy of the control volume. This energy is equal to the summation of the energy of the net influx to the control volume and energy due to the external forces. Assuming that no heat transfer is involved, the total energy of the system can be written as follows:

$$\left[\frac{d}{dt}\int_{V_\alpha(t)} \rho_\alpha(1/2v_\alpha^2 + u_\alpha)dV + \frac{d}{dt}\int_{V_\beta(t)} \rho_\alpha(1/2v_\beta^2 + u_\beta)dV\right]$$

$$+ \left[\int_{A_\alpha(t)} \rho_\alpha(1/2v_\alpha^2 + u_\alpha)(v_\alpha \cdot n_\alpha)dA + \int_{A_\beta(t)} \rho_\beta(1/2v_\beta^2 + u_\beta)(v_\beta \cdot n_\beta)dA\right]$$

$$- \left[\int_{V_\alpha} \rho_\alpha f \cdot v_\alpha dV + \int_{V_\beta} \rho_\beta f \cdot v_\beta dV + \int_{V_\alpha} (n_\alpha \cdot T_\alpha) \cdot v_\alpha dA + \int_{V_\beta} (n_\beta \cdot T_\beta) \cdot v_\beta dA\right]$$

$$= 0 \qquad (4.9)$$

where u = internal energy.

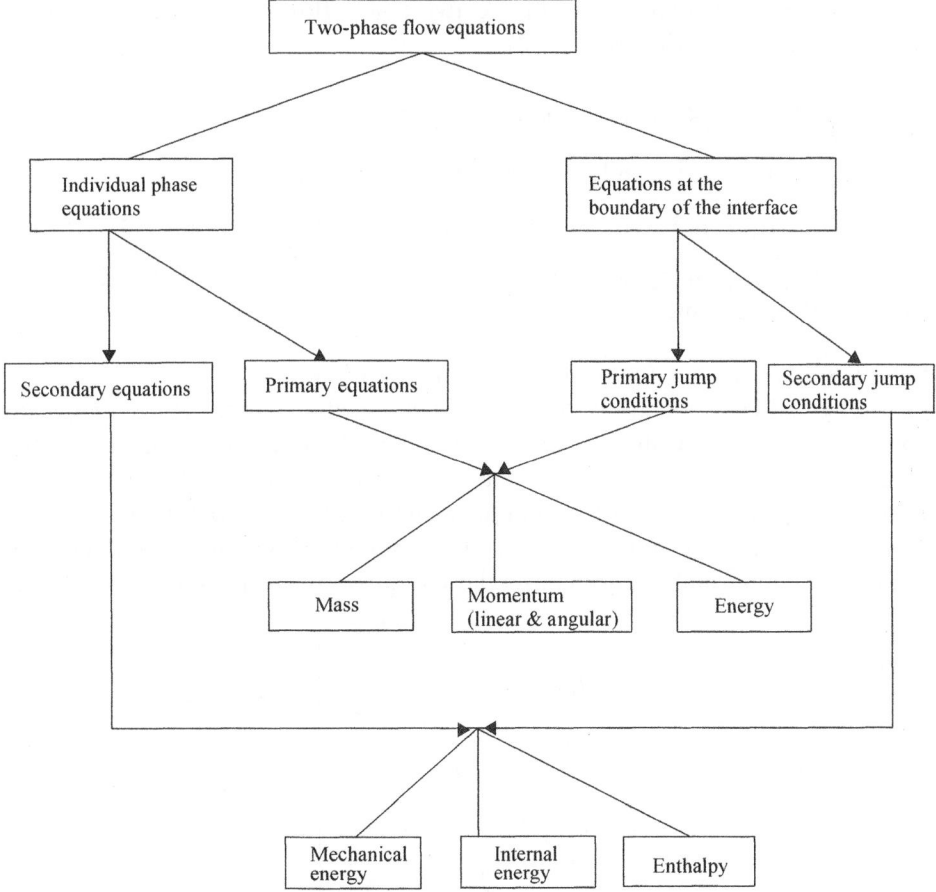

Figure 4.1. General formulation of governing equations for two-phase flow analysis.

Phase equations

Once the general equations are developed, the next step is to develop the equations for each phase. As shown in Fig. 4.1, the primary and secondary theories involving each phase are derived from the above equations, based on Gauss and Leibniz theorems.

For two-phase flow, the mass conservation theory is represented by

$$\frac{\partial(\rho_\alpha + \rho_\beta)}{\partial t} + \nabla \cdot (\rho_\alpha v_\alpha + \rho_\beta v_\beta) = 0 \tag{4.10}$$

From the momentum conservation theory, the following equation can be written:

$$\frac{\partial(\rho_\alpha v_\alpha + \rho_\beta v_\beta)}{\partial t} + \nabla \cdot (\rho_\alpha v_\alpha^2 + \rho_\beta v_\beta^2 - T_\alpha - T_\beta) - (\rho_\alpha + \rho_\beta)f = 0 \tag{4.11}$$

Considering the total energy of the two-phase flow system, the governing equation is given by (Hetsroni, 1982)

$$\frac{\partial}{\partial t}[(\rho_\alpha + \rho_\beta)(1/2(v_\alpha^2 + v_\beta^2) + u_\alpha + u_\beta)]$$
$$+ \Delta \cdot [(\rho_\alpha + \rho_\beta)(1/2(v_\alpha^2 + v_\beta^2) + u_\alpha + u_\beta)(v_\alpha + v_\beta)]$$
$$- (\rho_\alpha + \rho_\beta)f \cdot (v_\alpha + v_\beta) - \nabla \cdot [(T_\alpha + T_\beta) \cdot (v_\alpha + v_\beta)] = 0 \qquad (4.12)$$

By combining primary equations, secondary phase equations are developed for mechanical energy, internal energy and enthalpy.

Equations at the discontinuity boundary between phases
Basic physical laws of conservation of mass, momentum, energy and enthalpy are imperative to describe the characteristics of the discontinuity boundary or the interface. If the interface between phase α and phase β (Fig. 4.2) moves at a velocity of v_d, the mass transfer from the phase α via the discontinuity should be equal to the mass transfer from the phase β via the discontinuity surface. The mass, momentum and energy conservation equations are thereby expressed as

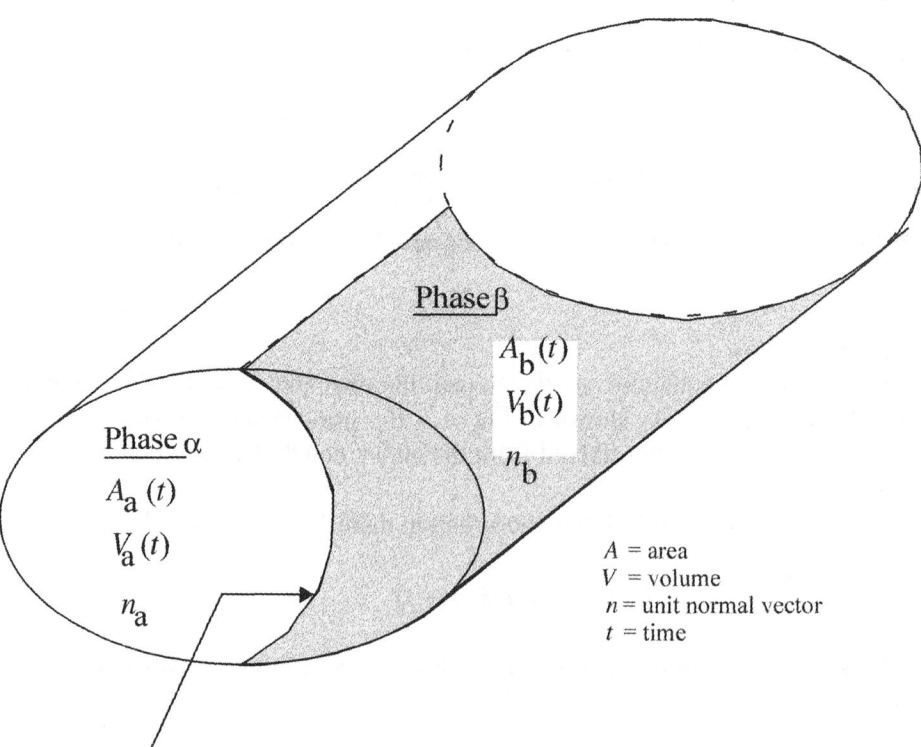

Figure 4.2. Two-phase flow in a given control volume.

follows:

$$\rho_\alpha(v_\alpha - v_d) \cdot n_\alpha + \rho_\beta(v_\beta - v_d) \cdot n_\beta = 0 \tag{4.13}$$

The momentum at the interface is due to the effect of velocity of each phase and surface forces, as modelled by

$$[\rho_\alpha(v_\alpha - v_d) \cdot n_\alpha]v_\alpha + [\rho_\beta(v_\beta - v_d) \cdot n_\beta]v_\beta - n_\alpha \cdot T_\alpha - n_\beta \cdot T_\beta = 0 \tag{4.14}$$

Kinematics energy, internal energy, surface forces and heat transfer contribute to the total energy along the interface between each phase, as expressed below:

$$[\rho_\alpha(v_\alpha - v_d) \cdot n_\alpha][1/2v_\alpha^2 + u_\alpha] + [\rho_\beta(v_\beta - v_d) \cdot n_\beta](1/2\,v_\beta^2 + u_\beta)$$
$$-(n_\alpha \cdot T_\alpha) \cdot v_\alpha - (v_\beta . T_\beta) \cdot v_\beta = 0 \tag{4.15}$$

4.4 DETERMINATION OF FLOW REGIMES

Accurate determination of flow structures is very important in developing mathematical models for multiphase flow analysis. The knowledge of the flow regimes is important in the study of transient flows and steady-state flows, in order to incorporate the flow parameters such as changes of fluid properties and wall shear stresses in various numerical models. The inlet pressures of each phase, their physical properties, interactions between the phases and the geometries of flow paths determine the flow patterns in a pipe or channel or in a rock joint. Common flow patterns that occur in pipe flows can be categorized as stratified, bubble, droplet, annular, complex, plug and churn. Two-phase flow can be in the form of gas–liquid or gas–solid or liquid–liquid or liquid–solid. Out of these, the most complex form is the gas–liquid because of the complex interaction between each phase. For example, gas can dissolve inside the body of liquid depending on the temperature and pressure changes. The gas–liquid flows come across in various applications including in human bodies (e.g. blood + O_2), boilers, nuclear reactors, underground mines, and oil recovery process. Also, natural gas reservoirs with two-phase flow of gas and water are located in tight rocks.

In the past, various approaches have been employed to identify flow structures as listed below (Fiori & Bergles, 1966; Hewitt, 1978; Fourar & Bories, 1995):

(1) Visual observations supported by photographic method and
(2) Selected fluid flow parameters such as superficial velocities.

From the work carried out in two-phase flow analysis through pipes, flow regimes can be mainly divided into two groups (Fig. 4.3):

(a) Continuous two-phase flow and
(b) Discontinuous two-phase flow.

In continuous flow, both phases prevail all along the flow path, whether stratified or annular. Stratified flow is the simplest flow pattern in multiphase flows. In such stratified flows, the main difficulty is in defining the interface level between the phases, as the interface location is time dependent and subjected to changes,

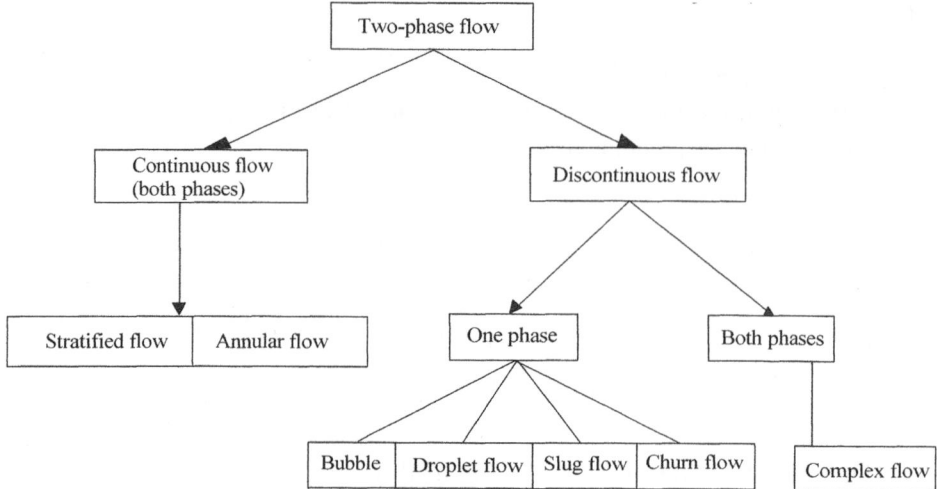

Figure 4.3. Common two-phase flow regimes.

depending upon various external conditions such as flow path geometry, fluid pressure and fluid properties. In discontinuous flow, both or one phase is discontinuous (e.g., bubble, droplet, slug and complex flows). Various types of flow patterns are shown in Fig. 4.4. These flow patterns may develop one or combined effects of the following:

(1) Pressure/velocity of each phase
(2) Viscosity of gas and liquid
(3) Density difference and buoyancy
(4) Flow path geometry and
(5) Surface tension and surface contamination.

4.4.1   Visual observations supported by photographic method

This method is only possible when flow takes place through transparent materials like glass or plastic. Many attempts of the visual observation and photographic methods can be found in the literature (Hewitt & Lovegrove, 1969; Arnold & Hewitt, 1967; Delhaye, 1979; Fourar & Bories, 1995). The experimental work on identification of gas–water flow pattern in a narrow channels using combined visual observation and photographic method was carried out by Fourar and Bories (1995). Water and air was driven through two glass plates 1 m long and 0.5 m wide separated by 1 mm, and the flow pattern was observed through the glass plates and photographed using a high speed camera (Fig. 4.5). The fracture was initially saturated with water and then the air phase was introduced to the fracture. Fourar and Bories (1995) observed different flow patterns for various inlet air pressures, as shown in Fig. 4.6. These flow patterns vary from bubble flow to droplet flow. Droplet flow and annular flow usually occur at elevated inlet gas pressures, whereas bubble flow may develop because of the

(1) Separated Flows

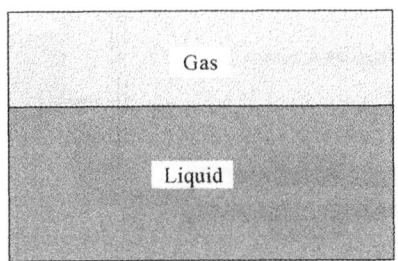

Film flow

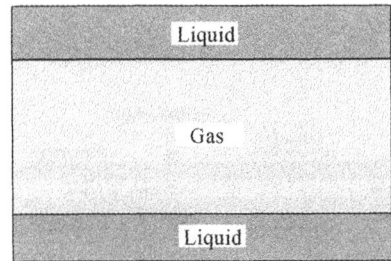

Annular flow - Gas core and liquid film

(2) Mixed Flows

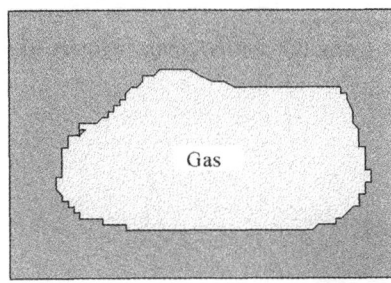

Slug flow - Gas pocket in liquid

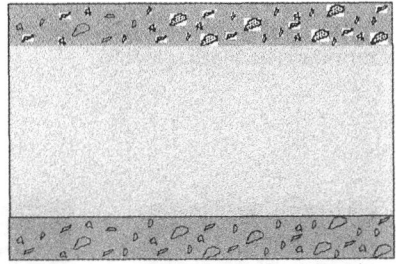

Bubbly annular flow - Gas bubbles in liquid film with gas core

(3) Dispersed flows

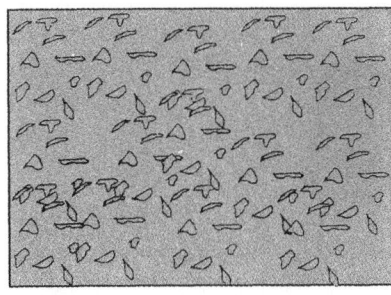

Bubbly flow - Gas bubbles in liquid

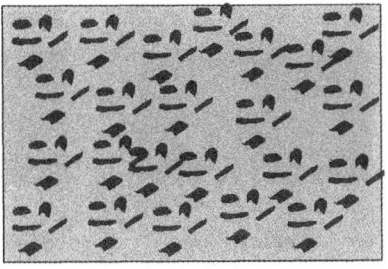

Droplet flow - Liquid droplets in gas

Figure 4.4. Different flow modes of two-phase liquid and gas interaction.

high liquid pressures compared to gas pressures. At higher gas inlet pressures, gas occupies the main part of the fracture and the liquid may flow as droplets or as an unstable film. Mishima and Hibiki (1996) studied two-phase flow through small diameter (1.05–4.08 mm) vertical tubes, and they observed the following flow patterns (Fig. 4.7).

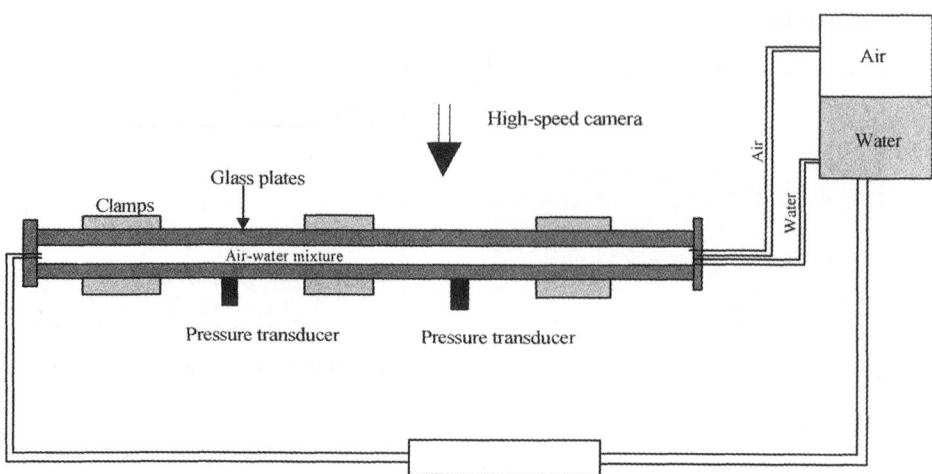

Figure 4.5. Flow pattern observation technique using in a narrow glass channel (modified after Fourar & Bories, 1995).

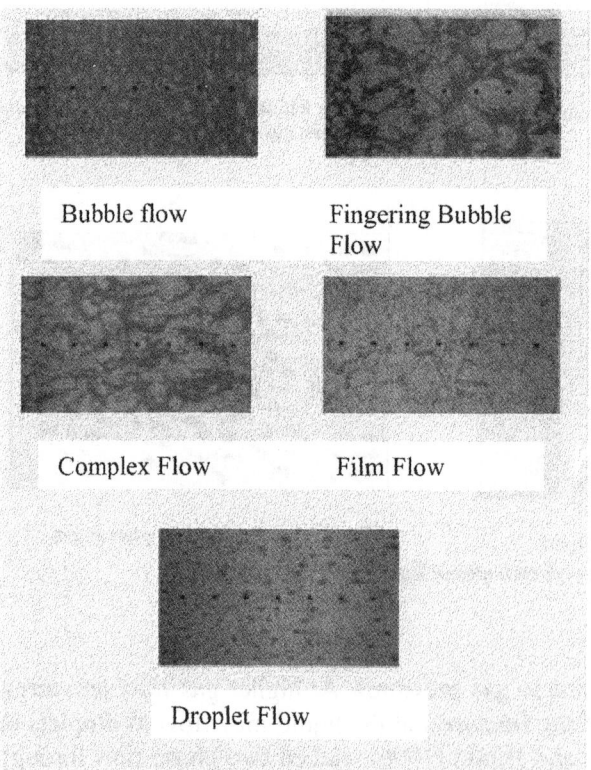

Figure 4.6. Flow patterns observed in a narrow glass channel. The dark colour shows the liquid and the light colour shows the gas (after Fourar & Bories, 1995).

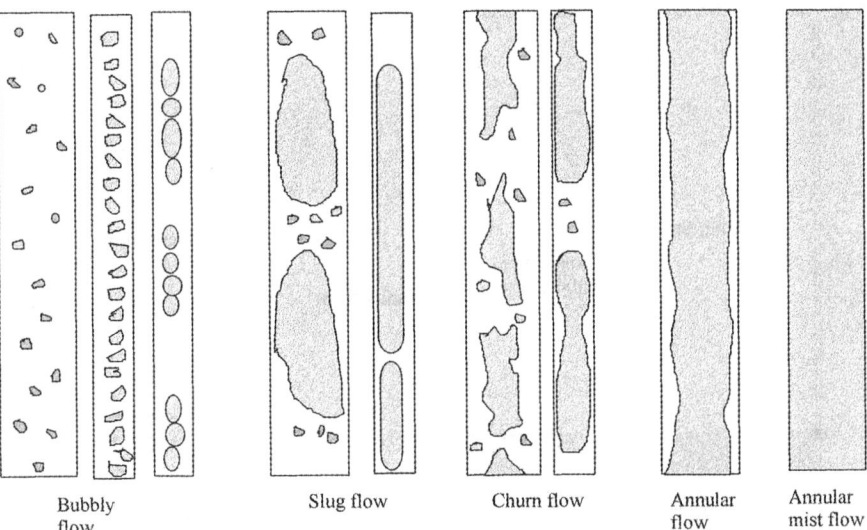

Figure 4.7. Flow patterns observed in a small vertical tube (modified after Mishima & Hibiki, 1996).

4.4.2 Flow pattern based on fluid flow parameters

The technique based on the study of the fluid flow parameters is an indirect method of analysing the flow pattern of gas–liquid two-phase flow system. This technique is particularly suitable when the flow is not visible or when the flow takes place at high-speed. In multiphase flow analysis through pipes, the common procedure is to plot the liquid superficial velocity against the gas superficial velocity. Such a plot on water–gas flow through horizontal glass plates is shown in Fig. 4.8 (Fourar & Bories, 1995). Golan and Stenning (1969) reported a flow regime map for a vertical pipe, as shown in Fig. 4.9. They observed annular, oscillatory and slug flow patterns for elevated gas and liquid superficial velocities. Both plots show that bubble flow, in which liquid flow is dominant, occurs at low gas velocities. In contrast, annular flow prefers to develop at elevated gas velocities.

4.5 POSSIBLE FLOW STRUCTURES IN A SINGLE ROCK JOINT

The difficulty involved in determining flow pattern through rock is that the flow is neither visible nor accessible. Unlike in flow through transparent pipes, no direct observations or photographic techniques can be employed. Nevertheless, the photographic method may be used to map the flow pattern in artificial fractures made out of transparent plastics. For example, a natural rock joint surface may first be transformed to a transparent plastic surface, subsequently, water and air flow are driven through the plastic joint to observe the flow type.

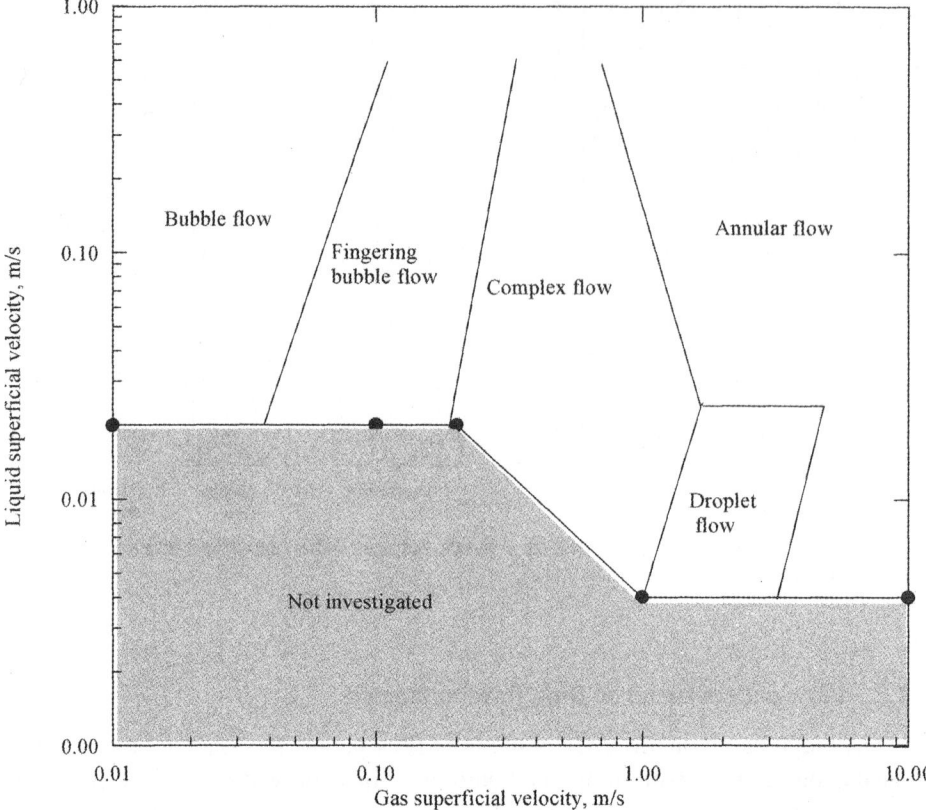

Figure 4.8. Possible water-gas flow patterns (data from Fourar & Bories, 1995).

Extensive laboratory work on water–gas flow through fractured rocks was conducted using the newly designed Two-Phase, High-Pressure Triaxial Apparatus (TPHPTA, described earlier in Ch. 2). The test procedure is as follows. The specimen is first saturated with water, and then the air phase is forced through the specimen. Once the water and air mixture passes through the dreschel bottle, airflow rates and water flow quantities are recorded by the film flow meter and electronic weighing scale, respectively. These individual flow rates are used to calculate the superficial velocity (v_α) of each phase, using Eqn. 4.16. It is important to note here that the effect of gravity on fluid flow is negligible when compared to the magnitude of applied fluid pressure.

$$v_\alpha = \frac{k_\alpha K_{r\alpha}}{\mu_\alpha}\left(\frac{dp}{dx}\right)_\alpha = \left(\frac{q_\alpha}{A_\alpha}\right) \tag{4.16}$$

where α = phase, μ = dynamic viscosity, k = intrinsic permeability, K = relative permeability, dp = pressure difference along the joint length, dx, q = flow rate, and A = area perpendicular to the flow.

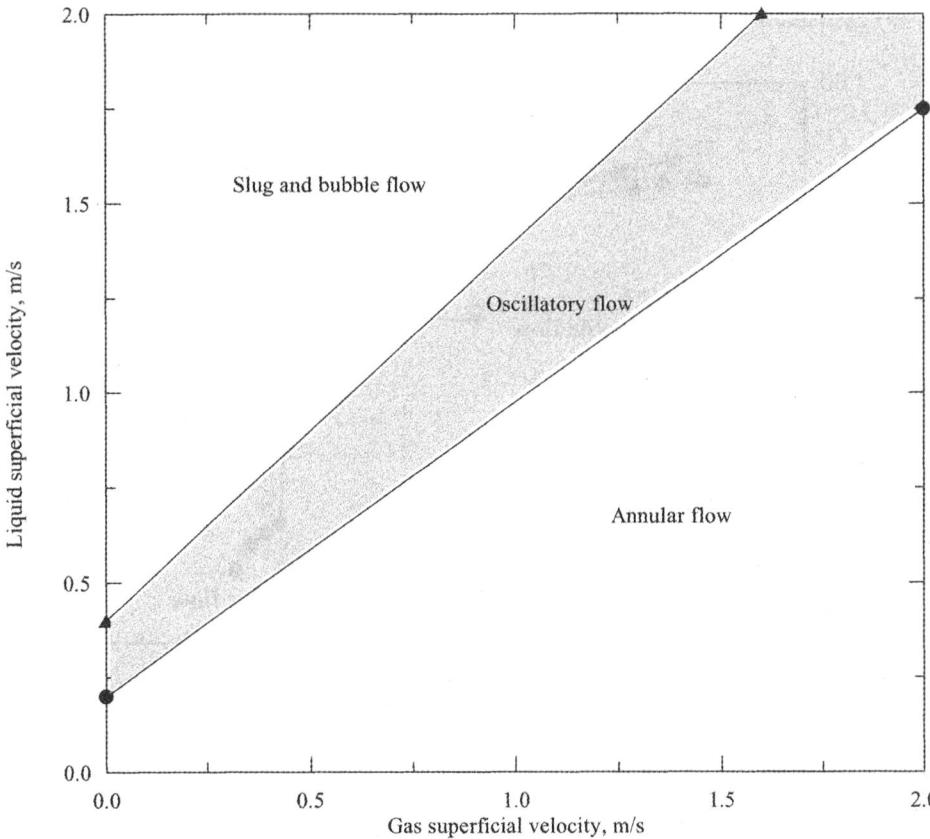

Figure 4.9. Flow patterns in vertical pipe (data from Golan & Stenning, 1969).

Using this technique, the flow patterns within natural rock fractures were studied for different boundary conditions including confining pressure and inlet fluid pressures. The typical flow patterns are shown in Fig. 4.10. For the results shown in Fig. 4.10a and b, the inlet water pressure was maintained at a given constant value of 0.25 MPa and the air pressure was increased gradually from zero until the specimen was fully saturated with air. For the results shown in Fig. 4.10c and d, the inlet air pressure (p_a) and water pressure (p_w) were made to be equal ($p_a = p_w$). A similar flow study for water–gas flow through a pipe was recorded by Golan and Stenning (1969). Although these plots do not clearly distinguish the different flow regimes, they indicate the changes from bubble flow patterns to annular or complex flow regimes.

One of the possible flow mechanisms, which can occur within a joint, is explained below. A simplified flow chart for the formation of different flow regimes is shown in Fig. 4.11, and the corresponding flow regimes are also represented in Fig. 4.12. When the joint is fully saturated with water, it is assumed that the joint has no air (Fig. 4.12a). With the subsequent injection of the air-phase,

(a) Specimen 1, $P_a \neq P_w$

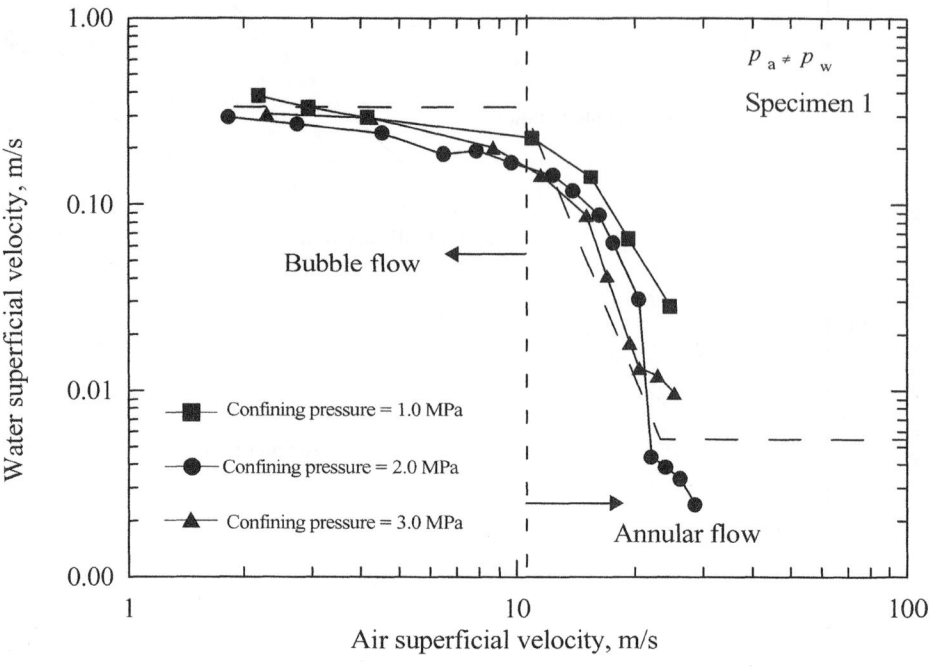

(b) Specimen 2, $P_a \neq P_w$

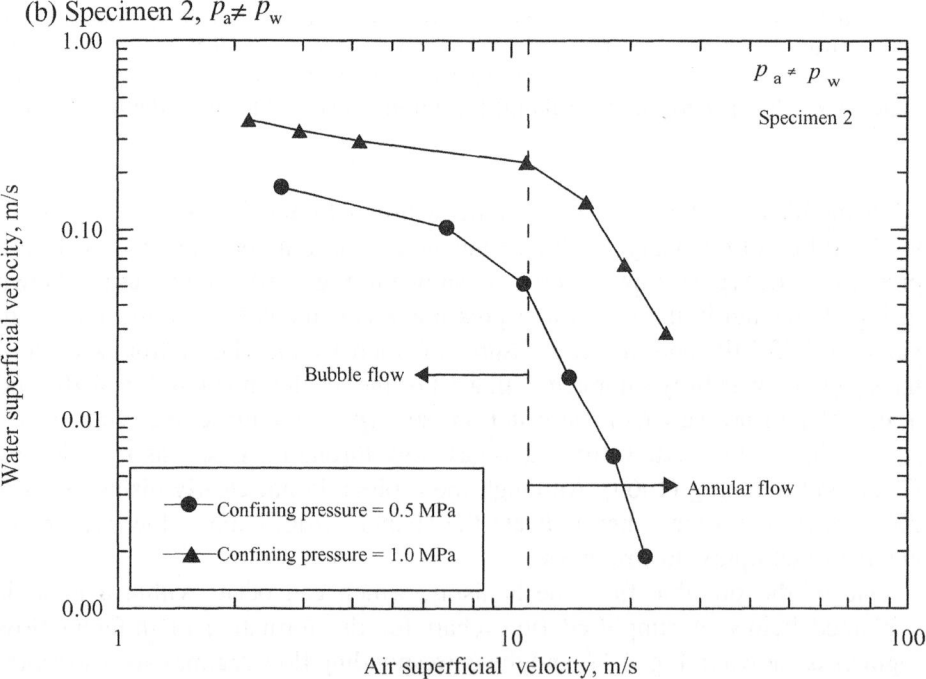

Figure 4.10a,b. Possible flow types in a natural single rock joint for different capillary pressures.

(c) Specimen 1, $P_a = P_w$

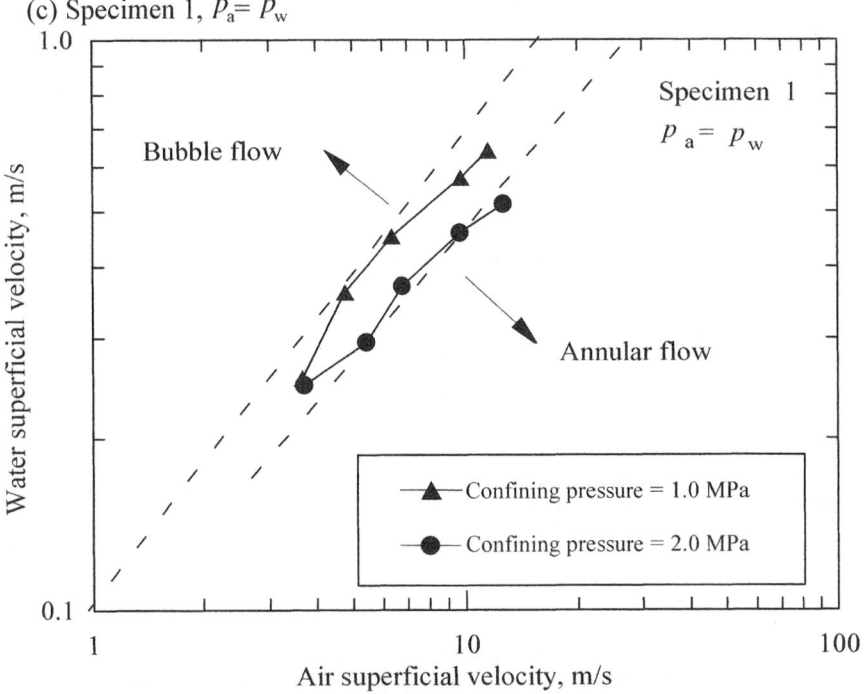

(d) Specimen 2, $P_a = P_w$

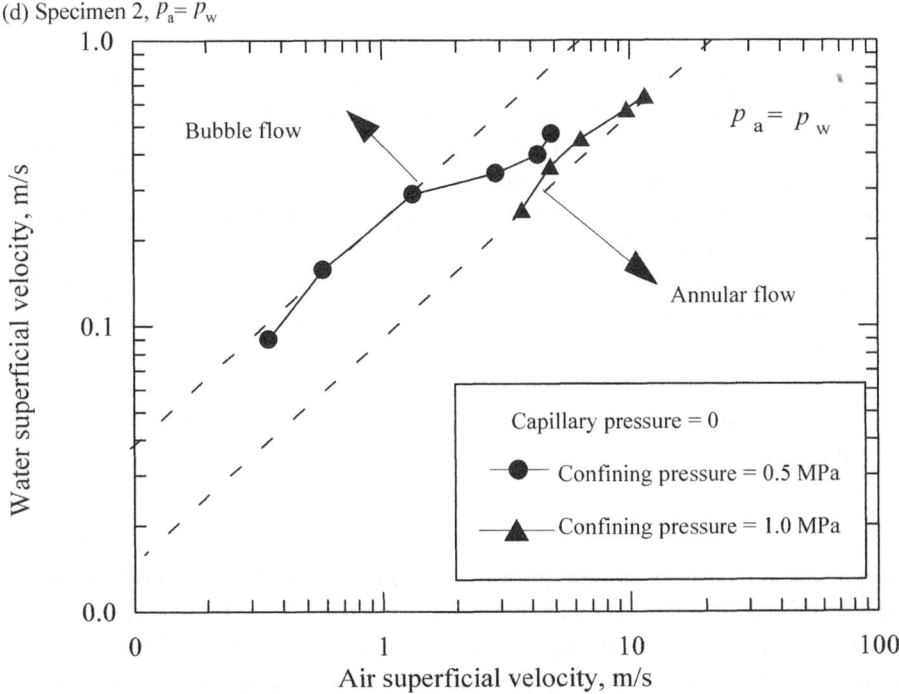

Figure 4.10c,d. Possible flow types in a natural single rock joint at zero capillary pressure.

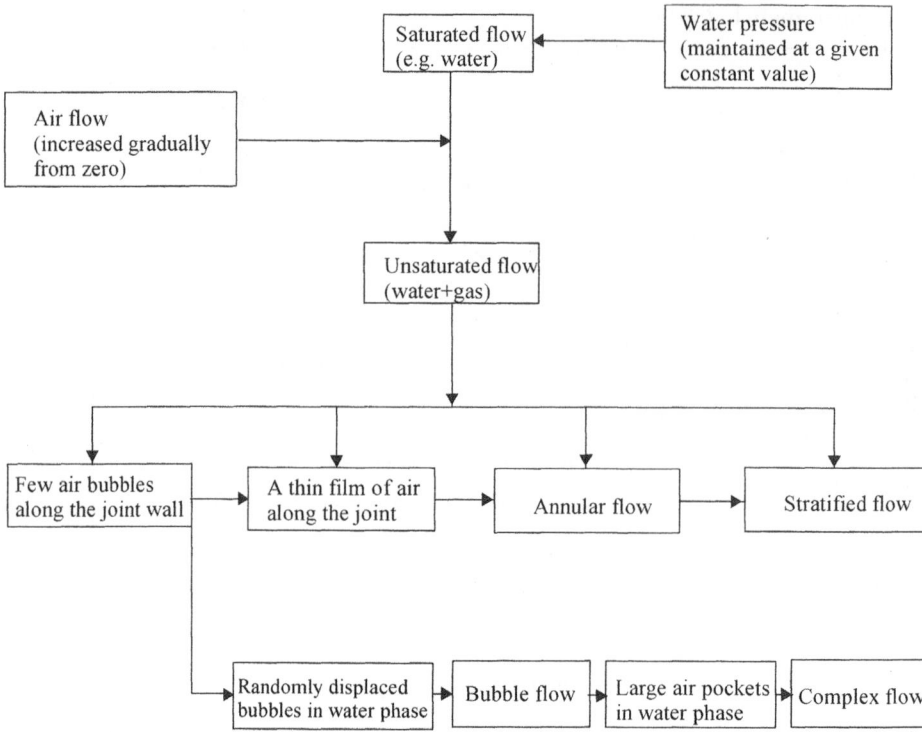

Figure 4.11. Simplified flow chart for formation of flow regimes in a joint.

tiny air bubbles enter the joint and stay along the joint walls (Fig. 4.12b). Further increase in inlet air pressure results in the development of a string of larger bubbles along the joint walls (Fig. 4.12c). The continuous flow of air begins once a film of air develops along the joint walls (Fig. 4.12d).

For elevated capillary pressures ($p_a \gg p_w$), the initial flow pattern may be smooth-stratified or wavy stratified (Fig. 4.12e), and ultimately, single-phase air-flow may result once the water phase is totally replaced by air (Fig. 4.12f). Possible flow patterns in a single rock joint are shown in Fig. 4.13.

4.6 THEORETICAL ASPECTS OF TWO-PHASE STRATIFIED FLOWS: WATER AND AIR

4.6.1 Introduction

The analysis is based on a single joint filled with water and air as stratified layers, which is the simplest flow pattern that can occur (Fig. 4.11). Unlike in two-phase flow through pipes, flows through rocks involve additional variables including continuous deformation of joint fractures associated with changes in ground

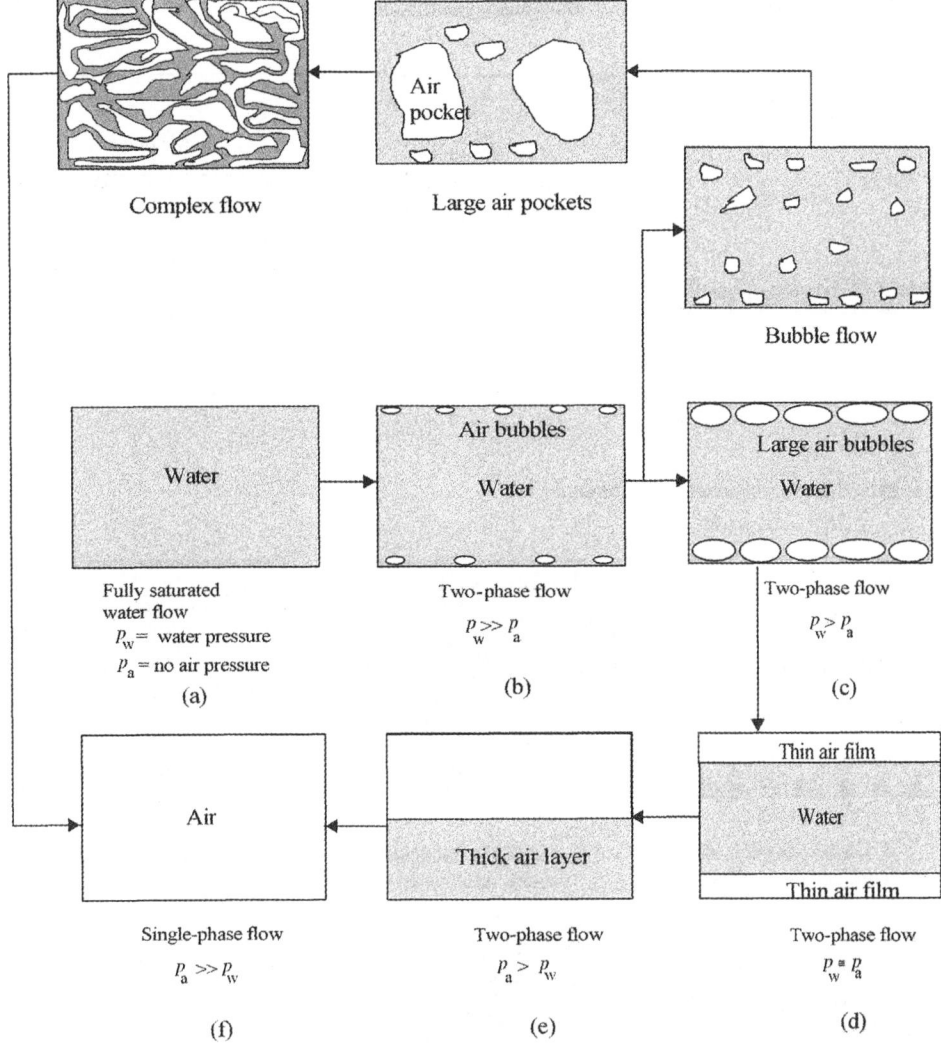

Figure 4.12. Possible flow mechanisms within a joint.

stresses and fluid pressures. Therefore, the use of a simplified stratified flow model to simulate complex flows in rock joints is an initial step to understand the fundamental mechanisms of two-phase flow. The stratified flow is characterized by the liquid flowing as a layer at the bottom section of the discontinuity with gas flowing above it. This kind of flow generally occurs at low gas velocities. An increase in gas flow volume leads to the generation of waves on the gas–liquid interface, giving a stratified but undulating (wavy) flow. Further increases of gas velocity will form different kinds of flow structures, including mixed flow with no continuous interface.

(a) Possible stratified flow patterns (both phases continuous)

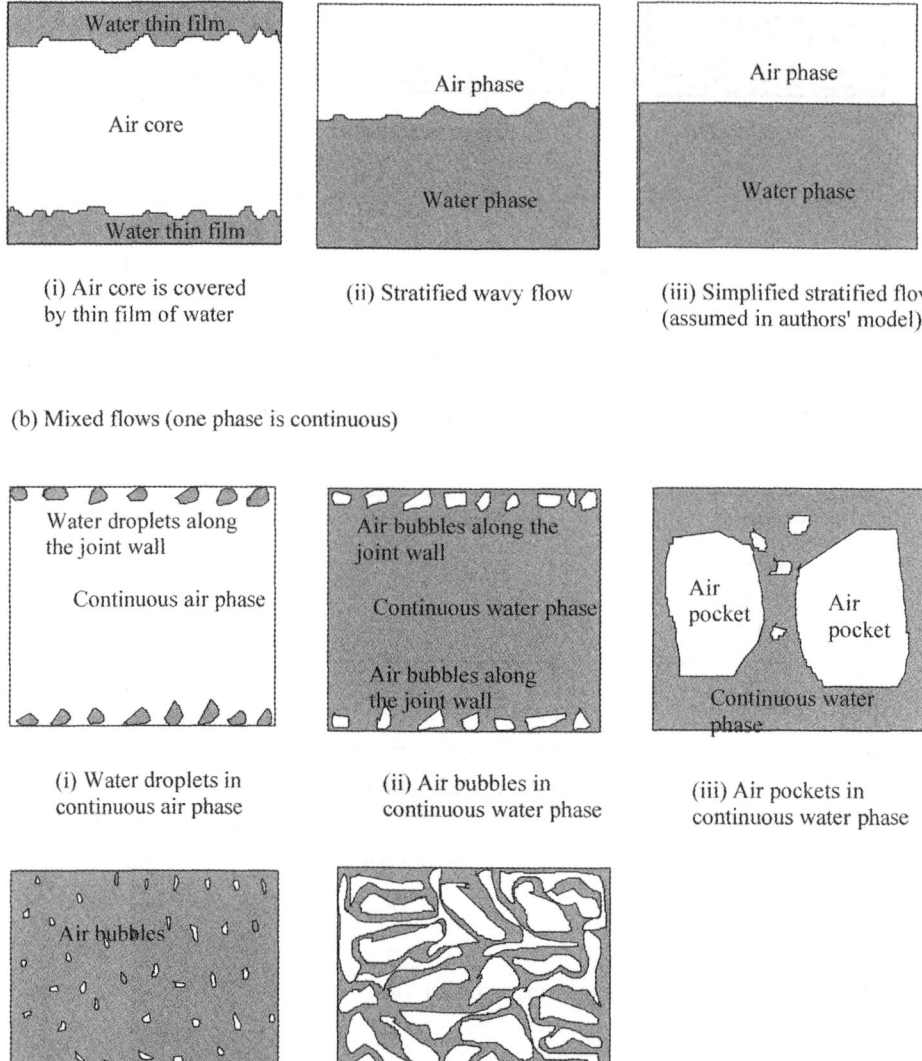

(i) Air core is covered by thin film of water

(ii) Stratified wavy flow

(iii) Simplified stratified flow (assumed in authors' model)

(b) Mixed flows (one phase is continuous)

(i) Water droplets in continuous air phase

(ii) Air bubbles in continuous water phase

(iii) Air pockets in continuous water phase

(iv) Bubble flow

(v) Complex flow

Figure 4.13. Possible flow patterns in a typical rock joint.

4.6.2   Governing equations for two-phase flow

The joint shown in Fig. 4.14, carrying gas and water as layers is initially subjected to confining and vertical stresses. Due to increment of stress with time, the top surface of the joint deforms by a certain amount, and consequently, the

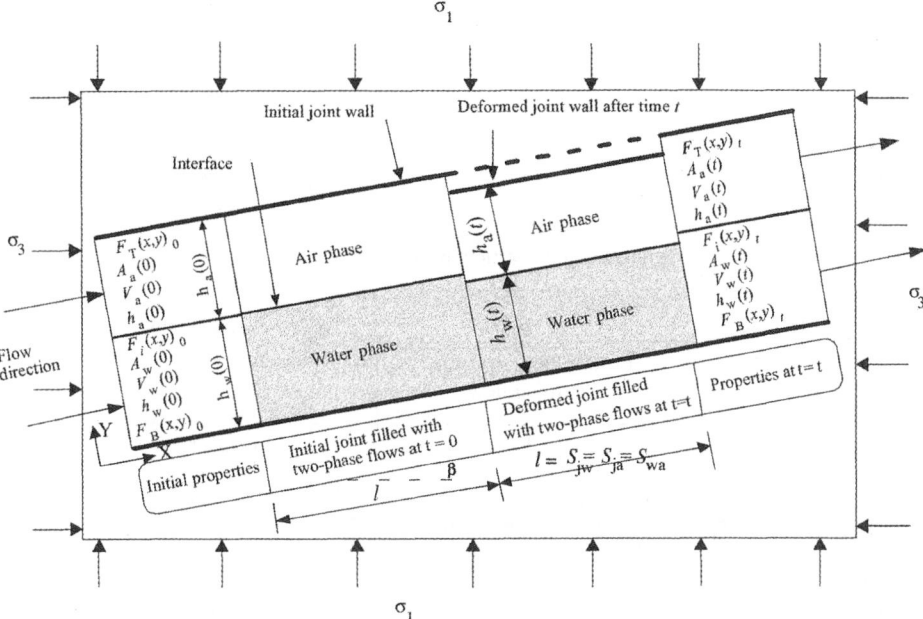

Figure 4.14. A typical inclined joint filled with two-phase flows indicating change of interface and deformation of joint wall associated with change of stress level.

interface level also alters. The flow properties of air and water are time dependent due to the deformation of the joint, as indicated by a function of time t as shown in Fig. 4.14. The mechanical deformation of a joint wall due to the changes in external forces (gravitational stress, e.g. σ_1 and σ_3) and the fluid pressure itself are combined in the model, with factors such as compressibility of fluids, solubility of air in water and phase change of fluids are known to influence the behaviour of the interface between air and water. The general procedure used to analyse the two-phase flow is described by the flow chart given in Fig. 4.15.

The application of momentum conservation in the flow along a given rock joint yields the following governing equations which have been extended by the authors from the original equations proposed by Graham (1969).

For the water phase,

$$(\rho_w v_w A_w)\frac{dv_w}{dx} - (\rho_w A_w)\frac{dv_w}{dt} - (A_w)\left(\frac{dp}{dx}\right)_w - (\rho_w g \sin\beta A_w)$$

$$+ (F_{wa}S_{wa}) - (F_{jw}S_{jw}) - (v_a - v_w)(1 - \eta)\,M\,\frac{dR}{dx}$$

$$+ [(\sigma_1 l \sin\beta\cos\beta - \sigma_3 l\cos\beta\sin\beta) + \sigma_3 h_w \cos^2\beta] = 0 \qquad (4.17)$$

where a = the subscript represents air phase, w = the subscript represents water phase, A = area of the cross-section of the fluid (m²), l = length of fluid element

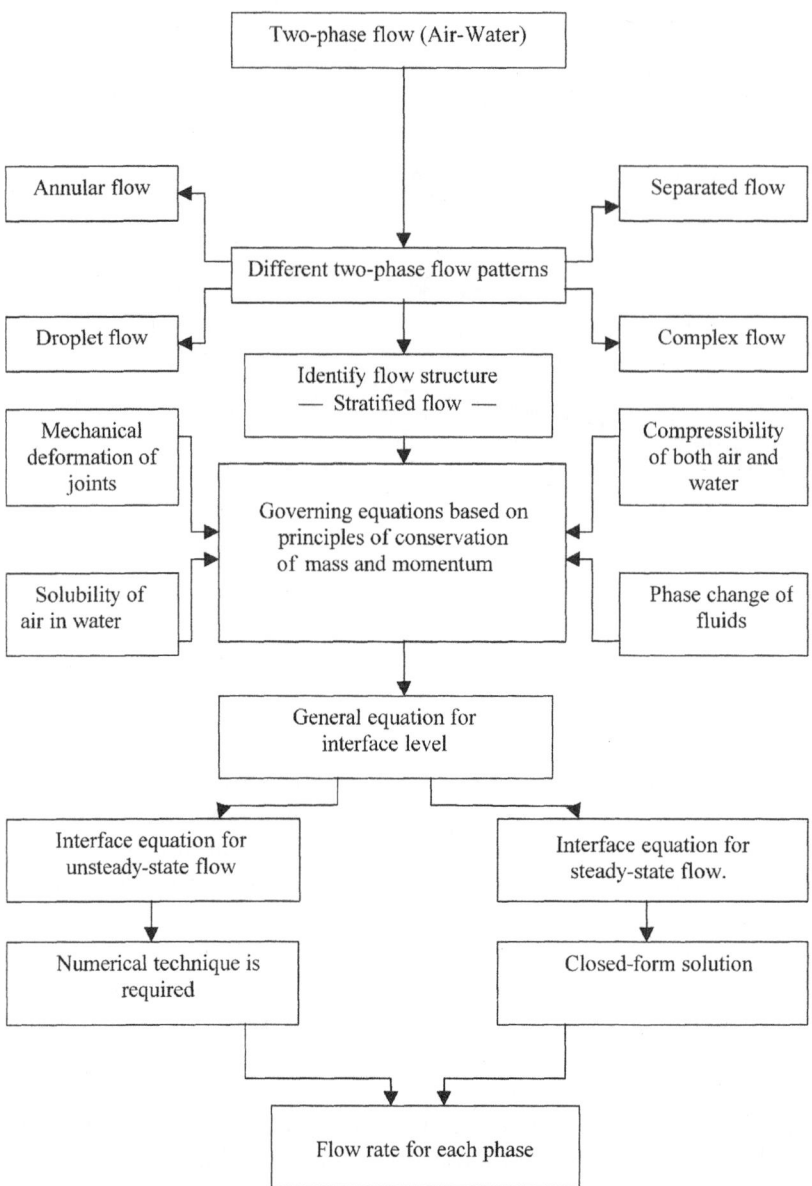

Figure 4.15. Simplified analysis of two-phase stratified flow.

considered (m), ρ = density of fluid (kg/m^3), v = velocity of fluid at time t (m/s), S_{wa} = perimeter of the interface between the phases (m), S_{jw} = perimeter of bottom joint wall (m) and, $F_{wa} = f_I(v_a - v_b)^2\rho_a/2$ = shear stress acting on the water–air interface (N/m^2). It is convenient to express the interfacial and joint wall shear stress in terms of friction factors (Taitel & Dukler, 1976), where f_I is the friction factor between two phases.

$F_{jw} = f_w \rho_w V_w^2/2$ = shear stress acting on the wall (N/m²), where f_w is the friction factor between the wall and water

σ_1 = vertical stress applied on the discontinuity (N/m²)

σ_3 = horizontal stress applied on the discontinuity (N/m²)

M = total mass rate (kg/s) = $(Q_a \rho_a + Q_w \rho_w)$

Q_a = volumetric flow rate of air-phase (m³/s)

Q_w = volumetric flow rate of water-phase (m³/s)

R = air fraction of the total mass flow across the interface = $M_1/M = Q_a \rho_a/(Q_a \rho_a + Q_w \rho_w)$

η = fraction of the force taken by air phase associated with phase change

β = orientation of the joint relative to the horizontal (degrees)

p = fluid pressure inside the joint (N/m²)

x = joint length (m)

Similarly, for the air-phase

$$(\rho_a v_a A_a)\frac{dv_a}{dx} - (\rho_a A_a)\frac{dv_a}{dt} - (A_a)\left(\frac{dp}{dx}\right)_a - (\rho_a g \sin \beta A_a)$$

$$-(F_{wa}S_{wa}) - (F_{ja}S_{ja}) - (V_a - V_w)\eta M \frac{dR}{dx}$$

$$+(\sigma_3 h_a \cos \beta \sin \beta - \sigma_1 l \sin \beta \cos \beta + \sigma_3 l \cos^2 \beta) = 0 \qquad (4.18)$$

where S_{ja} = perimeter of the top joint wall (m) and $F_{ja} = f_a \rho_a v_a^2/2$ = shear stress acting on the wall (N/m²), where f_a is the friction factor between wall and air. The other notations are similar to water phase notations with the subscript "a" representing the air-phase.

The pressure drop dp/dx needs to be eliminated from the above two equations in order to attain a solution for the interface level. The pressure drop along the joint length dp/dx is given by the following equation, after a mathematical rearrangement of Eqn. 4.17.

$$\left(\frac{dp}{dx}\right)_w = (\rho_w v_w)\frac{dv_w}{dx} - (\rho_w)\frac{dv_w}{dt} - (\rho_w g \sin \beta)$$

$$+\frac{(F_{wa}S_{wa})}{A_w} - \frac{(F_{jw}S_{jw})}{A_w} - \frac{1}{A_w}(v_a - v_w)(1 - \eta)M\frac{dR}{dx}$$

$$+\frac{1}{A_w}\left[l \sigma_1 \sin \beta \cos \beta - l \sigma_3 \cos \beta \sin \beta + \sigma_3 h_w \cos^2 \beta\right] \qquad (4.19)$$

The cross-sectional area of the water phase may be written in terms of the phase level and the length of wetted perimeter of the joint wall. The wetted perimeter of the interface and the joint walls (i.e. S_{wa}, S_{ja} and S_{jw}) can be assumed to be same, since the joint aperture is very small compared to the length of the wetted perimeter. Therefore, the pressure drop may be re-written as:

$$\left(\frac{dp}{dx}\right)_w = (\rho_w v_w)\frac{dv_w}{dx} - (\rho_w)\frac{dv_w}{dt} - (\rho_w g \sin \beta) + \frac{(F_{wa})}{h_w(t)}$$

$$- \frac{(F_{jw})}{h_w(t)} - \frac{1}{l\,h_w(t)}(v_a - v_w)(1 - \eta)\,M\frac{dR}{dx}$$

$$+ \frac{1}{h_w(t)}(\sigma_1 \sin \beta \cos \beta - \sigma_3 \cos \beta \sin \beta) + \frac{1}{l}\sigma_3 \cos^2 \beta \qquad (4.20)$$

When capillary forces are negligible (compared to viscous forces), the pressure-gradient in both phases is equal $[ie.(dp/dx)_w = (dp/dx)_a = (dp/dx)]$. The pressure drop (dp/dx) needs to be eliminated from the above two equations in order to attain a solution for the interface level. Having written a similar expression (as Eqn. 4.20) to the air-phase, the following equation can be obtained after eliminating the (dp/dx) term to give:

$$\frac{1}{h_w(t)}\left[\sigma_1 \sin \beta \cos \beta - \sigma_3 \cos \beta \sin \beta + F_{wa} - F_{jw}\right.$$

$$\left.- \frac{1}{l}(v_a - v_w)(1 - \eta)M\frac{dR}{dx}\right] + \frac{1}{h_a(t)}\left[\sigma_1 \sin \beta \cos \beta\right.$$

$$\left.- \sigma_3 \cos \beta \sin \beta + F_{wa} + F_{ja} + \frac{1}{l}(v_a - v_w)\eta\,M\frac{dR}{dx}\right]$$

$$= \rho_w v_w \frac{dv_w}{dx} - \rho_a v_a \frac{dv_a}{dx} + \rho_w \frac{dv_w}{dt} - \rho_a \frac{dv_a}{dt} - g\sin \beta[\rho_a - \rho_w] \qquad (4.21)$$

The mechanical and hydraulic behaviour of discontinuities in rocks depend strongly on the topography of the joint surfaces and the degree of correlation between them. The topography of joint walls has been studied using various techniques including profilometers (Pyrak-Notle et al., 1992; Brown et al., 1987). Brown (1987) studied the effects of surface roughness of rock joints on fluid flow along discontinuities using the Reynolds equation and a fractal model of surface topology. The effects of surface roughness on flow were discussed earlier in Ch. 3. The topography of fracture surfaces are quantitatively described here by specifying the location of the top and bottom fracture surfaces by $F_T(x, y)_t$ and $F_B(x, y)_t$ respectively, relative to some co-ordinate system as shown in Fig. 4.15. Unlike in single-phase flow through discontinuities, for two-phase flow, the geometry of the interface between the two phases can be defined by $F_I(x, y)_t$ relative to the same co-ordinate system. The initial surfaces of top and bottom joint walls profiles are taken as $F_T(x, y)_0$, and $F_B(x, y)_0$, respectively. When they are subjected to deformation associated with external stresses and fluid pressure after time t, they are assumed to take the form of $F_T(x, y)_t$, and $F_B(x, y)_t$, respectively. The interface between the two phases is assumed to be $F_I(x, y)_0$, and $F_I(x, y)_t$ at the beginning and after time t, respectively.

The water-phase level (height), $h_w(t)$, can be represented as

$$h_w(t) = F_I(x, y)_t - F_B(x, y)_t \qquad (4.22)$$

If ξ_{wc} is the change of interface level between two phases, due to compressibility of water, then,

$$h_w(t) = F_I(x, y)_0 - F_B(x, y)_0 - \xi_{wc}, \quad \text{or}$$
$$h_w(t) = F_I(x, y)_0 - F_B(x, y, \Delta_B)_t \qquad (4.23)$$

where $F_B(x, y, \Delta_B)_t$ is given by the expression, $F_B(x, y)_0 + \xi_{wc}$.

In the same manner, the air-phase level, $h_a(t)$, at time t is given by:

$$h_a(t) = F_T(x, y)_t - F_I(x, y)_t \qquad (4.24)$$

Factors which control the air-phase level are (a) mechanical deformation of joint, (b) compressibility of air, (c) rate of solubility of air in water and (d) effects of change of fluid properties and temperature. Hence, Eqn. 4.24 can be linked to the initial condition by

$$h_a(t) = F_T(x, y)_0 - F_I(x, y)_0 - \Delta_T \qquad (4.25)$$

where Δ_T is the deformation of wetted wall in contact with the air-phase.

The total deformation, Δ_T, includes the effects of compressibility of water (ξ_{wc}), compressibility of air (ξ_{ac}), solubility of air in water (ξ_{ad}), and the elastic deformation of the joint wall (δ_n) on the air-phase level $h_a(t)$. Hence,

$$\Delta_T = \xi_{ac} + \xi_{ad} + \delta_n - \xi_{wc} \qquad (4.25a)$$

The evaluation of the functions ξ_{ac}, ξ_{ad}, δ_n and ξ_{wc} will be discussed later.

If $F_T(x, y, \Delta_T)_t$ is represented by the expression $[F_T(x, y)_0 - \Delta_T]$, Eqn. 4.25 can now be re-written as:

$$h_a(t) = F_T(x, y, \Delta_T)_t - F_I(x, y)_0 \qquad (4.26)$$

Substituting Eqns. 4.23 and 4.26 into the Eqn. 4.21 yields

$$\begin{aligned}
&\Big\{ \sigma_1 \sin\beta \cos\beta \, [F(x, y, \Delta)] + F_{wa}[F(x, y, \Delta)] \\
&\quad - [F_{jw} + \sigma_3\cos\beta \sin\beta] F_T(x, y, \Delta_T) + [\sigma_3\cos\beta \sin\beta - F_{ja}] F_B(x, y, \Delta_B) \\
&\quad - \frac{1}{l}(v_a - v_w)(1 - \eta) M \frac{dR}{dx} [(1 - \eta)F_T(x, y, \Delta_T) + \eta \, F_B(x, y, \Delta_B)] \\
&\quad + F_I(x, y)_0 \Big[F_{jw} + F_{ja} + \frac{1}{l}(v_a - v_w)M \frac{dR}{dx} \Big] \Big\} \\
&= \{ F_I(x, y)_0 [F_T(x, y\, \Delta_T) + F_B(x, y, \Delta_B)] - F_I^2(x, y)_0 \\
&\quad - F_T(x, y, \Delta_T) F_B(x, y, \Delta_B) \} \{ A - g \sin\beta \, (\rho_a - \rho_w) \} \qquad (4.27)
\end{aligned}$$

where

$$F(x, y, \Delta) = F_T(x, y, \Delta_T) - F_B(x, y, \Delta_B) \tag{4.27a}$$

The above expression may be presented in a simplified form as

$$\left[F_I(x, y)_0 \sum_{k=1}^{2} \Delta_k - F_I^2(x, y)_0 - \Delta_1 \Delta_2 \right][A]$$

$$= F(x, y, \Delta)[\sigma_1 \sin \beta \cos \beta - \sigma_3 \cos \beta \sin \beta + F_{wa}]$$

$$- \left[\sum_{\substack{k=1 \\ i=a,w}}^{2} F_{Ji} \Delta_k \right] - \sum_{\substack{N=1-\eta,\eta \\ k=1}}^{(1-\eta)} CN\Delta_k + F_I(x, y)_0 \left[\sum_{i=a,w} F_{Ji} + C \right] \tag{4.28}$$

$$A = [B_1 - g \sin \beta (\rho_a - \rho_w)] \tag{4.28a}$$

where $B_1 = [\rho_w v_w dv_w/dx - \rho_a v_a dv_a/dx + \rho_w dv_w/dt - \rho_a dv_a/dt] =$ force per unit area associated with unsteady effects of flow. $\tag{4.28b}$

$$C = \left[\frac{1}{l} (v_a - v_w)M \frac{dR}{dx} \right] \tag{4.28c}$$

$$\Delta_1 = F_T(x, y, \Delta_T), \text{ and} \tag{4.28d}$$

$$\Delta_2 = F_B(x, y, \Delta_B) \tag{4.28e}$$

Alternately, Eqn. 4.28 may be presented as

$$F_I(x, y)_0 \left[A \sum_{k=1}^{2} \Delta_k - \sum_{i=a, w} F_{Ji} - C \right] - A F_I^2(x, y)_0$$

$$= F(x, y, \Delta)[\sigma_1 \sin \beta \cos \beta - \sigma_3 \cos \beta \sin \beta + F_{wa}]$$

$$- \left[\sum_{\substack{k=1 \\ i=a, w}}^{2} F_{Ji} \Delta_k \right] - \sum_{\substack{N=1-\eta, \eta \\ k=1}}^{2} CN\Delta_k + A\Delta_1 \Delta_2 \tag{4.29}$$

Equation 4.29 can also be simplified to write the main governing equation as follows:

$$F_I(x, y)_0 \Delta_3 - A F_I^2(x, y)_0 - D = 0 \tag{4.30}$$

where

$$\Delta_3 = \left[A \sum_{k=1}^{2} \Delta_k - \sum_{i=a, w} F_{Ji} - C \right], \tag{4.30a}$$

and

$$D = F(x, y, \Delta)[\sigma_1 \sin \beta \cos \beta - \sigma_3 \cos \beta \sin \beta + F_{wa}]$$

$$- \left[\sum_{\substack{k=1 \\ i=a, w}}^{2} F_{Ji} \Delta_k \right] - \sum_{\substack{k=1 \\ N=1-\eta, \eta}}^{2} CN\Delta_k + A\Delta_1 \Delta_2 \tag{4.30b}$$

Equation 4.30 in quadratic form, provides a definite solution to represent the interface level, which facilitates the computation of both water and air levels at any time, incorporating the water and air levels given earlier by Eqns. 4.23 and 4.26.

In two-phase flow, the main difficulty is the location of the interface, which if known will enable the evaluation of other parameters such as the flow quantity of each phase at a given time. Although the solution of the governing Eqn. 4.30 is somewhat cumbersome, the apparent complexity will be lessened under steady-state flow conditions, on which most practical engineering solutions are based. For steady-state flow, the acceleration term (B_1) in Eqn. 4.28b vanishes. Moreover, the gravity effects due to air can be neglected, noting that the density of air at atmospheric pressure (1.23 kg/m^3 at 15° C) is insignificant compared to that of water density (1000 kg/m^3 at 15° C). Further simplification will provide a new expression for the term A in Eqn. 4.28a, as given by

$$A = [\rho_w \, g \sin \beta] \tag{4.31}$$

By assuming that no sudden temperature change occurs, it is then reasonable to neglect any change of properties within the phases. This will make the term (dR/dx) associated with phase change to be zero in Eqns. 4.18–4.21 and 4.27–4.30.

A simple form of expression for the interface level could be found, if one would ignore the gravity effects of fluid. Such assumptions are not unreasonable given that, when computing flow rates using Darcy's law it is customary to ignore the gravity effects in conventional soil mechanics. The effects of horizontal stress (σ_3) acting on the face of a horizontal joint segment can be neglected, as the effective area of the face on which the horizontal stress acts, is insignificant. Considering the same assumptions, for the interface, Eqn. 4.30 will take the following simplified form, for steady-state flow through a horizontal joint:

$$F(x, y, \Delta)[F_{wa}] - \sum_{\substack{k=1 \\ i=a,w}}^{2} F_{Ji} \, \Delta_k + F_I(x, y)_0 \sum_{i=a,w} F_{Ji} = 0 \tag{4.32}$$

$$F_I(x, y)_0 = \frac{\displaystyle\sum_{\substack{k=1 \\ i=a,w}}^{2} F_{Ji} \, \Delta_k - F(x, y, \Delta) [F_{wa}]}{\displaystyle\sum_{i=a,w} F_{Ji}} \tag{4.33}$$

$$F_I(x, y)_0 = \frac{F_{Ja} F_B(x, y, \Delta_B) + F_{Jw} F_T(x, y, \Delta) - F_{wa}(x, y, \Delta)}{F_{Ja} + F_{Jw}} \tag{4.34}$$

where $F(x, y, \Delta) = F_T(x, y, \Delta_T) - F_B(x, y, \Delta_B)$, $\Delta_2 = F_T(x, y, \Delta_T)$, and $\Delta_1 = F_B(x, y, \Delta_B)$.

$F_{Ji} = F_{Ji}$ where i takes subscripts 'a' and 'w' for air and water, respectively. For the conditions of steady-state flow through a horizontal joint and with the assumptions that the effect of horizontal stress on joint segment is insignificant, the pressure drop $(\mathrm{d}p/\mathrm{d}x)$ given by general Eqn. 4.20 can now be simplified to

$$\frac{\mathrm{d}p}{\mathrm{d}x} = \frac{F_{\mathrm{wa}}}{h_{\mathrm{w}}(t)} - \frac{F_{\mathrm{jw}}}{h_{\mathrm{w}}(t)}$$

Equation 4.34 represents the surface of discontinuity between phases for steady-state flow through a single horizontal joint. The above solution confirms that the following parameters govern the position of the interface level for two-phase flow through a horizontal discontinuity:

(1) External stress (e.g. gravitational stress, fluid pressure);
(2) Surface topography of joint walls;
(3) Initial discontinuity aperture;
(4) Compressibilities of the fluids and solubility of air in water; and
(5) Pressure causing fluid flow.

4.7 EFFECTS OF SOLUBILITY, COMPRESSIBILITY AND DEFORMATION ON FLOW RATES

This section is focused on developing a mathematical expression for flow rate of each phase. The deformation of a joint associated with the compressibilities of air and water, solubility of air in water, fluid pressure itself and external forces such as gravitational stresses are discussed below. The changes in the properties of both phases and their effects on the flow rates are also discussed.

4.7.1 Effects of solubility

The water phase initially has an amount of dissolved air, which is in equilibrium with the free air. As a result of pressure increase in the air-phase due to the mechanical deformation of joint walls, the properties governing the flow such as density and fluid pressure will change. The temperature variation will also influence these properties. Because of the pressure difference between the air-phase and the water-phase, some air (in air-phase) will dissolve in the water-phase according to Henry's law described below. This process will continue until a new equilibrium is attained (i.e. new pressure in the air-phase is the same as the new pressure of dissolved air in the water-phase). The dissolved quantity of air in water can be predicted by Henry's law assuming the air to be an ideal gas: $p = mc$, where p is the partial pressure in the air phase (atm), c is the concentration of soluble component in liquid (lb moles/cuft) and m is the Henry's constant (atm cuft/mole). The mass transfer from the air phase to water phase takes place in two stages: (a) air phase to interface and (b) interface to water phase. The solubility of air in water at the equilibrium state is best described by the ideal gas law and Henry's law at given pressure and temperature conditions

(Fredlund & Rahardjo, 1993):

$$V_{dt} = \left(\frac{M_d}{p_a}\right)_t \left(\frac{RT}{W_a}\right)_t \tag{4.35}$$

In the above equation, V_{dt} is the volume of dissolved air in water at time t, M_d and p_a are the mass of dissolved air in water and absolute pressure of the air, respectively. W_a is the molecular mass of air (kg/kmol), R is the universal gas constant (8.314 J/mol·K) and T is the absolute temperature (K) ($T = 273 + t°$) where $t°$ is the temperature in °C.

For constant temperature, the ratio of the mass of dissolved air to absolute air pressure can be given as follows (Universal gas law):

$$\frac{M_{d0}}{p_0} = \frac{M_{dt}}{p_{at}} \tag{4.36}$$

where the subscripts "0" and "t" represent the initial conditions ($t = 0$) and at any given time, t. At constant temperature, the volume of dissolved air in water-phase takes a constant value at different pressures. The change of equivalent air phase level (ξ_{ad}) associated with dissolved air may be calculated for a given joint length (l) per unit width, as follows:

$$\xi_{ad} = \frac{V_{dt}}{l} \tag{4.37}$$

Air can dissolve in water and can occupy approximately 2% by volume of water (Dorsey, 1940). The coefficient of solubility of a gas in liquid is defined as the volume of the gas which is measured at the same phase conditions contained in a unit volume of the saturated solution. The coefficient of solubility and volumetric coefficient of solubility of different gases for various temperatures have been discussed by Dorsey (1940). At an atmospheric pressure of 1, Fig. 4.16 shows the solubility of air in water at various temperatures. This shows a decrease in solubility with increase in temperature. At elevated temperatures, the rate of solubility is very small. The effect of pressure on the solubility of some gases in water is illustrated in Fig. 4.17.

4.7.2   Effects of compressibility of air

The compressibility of a fluid is a measure of the change in density due to the specified change in pressure. Generally, pressure change will induce density changes, which will influence other flow parameters, such as the flow rate and velocity. A knowledge of the compressible fluid flow theory is required in various engineering applications, such as gas turbines, steam turbines, combustion chambers and natural transmission pipe lines. More significantly, the compressibility of gas is an essential consideration in coupled water–gas flow through rock joints.

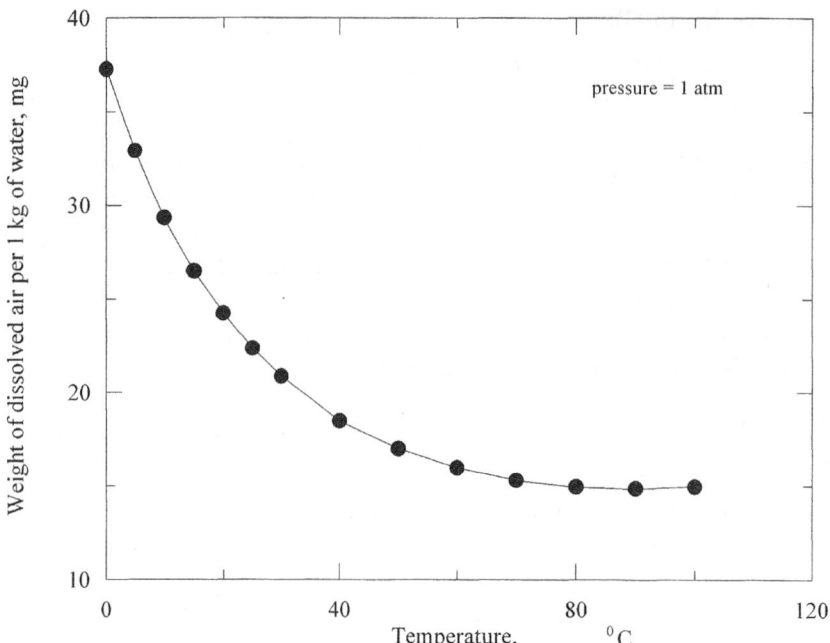

Figure 4.16. Solubility of air in water at different temperatures (data from Dorsey, 1940).

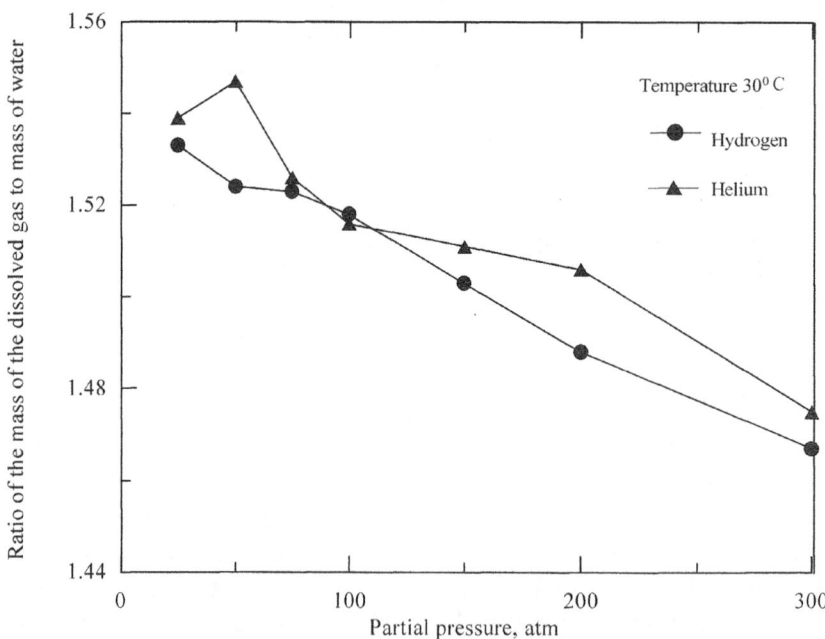

Figure 4.17. Effect of pressure on the solubility of gases in water (data from Dorsey, 1940).

In a simplified compressible flow theory, the following assumptions are made:
(1) The gas is a perfect fluid (i.e., $PV = mRT$);
(2) Gas is a continuous phase;
(3) Gravity effects on flow are negligible; and
(4) No chemical changes occur between the phases.

Under such conditions, the general equations for gas flow are based on (a) continuity equation, (b) momentum equation and (c) law of thermodynamics. Compared to water, gases are highly compressible, and the changes in gas density are directly related to the changes in the pressure and temperature through Charles' equation:

$$p = \rho R T \tag{4.38}$$

where p is the absolute the pressure, ρ is the density of gas, R and T are the gas constant and the absolute temperature, respectively.

The compressibility of air at a constant temperature with respect to pressure is described by (Fredlund & Rahardjo, 1993)

$$C_a = -\frac{1}{V_a}\left(\frac{dV_a}{dp_a}\right)_t \tag{4.39}$$

where C_a = compressibility of air (m^2/N), V_a = volume of air, p_a = pressure of air, t is time and dp_a = change of air pressure.

For a joint length, l, the change of equivalent air-phase level due to compressibility of air is given by

$$\xi_{ac} = \frac{C_a V_a dp_a}{l} \tag{4.40}$$

The effect of ξ_{ac} on the air-phase level $h_a(t)$ was discussed earlier via Eqn. 4.25a.

4.7.3 Effects of compressibility of water

The compressibility of water due to the deformation of joint walls is discussed in this section. Water has a very low value of compressibility [4.58×10^{-7} (1/kPa)] compared to gas [4.94×10^{-3} (1/kPa)]. A property that is commonly used to characterise the compressibility of water is the bulk modulus [$(K = dp/(dv/V)]$. For the analysis presented here, the term compressibility coefficient (C_w) is adopted, where $C_w = 1/K$.

Similar to Eqn. 4.40, the change of equivalent water-phase level due to the compressibility effects of water is given by

$$\xi_{wc} = \frac{C_w V_w dp_w}{l} \tag{4.41}$$

The effect of ξ_{wc} on the water-phase level $h_w(t)$ and air-phase $h_a(t)$ was introduced earlier through Eqns. 4.23 and 4.25a, respectively. The accuracy of Eqn. 4.41

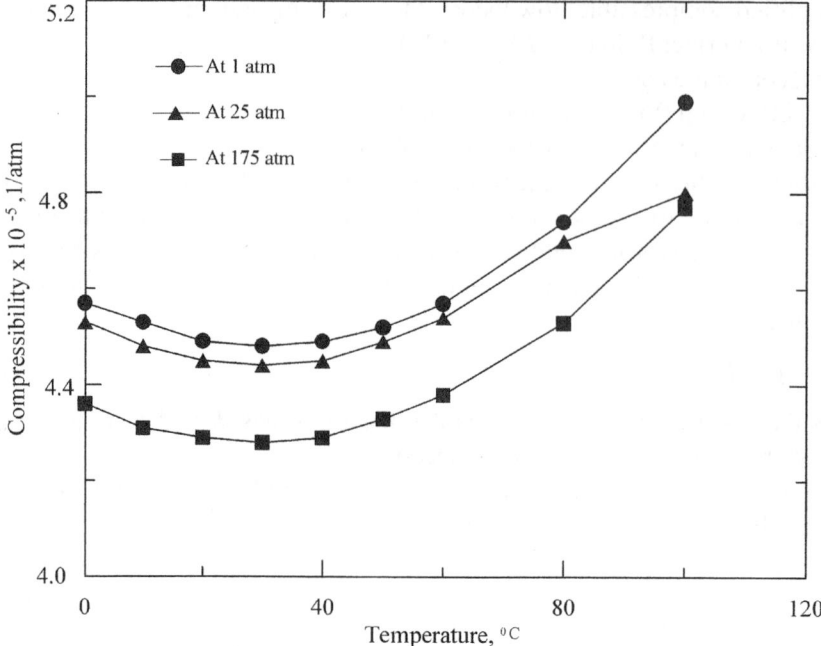

Figure 4.18. Effect of temperature on compressibility of water at different pressures (data from Dorsey, 1940).

may be less valid for air-dissolved water, which inevitably occurs in real discontinuities. However, Dorsey (1940) has shown that there is no significant difference between the compressibility of air-free water and that of air-saturated water. Figure 4.18 presents the effect of temperature on the compressibility of water for different pressures.

4.7.4 Influence of water and air density

In two-phase flow, on one hand, the pressure changes caused by deformation of joint walls or inlet velocity of air will generally induce density changes, which in turn will affect the flow rate. On the other hand, the temperature changes in the flow, which arise due to the kinetic energy changes associated with the velocity changes will also influence the flow quantity. In the following analysis, the effects of temperature changes are omitted for simplicity, in which case, the changed density of air-phase at time t is determined by (Fredlund & Rahardjo, 1993):

$$\rho_a(t) = \frac{p_a(t)}{p_a(0)} \rho_a(0) \tag{4.42}$$

where $\rho_a(t)$ = final density of air corresponding to final pressure $p_a(t)$, $p_a(0)$ = initial pressure of air in air-phase and $\rho_a(0)$ = initial density of air at $t = 0$.

As a result of dissolved air in water, the water-phase will be characterized by the following density term:

$$\rho_{\mathrm{w}}(t) = \frac{M_{\mathrm{w}}(0) + M_{\mathrm{da}}(t)}{V_{\mathrm{w}}(t)} \qquad (4.43)$$

where $M_{\mathrm{w}}(0)$ = initial mass of water, $M_{\mathrm{w}}(t)$ = final mass of dissolved air in water and $V_{\mathrm{w}}(t)$ = final volume of water (i.e. water + dissolved air).

Equation 4.43 may be presented in a different way by incorporating the Boyle's law, and assuming that the final volume of water and initial volume will be the same, hence,

$$\rho_{\mathrm{w}}(t) = \rho_{\mathrm{w}}(0) + \left[\frac{V_{\mathrm{da}}(t)}{V_{\mathrm{w}}(0)}\right]\left[\frac{p_{\mathrm{da}}(t)}{p_{\mathrm{a}}(0)}\right][\rho_{\mathrm{a}}(0)] \qquad (4.44)$$

The deviation of the density of water from standard conditions (1 atm, 20° C) to extreme conditions such as high pressure and low temperature (300 atm, 0° C), or high temperature and low pressure (100° C, 0 atm), can be ignored in most geotechnical applications. Therefore, in this analysis, the second term of Eqn. 4.44 has been neglected [i.e. $\rho_{\mathrm{w}}(t) = \rho_{\mathrm{w}}(0)$].

4.7.5 Deformation of joint due to external stress

The aperture variations of discontinuities due to stress changes are examined in this section. The rock material is considered as impermeable and the flow is assumed to be confined to the discontinuities. Moreover, the rock matrix is assumed to be isotropic and linear elastic, obeying Hooke's law. The aperture of the discontinuity, e_t, at any time is given by

$$e_t = e_0 \pm \delta_{\mathrm{n}} \qquad (4.45)$$

where e_0 = aperture at time $t = 0$, e_t = aperture at time t and δ_{n} = aperture increment during time interval t

In conventional rock mechanics, the normal and shear deformation components are given by

$$\delta_{\mathrm{n}} = \frac{1}{k_{\mathrm{n}}}[\sigma_1 \cos^2\beta + \sigma_3 \sin^2\beta] \qquad (4.46a)$$

and

$$\delta_\tau = \frac{1}{k_{\mathrm{s}}}[\sigma_3 \cos^2\beta - \sigma_1 \sin^2\beta] \qquad (4.46b)$$

Considering water pressure acting perpendicular to the joint surface, Eqn. 4.46a can be modified to give

$$\delta_{\mathrm{n}} = \frac{1}{k_{\mathrm{n}}}[\sigma_1 \cos^2\beta + \sigma_3 \sin^2\beta - p_{\mathrm{w}}] \qquad (4.46c)$$

and δ_τ remains unchanged, where σ_1 = vertical stress applied to the discontinuity; σ_3 = horizontal stress applied to the discontinuity; k_n = normal stiffness of discontinuity; k_s = shear stiffness of discontinuity; p_w = water pressure within the discontinuity; β = orientation of discontinuity relative to the horizontal; δ_n = normal aperture of discontinuity; and δ_τ = tangential displacement of discontinuity.

In reality, δ_n may be a function of both water and air pressures (p_w and p_a), but it is assumed for simplicity that a critical value for δ_n will be associated with p_w only. One may describe the elasto-plastic behaviour in the mechanical deformation of the rock joint by the Mohr–Coulomb theory. However, it is assumed that yielding of rock with respect to the Mohr–Coulomb failure envelope will not occur, and that elastic conditions will prevail.

Knowing the individual components, ξ_{ad}, ξ_{ac}, ξ_{wc} and δ_n from Eqns. 4.37, 4.40, 4.41 and 4.46, Eqn. 4.23 representing $h_w(t)$ and Eqn. 4.25 representing $h_a(t)$ can both be solved.

4.8   FLOW LAWS RELATED TO TWO-PHASE FLOW

Once the governing equations for interface level, and phase levels of air and water have been evaluated, the next task is to evaluate the flow rates of each phase through a given discontinuity. Unlike in single-phase flow, there can be more than one driving potential for the flow rate in two-phase flow. For example, air–water stratified flow can have two components: (a) for air the phase which is governed by the pressure gradient with negligible gravity effects, and (b) the water phase which is usually caused by hydraulic gradient determined on the basis of elevation head, pressure head and velocity head, all of which constitute the total pressure head.

Darcy's law is commonly applied to flow of water in saturated rock joints, which is described by

$$v_w = k_w\, i_w \qquad (4.47)$$

where v_w = flow rate of water, k_w = coefficient of permeability and i_w = hydraulic gradient. Darcy's law has also been applied to unsaturated rocks (Richard, 1931; Buckingham, 1907), with a variable coefficient of permeability. For a saturated discontinuity, k_w is the hydraulic conductivity of water, whereas, if the fracture is unsaturated, then the hydraulic conductivity of the fracture becomes a function of the frictional drag forces generated by the fluids acting along the joint walls.

Poiseuille's law is best suited for describing flow (both gas and air) through a single smooth fracture (parallel plate), which is represented mathematically by

$$v_t = \frac{e_t^2 g}{12\nu}\, \nabla \left(\frac{p}{\rho g} + z \right)_t \qquad (4.48)$$

where v_t = the average velocity vector of fluid inside the fracture, e_t = the aperture of discontinuity, g = the gravitation of acceleration, v = the kinematics viscosity, p = fluid pressure, z = elevation head and ∇ = partial differential operator.

The effects of gravity on fluid flow through a vertical fracture are negligible over the small length of the specimen (i.e. 120 mm) compared to the high applied inlet pressure. Therefore, the elevation head (z) is assumed to be negligible compared to the inlet fluid pressure. Equation 4.48 is then given by the simplified cubic law:

$$q_t = \frac{e_t^3 g}{12v}\left(\frac{dp}{dx}\right) \tag{4.49}$$

The aperture e in Eqn. 4.49 can be replaced by the height of water phase $h_w(t)$ to give:

$$q_w(t) = \frac{h_w^3(t)g}{12v_w}\left(\frac{dp}{dx}\right)_t \tag{4.50}$$

in which $h_w(t) = F_I(x, y)_0 - F_B(x, y)_0 - \xi_{wc}$.

In the same manner, for the air-phase

$$q_a(t) = \frac{h_a^3(t)g}{12v_a}\left(\frac{dp}{dx}\right)_t \tag{4.51}$$

where $h_a(t) = F_T(x, y)_0 - F_I(x, y)_0 - (\delta_n + \xi_{ad} + \xi_{ac} - \xi_{wc})$, and

$$\frac{dp}{dx} = \frac{F_{wa}}{h_w(t)} - \frac{F_{jw}}{h_w(t)}$$

For the given water and air pressure of 0.25 MPa inlet pressures, the phase heights of air and water for different joint apertures are shown in Fig. 4.19. The material properties of the rock and the fluids are presented in Table 4.1. Increased joint aperture results in more air entering into the joint than water. This causes an increase in the solubility of air in water and compressibility of air. A significant increase in the solubility and compressibility of air occurs at small apertures. An increase in the joint aperture does not increase the solubility and compressibility linearly, if the solubility and compressibility of fluid reach their maximum values. As expected, the phase heights of both water and air increase with the increase in joint apertures.

According to Fig. 4.20 flow rates follow the same trend as phase heights against joint apertures. However, the airflow rates are significantly higher than the water flow rates because, water has a greater viscosity than air. At larger apertures, more air enters the joint in comparison with water, resulting in increased airflow. It also follows that air will occupy most part of the joint when the joint has a larger opening. In other words, similar to multiphase flow in pipes, a higher airflow rate is usually expected for larger apertures in the case of rock joints also.

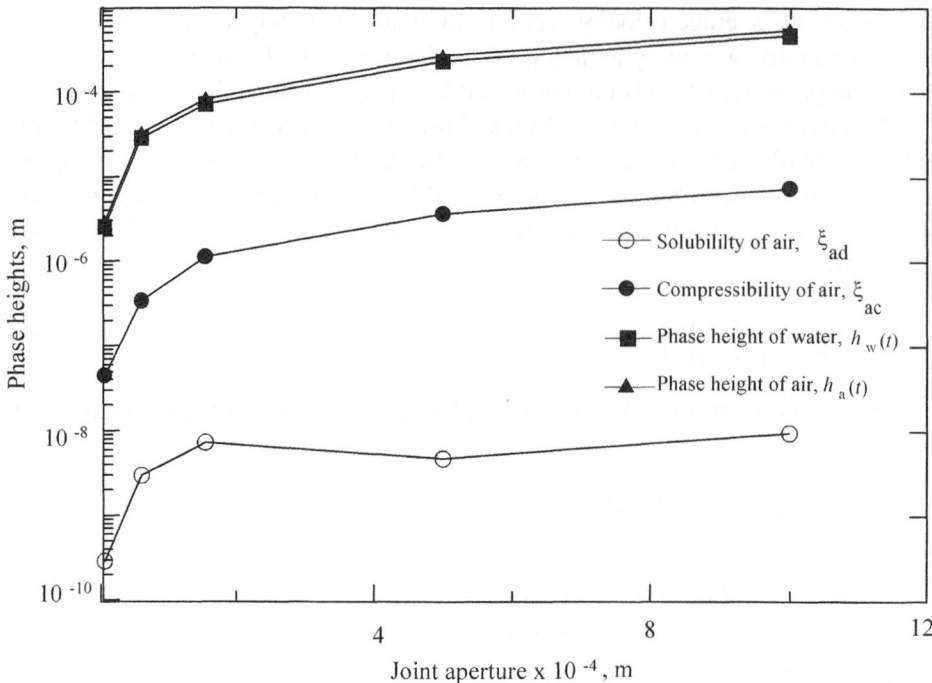

Figure 4.19. Phase heights of air and water.

Table 4.1. Typical input data for the analytical solution.

Initial aperture	0.0000062	m
Length	0.091	m
Density of the granite rock	2500	kg/m^3
Vertical stress	5×10^6	N/m^2
Density of air at 20° C	1.23	kg/m^3
Initial gas pressure	250000	N/m^2
Change pressure	1008000	N/m^2
Density of water	1000	kg/m^3
Joint normal Stiffness	2.8×10^{11}	N/m^2
Horizontal Stress	7480000	N/m^2
Water pressure in the joint	250000	N/m^2
Shear stiffness	9×10^{11}	N/m^2
Coefficient of compressibility of air, C_a	2.857143×10^{-6}	N/m^2
Vol. coefficient at 1 atm and 20° C	0.02918	
Water velocities	0.000018	m/s
Air velocities	0.0013	m/s
Interfacial friction factor	0.014	
Coefficient C for laminar flow	16	
Coefficient n for laminar flow	1	
Viscosity of water	0.00112	Pa·s
Viscosity of air	0.0000179	Pa·s

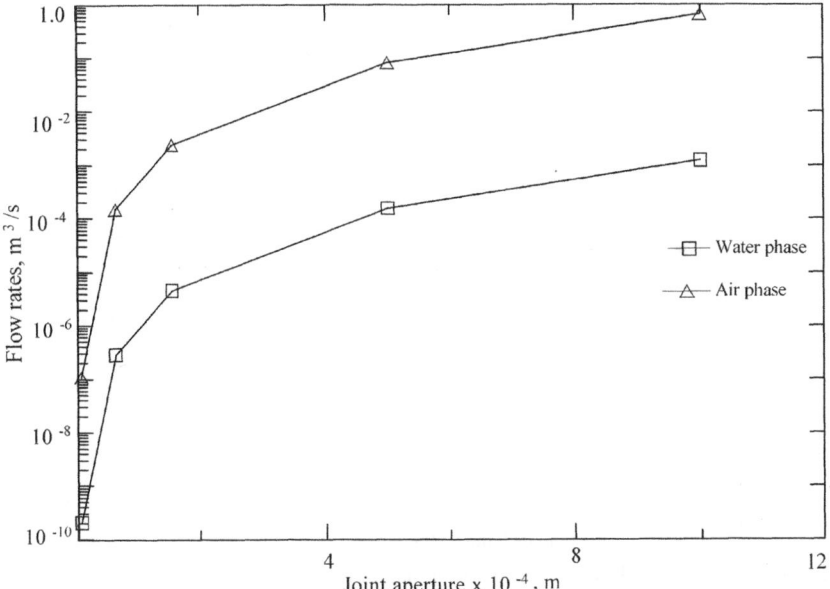

Figure 4.20. Two-phase flow rates for different apertures.

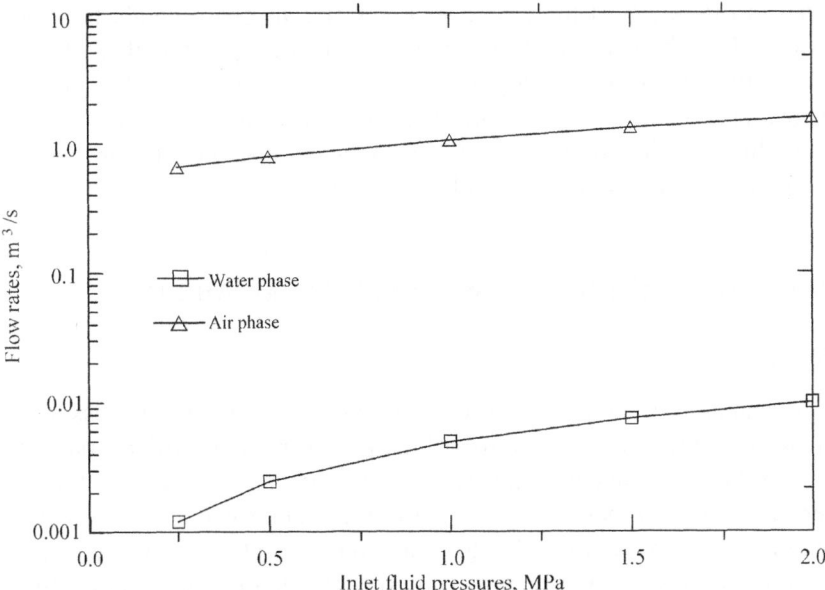

Figure 4.21. Effect of inlet fluid pressures on two-phase flow rates.

At a given confining pressure and axial stress, the two-phase flow rate against the inlet fluid pressure ratio is plotted in Fig. 4.21. The flow rates (vertical axis) are plotted on a log scale, in order to improve the clarity of the relatively small water flow rates in relation to the air flow rates. The corresponding phase heights

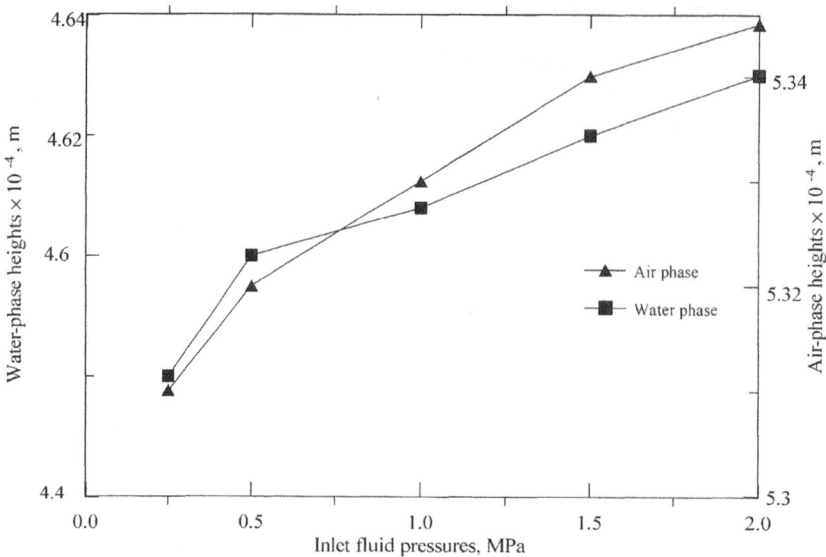

Figure 4.22. Effects of inlet fluid pressures on phase heights.

are presented in Fig. 4.22. At the crossover point, the air phase height becomes equal to water phase height, which indicates that fracture permeability of both phases are similar at 0.8 MPa inlet fluid pressure. It is obvious that airflow through a rock specimen is much greater than the water flow rate for the same boundary conditions. The comparison of the mathematical model with experimental results will be discussed in the following section.

4.9   LABORATORY STUDIES OF TWO-PHASE PERMEABILITY

4.9.1   Introduction

A number of available triaxial facilities can measure either the water pressure or air pressure or both within a fractured rock, but most of them are still incapable of measuring the relative permeability (air or water) of a fractured specimen. It is the relative permeability data that are most useful in the numerical analysis of flow through jointed rock mass. In order to study the two-phase flow behaviour through fractured rock specimens, a novel equipment, namely, Two-Phase, High-Pressure Triaxial Apparatus (TPHPTA) was employed. The features of commonly available single-phase triaxial equipment and those of TPHPTA for soil and rock testing were discussed earlier in Ch. 2. The following section concentrates on the effects of boundary conditions on flow, intrinsic and relative permeabilities, validity of Darcy's law and finally, validation of the mathematical model developed in the earlier part of this chapter.

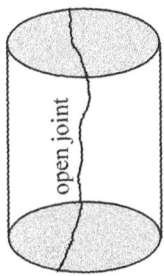

Artificial (tension) fracture

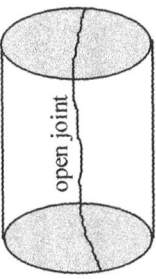

Natural fracture

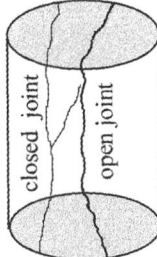

Naturally jointed rock specimen
with multiple fractures

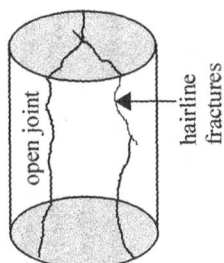

Naturally jointed rock specimen
with multiple fractures

Figure 4.23. Typical fractured rock specimens used for testing.

Test procedure for two-phase flow

Fractured granite rock specimens having diameters from 45 to 55 mm and length of twice the diameter with low matrix permeability (order of 10^{-19} m^2) were selected for the laboratory investigation. The tests were carried out on both artificially and naturally fractured specimens, having either a single fracture or multiple fracture networks. The test specimens were cored from large samples, using a small scale coring machine. Subsequently, artificial (tension) fractures were induced on the intact rock cores using the Brazilian test. Artificially fractured specimens have single fracture while naturally fractured specimens exhibit single or multiple fractures. Naturally fractured granite specimens were obtained from various mines in Australia. Some typical fractured specimens used for testing are shown in Fig. 4.23. The basic test procedure was the same as for single-phase flow illustrated in Ch. 3. Table 4.2 shows a combination of tests carried out using TPHPTA. As discussed in Ch. 3, prior to testing, all the fractured specimens were mapped using the digital co-ordinate profilometer to estimate the roughness of the fractured surfaces.

4.9.2   Effect of inlet fluid pressure

Flow rates through rock specimens were observed under both steady-state and transient conditions. Figure 4.24 indicates the change of flow rate against time, where a steady-state water flow rate has been attained after 30 minutes of applying

Table 4.2. Testing procedures for single-phase and two-phase flow measurements.

Stages	Cell pressure	Axial stress	Fluid pressures Water (p_w)	Air (p_a)	Comments
Single-phase flow analysis					
Stage 1	Constant	Constant	Variable	No air flow	Steady-state water flow, water pressure is changed
Stage 2	Variable	Constant	Constant	No air flow	Steady-state water flow, cell pressure is changed
Stage 3	Constant	Variable	Constant	No air flow	Steady-state water flow, axial stress is changed
Two-phase flow analysis (water + air)					
Stage 1 a	Constant	Constant	Constant	No air flow	Steady-state water flow is observed (joint is initially saturated with water)
b	Constant	Constant	Constant	Variable	Air pressure is gradually increased from zero. For $(p_w > p_a)$, steady state is observed for both water and air
c	Constant	Constant	Constant	Variable	Air pressure is gradually increased until air pressure equals water pressure. Capillary pressure becomes zero $(p_w = p_a)$. Steady-state two-phase flow is observed
d	Constant	Constant	Constant	Variable	Air pressure is gradually increased until no water flow is observed from the specimen. For different suction pressures $(p_a > p_w)$, steady-state two-phase flow is observed
Stage 2	Constant	Constant	Variable	Constant	Same as stages 1a, 1b, 1c and 1d
Stage 3	Variable	Constant	Constant	Constant	Cell pressure is changed at zero capillary pressure, $(p_w = p_a)$
Stage 4	Constant	Variable	Constant	Constant	Axial stress is changed at zero capillary pressure, $(p_w = p_a)$

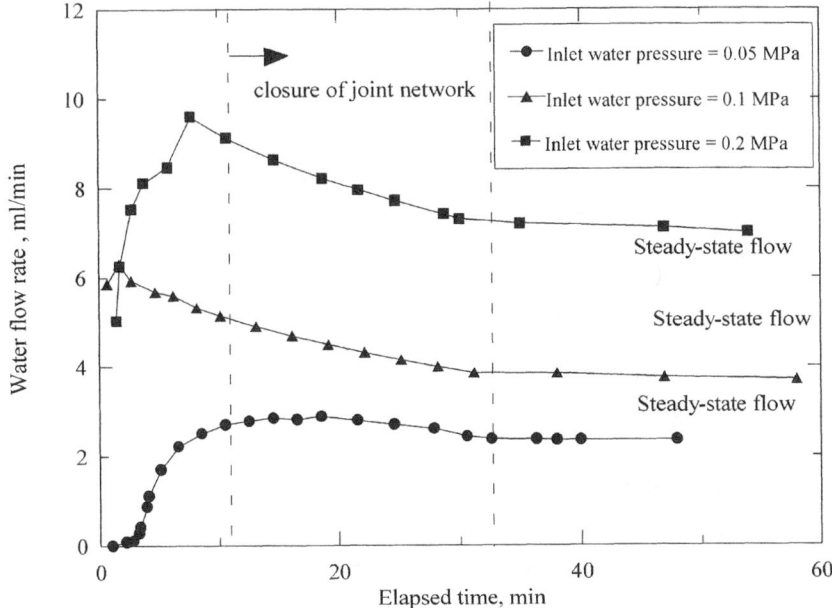

Figure 4.24. Transient conditions of water flow for different inlet pressures at confining pressure of 0.5 MPa.

the inlet fluid pressure. During the initial 10 minutes or so, the flow rate increased, probably due to increased fracture connectivity, facilitating the flow. The subsequent decrease in flow rate was due to the joint deformation associated with the confining pressure. Beyond 30 minutes or so, the flow rate became constant (steady-state) as the joints attained their residual apertures.

The observation of steady-state and transient flow in the specimens of two-phase flow were carried out in two distinct ways:
(1) Air flow through an initially water-saturated specimen and
(2) Water flow through an initially air-saturated specimen.

In (a), for given confining pressure and axial stress, the inlet water pressure was applied to the specimen and the steady-state water flow was monitored. Once the water flow became constant, the air phase was introduced, as shown in Fig. 4.25a. As an example, in Fig. 4.25b, the specimen was first saturated with water, at a pressure of 0.25 MPa, and then the air phase was introduced into the specimen. The inlet air pressure was gradually increased up to 0.25 MPa, and the corresponding flow rates of both the phases were observed. After 240 mins, water flow rate decreased to a constant value of 0.04 ml/min, but the airflow increased to a constant value of above 7 ml/min (note that the two flow rate scales are different for air and water). Therefore, under the above conditions, steady-state two-phase flow through the specimen occurred.

In the case of water flow through an initially saturated air specimen, the inlet air pressure was first applied to the specimen until the steady-state airflow was

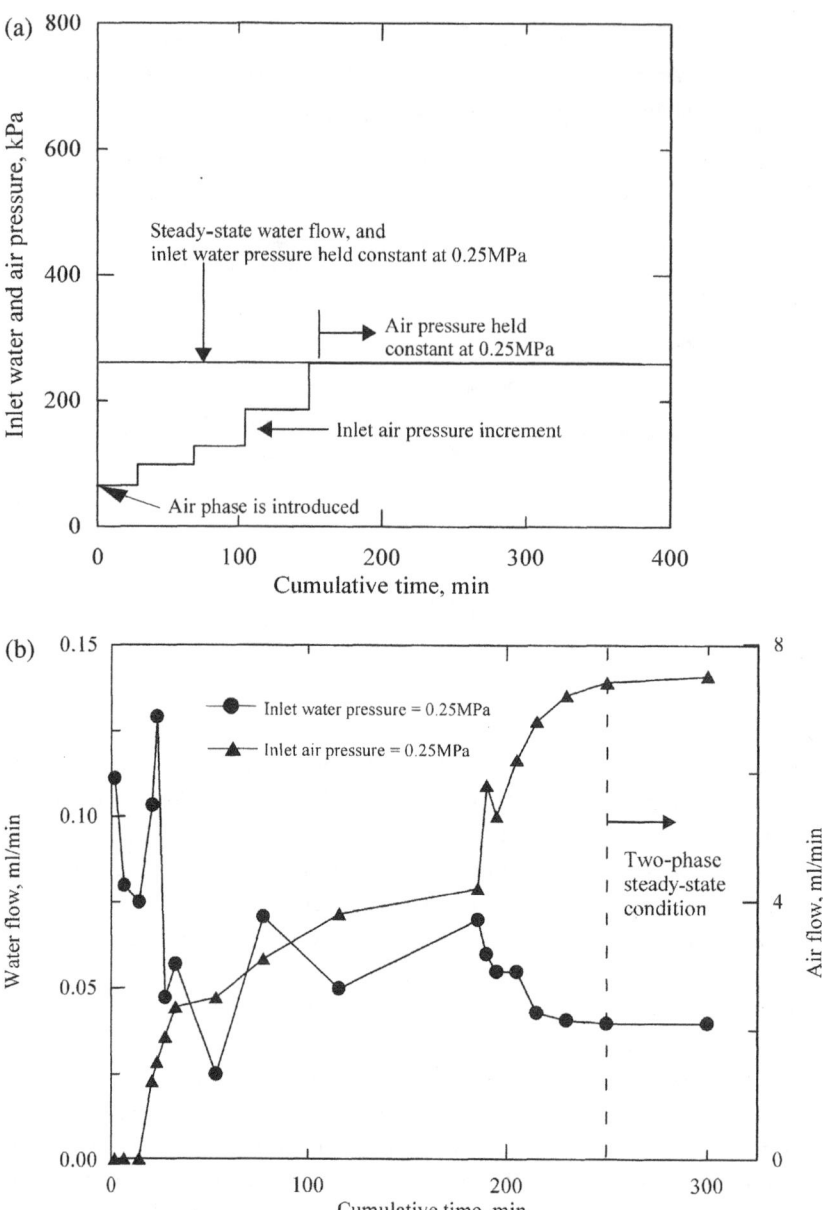

Figure 4.25(a) Applied water and air pressure to the specimen and (b) transient condition of two-phase flow at 2 MPa cell pressure and inlet water and air pressure of 0.25 MPa (initially the specimen is saturated with water).

observed. At this stage, the water phase was introduced to the specimen, however, no flow of water could be observed for a long time. Only after a period of 675 minutes, the water flow began to increase, while the airflow decreased. Comparison with Fig. 4.25b indicates that the response time for air flow through

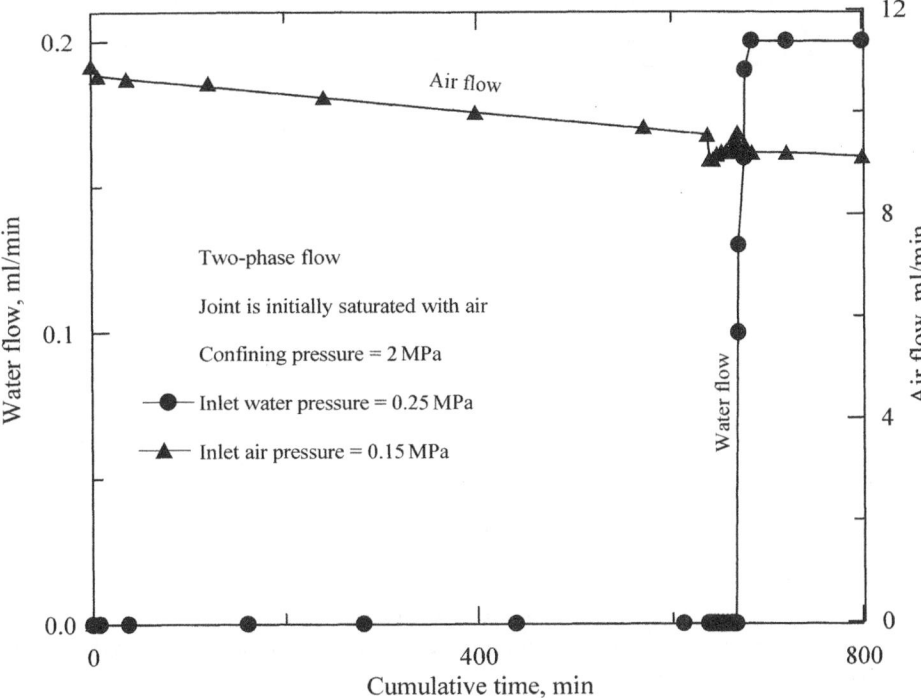

Figure 4.26. Transient two-phase flow for a specimen initially saturated with air.

the specimen is much shorter because the specimen was initially saturated with water. The time delay shown in Fig. 4.26 suggests that if the specimen is initially saturated with a low viscosity fluid such as air, the introduction of a higher viscosity fluid will not be able to penetrate the specimen for a relatively long period of time.

The comparison of the measured laboratory data with theoretical data is described in the following sections. The theoretical predictions for steady-state two-phase laminar flow through fractured rock specimen were carried out based on the mathematical model described in Section 4.6. The material properties of the rock and fluids are presented in Table 4.1. The heights of the air phase (Eqn. 4.25) and water phase (Eqn. 4.23) strongly depend on the normal deformation of the joint. For different normal stresses, the predicted normal deformation (Eqn. 4.46c) and the measured normal deformation are shown in Fig. 4.27. The Curve A shows the normal deformation based on Eqn. 4.46c, where the joint normal stiffness was calculated from uniaxial test data, in which the axial load was applied normal to the horizontal fracture surface, and the deformations of the mechanical aperture were measured for various axial loads. In contrast, for the Curve B also based on Eqn. 4.46c, the joint normal stiffness was determined from triaxial test data in which, the circumferential confining pressure (i.e. normal stress to the joint) and axial load were applied to the specimen and the deformation of the fracture

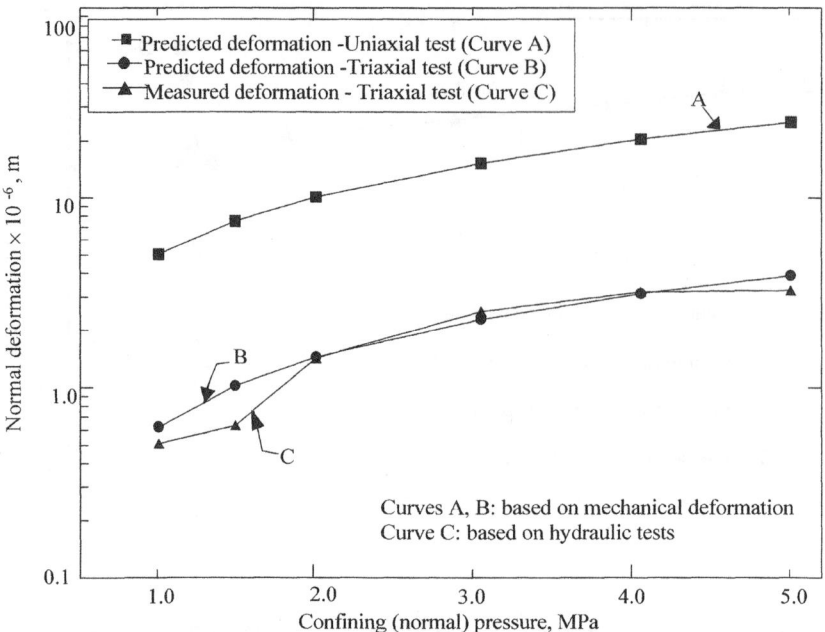

Figure 4.27. Measured and predicted normal deformation of jointed rock specimen for various normal stress levels.

for various confining pressure was measured by clip gauges. The data indicate a significant reduction in normal deformation in Curve B in comparison with Curve A. This is because, in the triaxial test, where a confining pressure is applied, the normal deformation is expected to be less. Also, from the measured fluid flow data, the hydraulic apertures of the joint could be back calculated for a given confining stress, using the cubic relation (Eqn. 4.49). Each data point plotted (Curve C) represents the average of at least three tests. The sudden drop in normal deformation at 1.5 MPa is probably due to some experimental error. In general, the Curve C based on cubic relation is in agreement with the triaxial data, while the normal deformation represented by Curve A is over-estimated. Therefore, the behaviour indicated by Curve B has been incorporated in the verification of the mathematical model.

Irrespective of single-phase or multiphase flow, increasing the inlet fluid pressure of one phase will result in an increased flow rate for the same phase, while decreasing the flow rate of the other. If the inlet fluid pressures of both the phases are increased simultaneously, then the relative fluid flow rate of each phase is governed by the properties of the individual phases. For given boundary conditions and fluid pressures, once the two-phase mixture was collected from the specimen and the individual phases separated, the respective flow rates were plotted against the inlet fluid pressures. For $p_a = p_w$, Fig. 4.28a–c illustrates the flow behaviour at three different confining pressures. At relatively low inlet fluid pressures, the flow rate varies linearly with the fluid pressure.

(a) Confining pressure 0.5 MPa and $P_a = P_w$

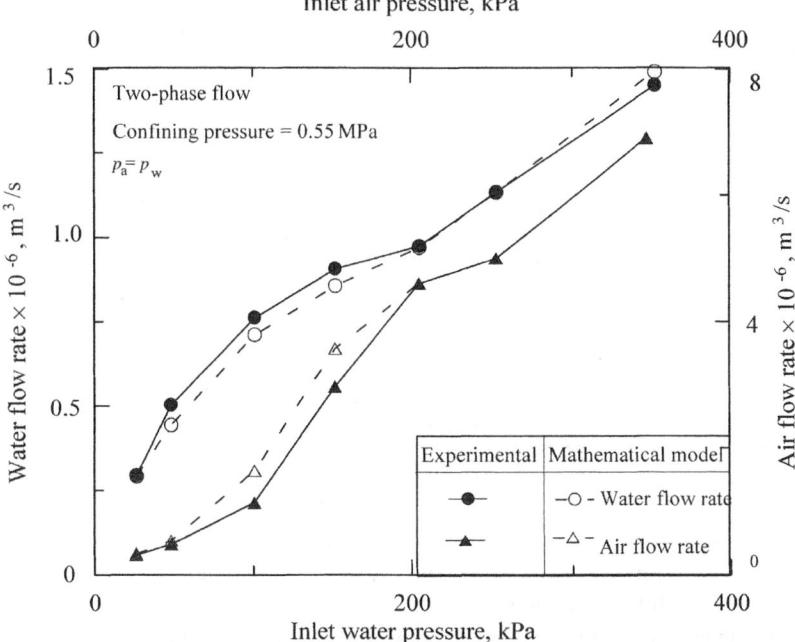

(b) Confining pressure 1.0 MPa and $P_a = P_w$

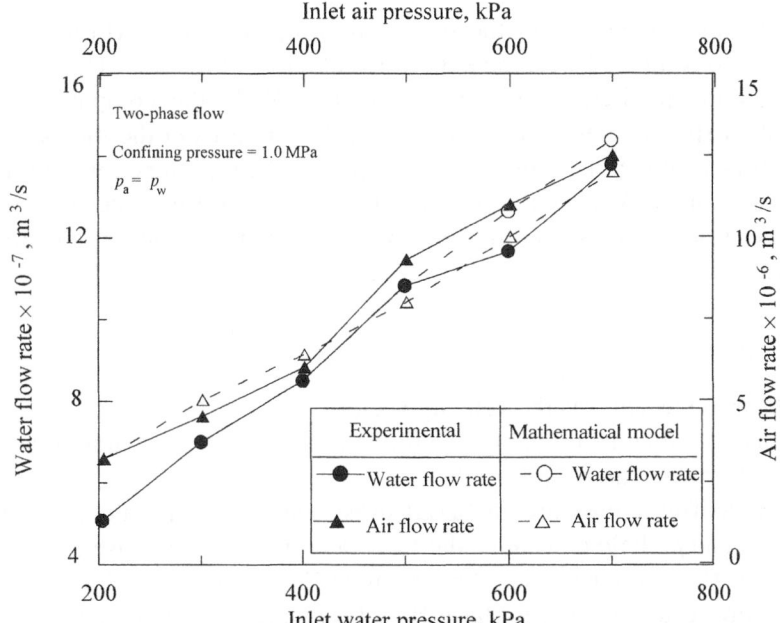

Figure 4.28a,b. Two-phase flow rates at different confining pressure and at $p_a = p_w$.

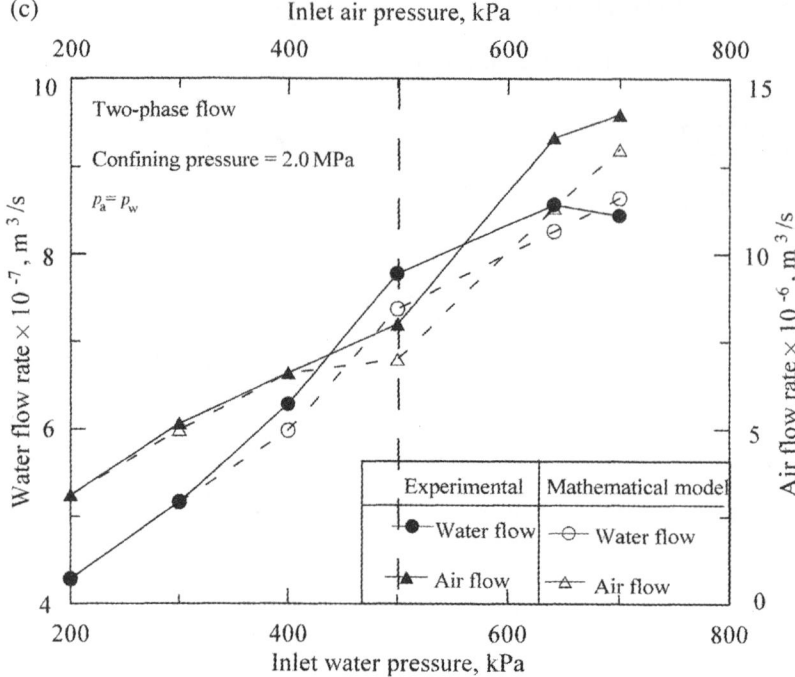

Figure 4.28c. Two-phase flow rates at 2.0 MPa confining pressure and at $p_a = p_w$.

The predictions obtained from the proposed two-phase model (Eqns. 4.50 and 4.51) are also compared with the experimental data in Fig. 4.28. The dotted lines represent the predicted flow by the model. While the comparison is acceptable, the experimental results show some deviation from the theoretical values, which is attributed to some air and water still being trapped within the pores of the specimen. At elevated inlet fluid pressure, the flow rates become less linear, because, Darcy's law ($q = av$) deviates from accuracy at elevated fluid pressures. Such non-linear flow can be simulated, using the following expression:

$$q = av + bv^2 \tag{4.52}$$

where a and b are constant, v is the velocity vector.

The non-linearity may be probably due to the formation of non-parallel laminar or turbulent flow within the joint.

For a given confining pressure and axial stress and increasing inlet air pressures ($p_a \neq p_w$, and p_w held constant), the flow rate of the water phase decreases, while the air flow rate increases. According to Fig. 4.29, further increase in inlet air pressure results in single-phase flow (i.e. water flow rate approaches zero).

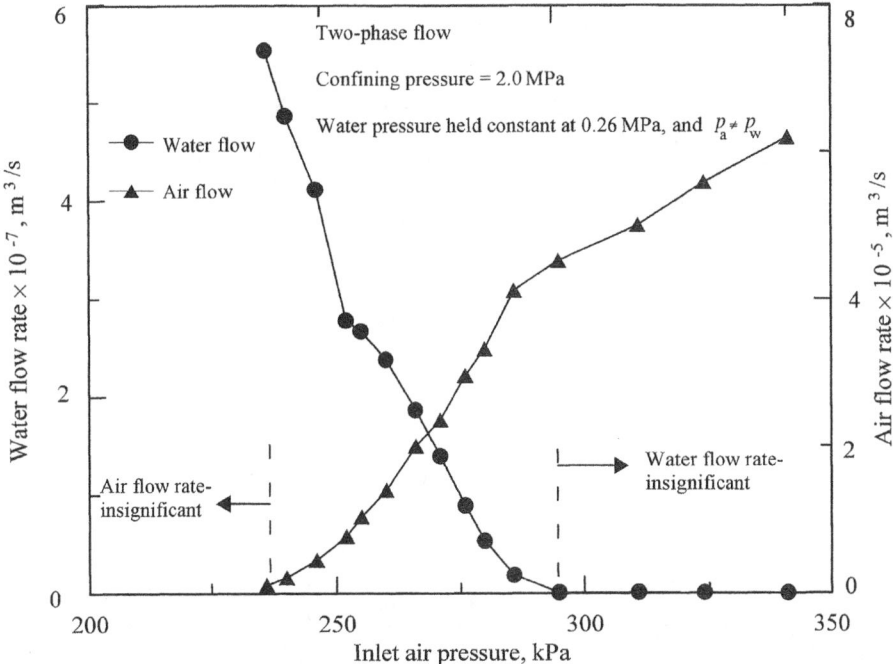

Figure 4.29. Two-phase flow rates at 2.0 MPa confining pressure and at $p_a \neq p_w$.

4.9.3   Effect of confining pressure and axial stress

As discussed in Ch. 3, for fully saturated water flow through rock joints, various studies have shown that the flow rates decrease with the increase in confining pressure (σ_3) due to the closure of apertures. Two-phase flow is also affected by the confining pressure in a way similar to single-phase flow. For given inlet water and air pressures, the effect of confining pressure on the permeability of two-phase flow is illustrated in Fig. 4.30, for $p_a = p_w$. The dotted lines represent the predicted flow from the proposed model (Eqns. 4.50 and 4.51). The measured flow rate is slightly smaller than the calculated values at low confining pressures which is probably due to some fluid being trapped in pores or along the joint walls as a thin layer of film. At low confining pressures, the air occupies the main volume of the joint, hence, water is expected to flow as an unstable film along the joint walls, thereby contributing to a significantly reduced water flow rate. As expected, fluid flow decreases with the increase in confining pressure, however, beyond a confining pressure of 6 MPa, the rate of change of permeability becomes marginal (Fig. 4.30). This is an indicative of joint apertures attaining their residual values.

Figure 4.31 shows that for constant inlet water pressures ($p_w = 125$, 200 and 250 kPa), the two-phase flow rate decreases with the increasing confining pressure (i.e., shift of curves to the right). For a constant confining pressure, when the

(a) $P_a = P_w = 0.25$ MPa

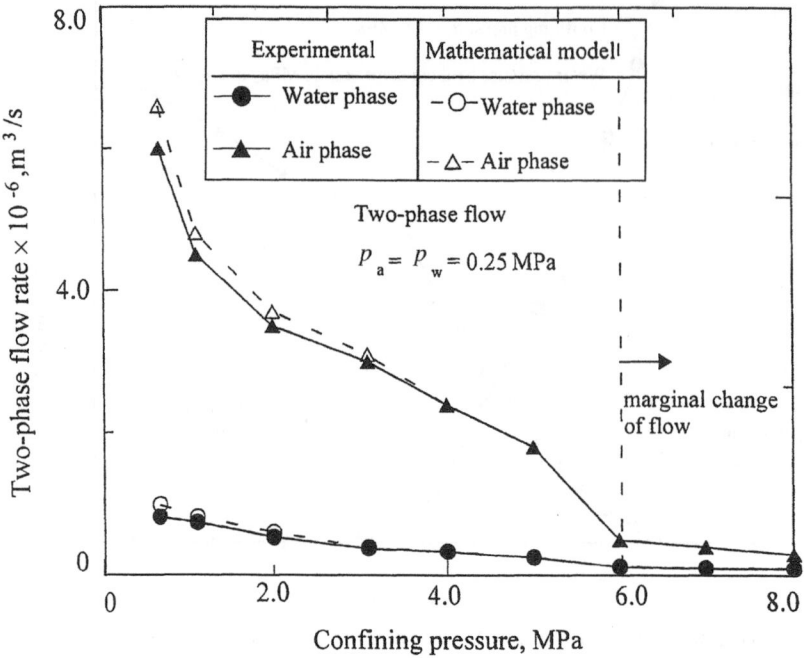

(b) $p_a = p_w = 0.125$ MPa

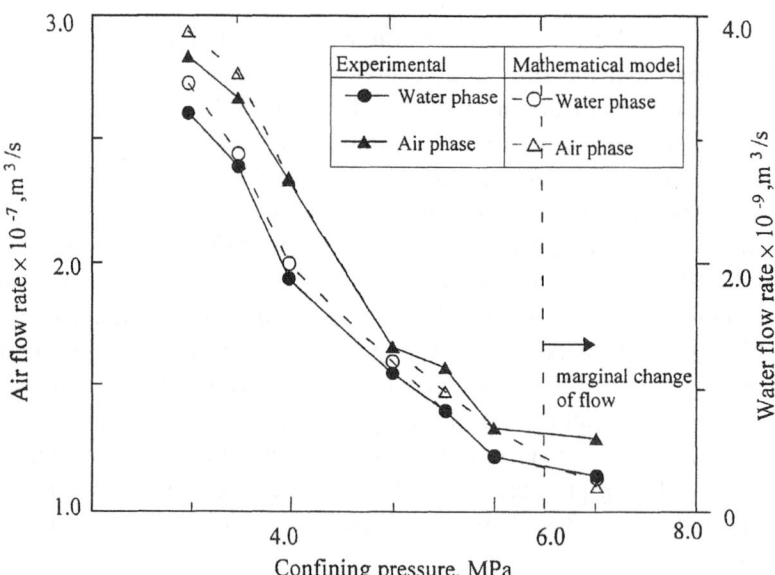

Figure 4.30. Effect of confining pressure on two-phase fracture permeability when $p_a = p_w$.

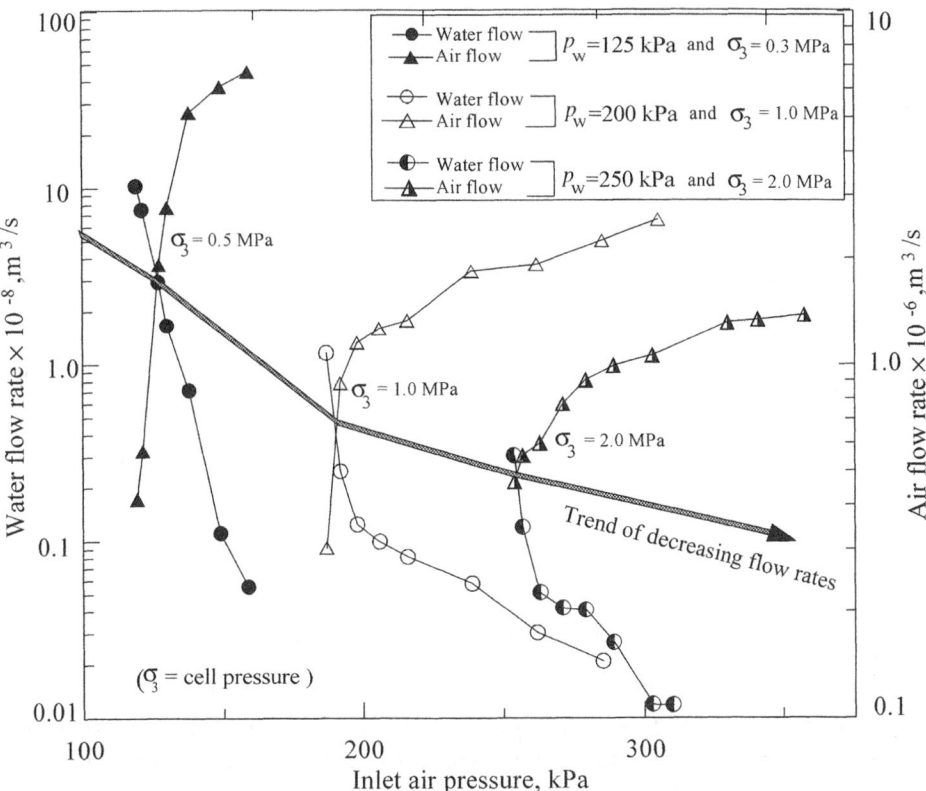

Figure 4.31. Comparison of two-phase flow at different cell pressures.

inlet air pressure (p_a) is increased, the airflow is expected to increase with an associated decrease in the water flow. Figure 4.31 illustrates that at increasing confining pressures, even with the increase in inlet air pressure, the airflow decreases. This verifies the dominant role of confining pressures.

Equivalent phase heights of water $h_w(t)$ (Eqn. 4.23) and air $h_a(t)$ (Eqn. 4.25), joint deformation due to normal stress, δ_n and the equivalent height components of air, ζ_{ad} and ζ_{ac} (based on solubility and compressibility, respectively) are plotted in Fig. 4.32. The total deformation of the joint wall is composed of these equivalent phase heights as represented earlier by Eqns. 4.22–4.25. Almost 95% of the magnitude of $h_a(t)$ and $h_w(t)$ is due to the normal joint deformation (δ_n), the rest being the combined effect of ζ_{ac} and ζ_{ad}. As expected, the contribution of the air compressibility term, ζ_{ac} is more significant than the solubility component ζ_{ad}. While ζ_{ac} amounts to about 4-5% of the value of δ_n, the magnitude of ζ_{ad} is not more than 0.01% of δ_n. At large confining pressures (>5 MPa), further decrease in δ_n may be marginal, once the joint aperture reaches its residual values. Beyond this stage, the role ζ_{ac} and ζ_{ad} will become increasingly pronounced.

Apart from the confining pressure, inlet fluid pressures and the degree of saturation, the deviator stress ($\sigma_1 - \sigma_3$) can also influence the permeability, strength

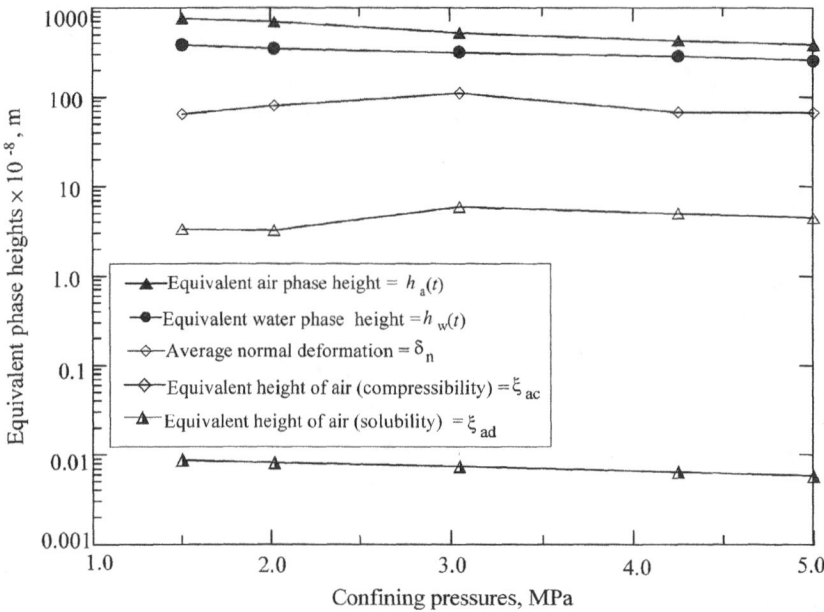

Figure 4.32. Equivalent phase heights of water and air in two-phase flow in a jointed granite specimen.

and deformation properties of the rock specimens. Continued increase in strain results in an increased flow through the specimen, either because of the formation of new fractures or dilation of existing fractures or both. Figure 4.33 illustrates the variation of flow rate with the deviator stress at a constant cell pressure of 1 MPa. The initial decrease in air and water flow is associated with the closure of joint apertures upon initial loading. However, with increased deviator stress, the air and water flow rates start to increase, probably due to the dilation of some existing fractures and the formation of new cracks. In Fig. 4.33, a sudden drop in water flow occurs at a deviator stress of 50 MPa, which is accompanied by an increase in airflow, as expected. This is because at these stress levels, airflow dominates when the dilation of joint takes place.

4.10   DETERMINATION OF FLOW TYPE WITHIN A ROCK JOINT

As illustrated in Section 4.9.2, further increase in the inlet fluid pressure does not increase the fluid flow rate linearly, and a non-linear flow relationship against inlet fluid pressure develops. The aim of this section is to demonstrate how the flow rate through real rock fractures under different boundary conditions assumes laminar or turbulent flow patterns. This is studied using the Reynolds number. The expression for estimating the Reynolds number in flow through pipes is given by

$$Re = \frac{vd}{\nu} \tag{4.53a}$$

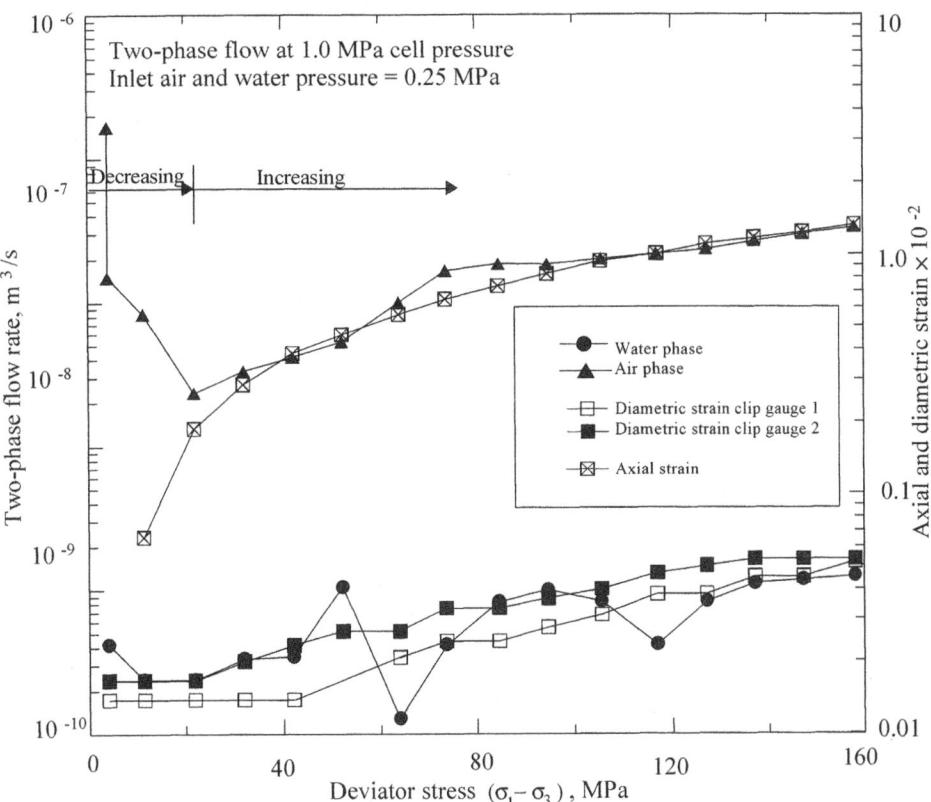

Figure 4.33. Relationship of two-phase flow and deformation of the specimen for different axial stress.

where v = average velocity, d = diameter of the pipe and v = kinematic viscosity of fluid. At a temperature $20°$ C, the kinematic viscosity of water and air are approximately 1.12×10^{-6} and 1.56×10^{-5} m²/sec, respectively.

If the hydraulic aperture is e, then the d is replaced by $2e$. For a single joint with a width w, and flow rate of q through the joint, the Reynolds number for the joint is expressed in terms of the flow rate, width of the joint and kinematic viscosity as

$$Re = \frac{2q}{vw} \qquad (4.53b)$$

(a) For single-phase flow: For a given confining pressure and axial stress, the calculated Reynolds numbers for single-phase flow using Eqn. 4.53b for typical specimens are indicated in Fig. 4.34. As expected in fluid flow through pipes, the increase in fluid pressure results in an increase in the Reynolds number. For single-phase water flow, the calculated Reynolds numbers are below 100, whereas, for the air phase, Reynolds numbers are as high as 500 because of the low viscosity of air (Fig. 4.34a). The Reynolds number depends mainly on the joint

(a) Air flow

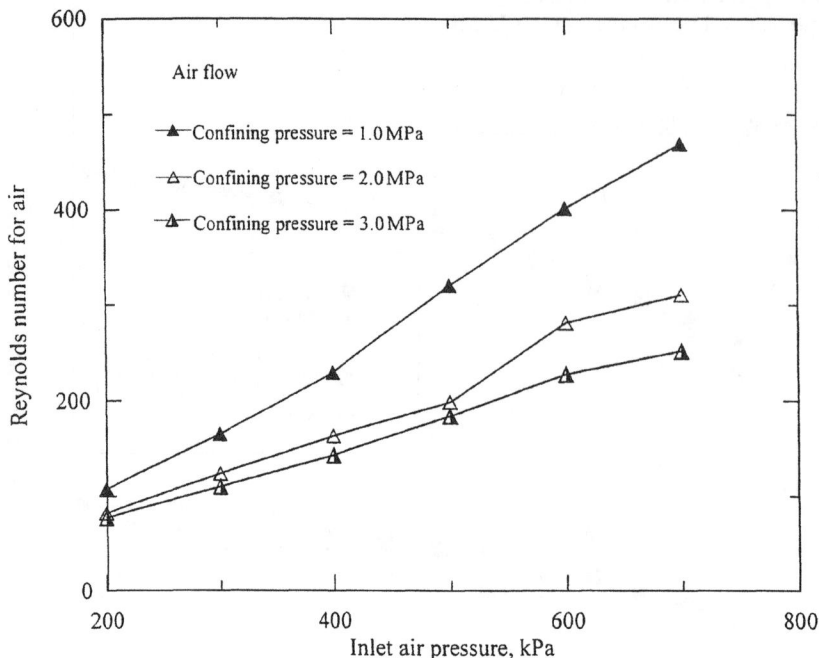

(b) Water flow

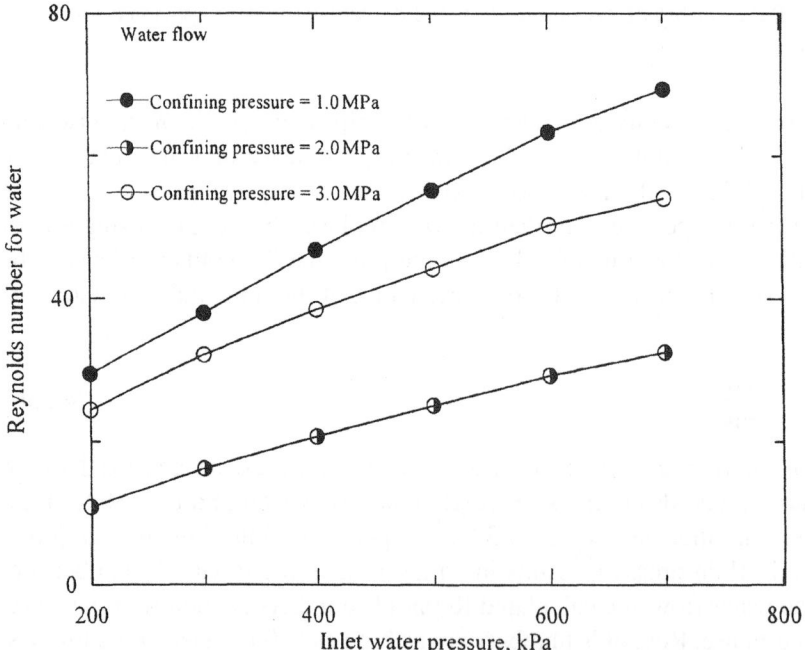

Figure 4.34. Average Reynolds number for water flow within the joint.

(a) Confining pressure 0.5 MPa, $P_a = P_w$

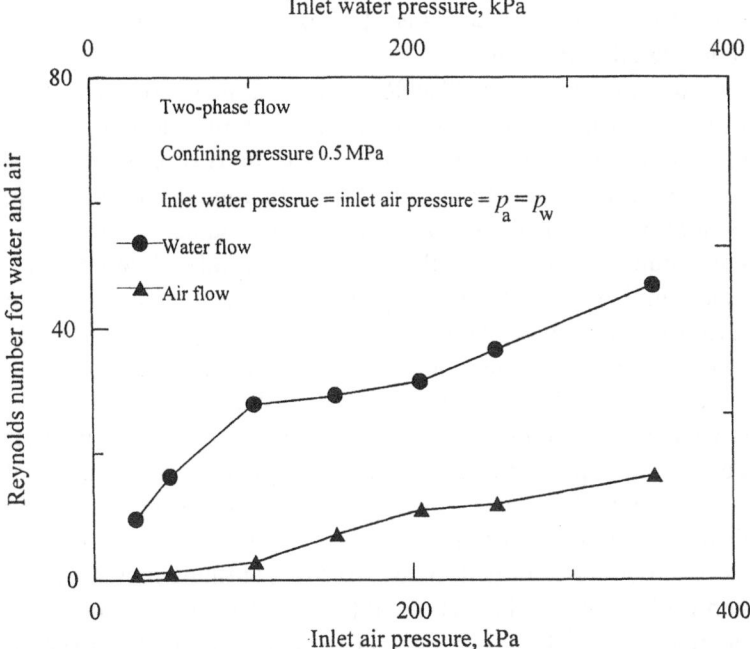

(b) Confining pressure 1.0, $P_a = P_w$

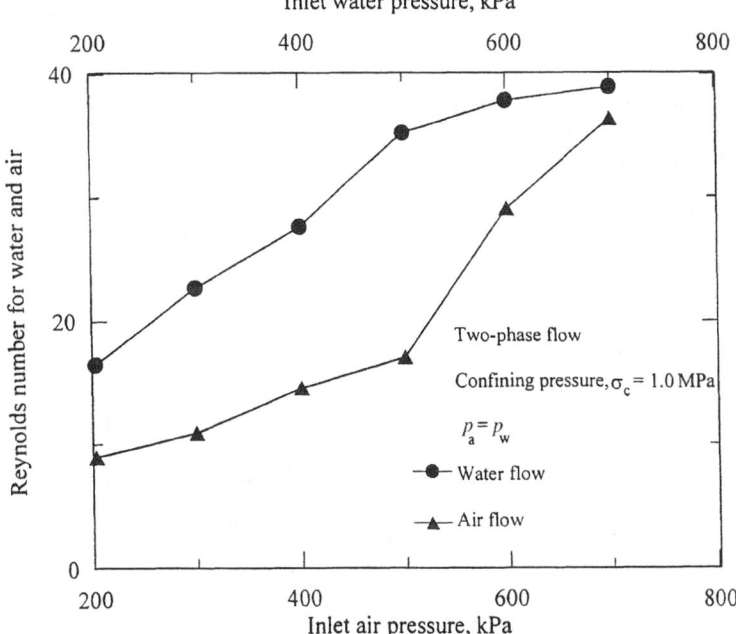

Figure 4.35. Average Reynolds number for two-phase flow for $p_a = p_w$.

surface roughness, the fluid properties and the inlet fluid pressure. The magnitude of critical Reynolds number representing the flow regimes between parallel walls is 1000 (Street et al., 1996). This indicates that the single-phase flow derived in this study can be considered as laminar.

(b)   For two-phase flow ($p_a = p_w$): The Reynolds numbers for two-phase flow were also estimated using Eqn. 4.53b, in which q is the individual flow rate of each phase. The estimated Reynolds numbers for the two-phase flow are shown in Fig. 4.35 for different confining pressures. In Fig. 4.35a and b the applied fluid pressures of each phase is approximately the same. As in single-phase flow, Reynolds numbers increase with the increase in inlet fluid pressure. However, Reynolds number for air in two-phase flow is smaller than the Reynolds number for the water phase. This is because, in two-phase flow, the difference between the flow rates of water and air is not as high as in single-phase flow, although the difference in the viscosities is considerable.

Figure 4.36 shows the corresponding Reynolds numbers for different inlet air pressures when the inlet water pressure was held constant and not equal to the inlet air pressure ($p_a \neq p_w$). Reynolds numbers corresponding to the air-phase increase considerably with the increasing inlet air pressure, whereas the corresponding Reynolds numbers based on water flow decrease. At low confining pressures (i.e. when the joint aperture is relatively larger), the difference between air and water phase Reynolds numbers is much higher. For very small joint apertures (i.e. at

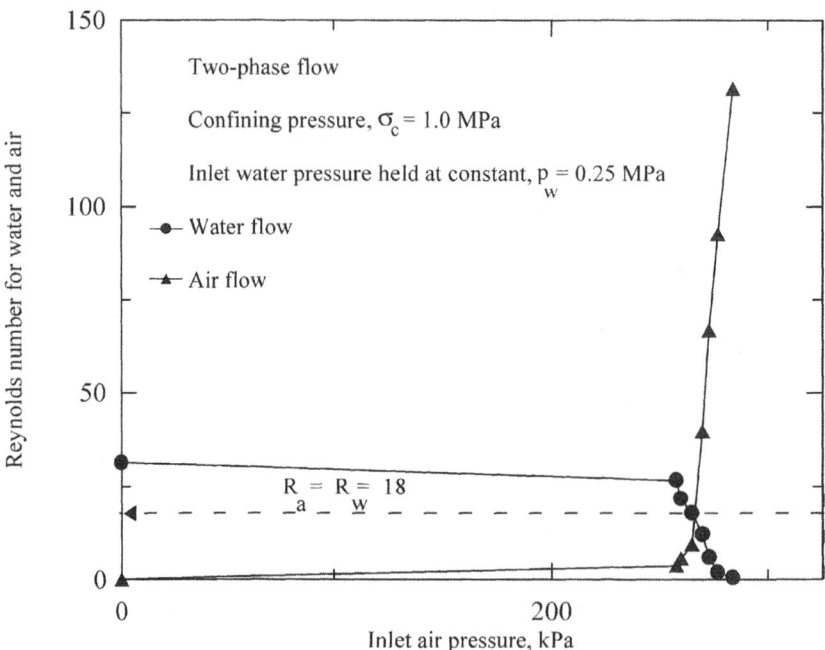

Figure 4.36. Average Reynolds number for two-phase flow when $p_a \neq p_w$.

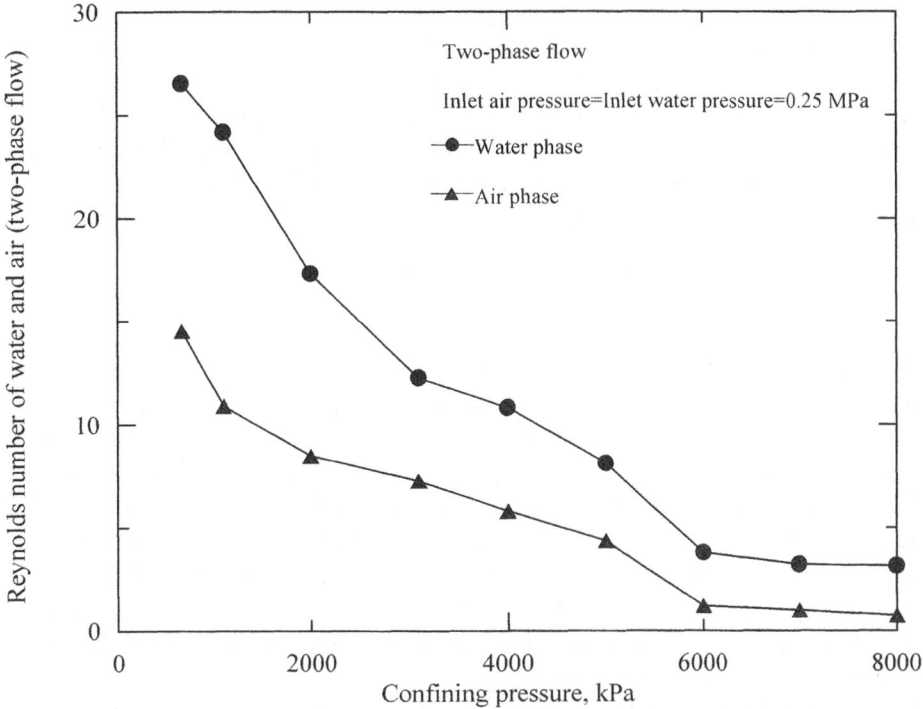

Figure 4.37. Effect of confining pressure on Reynolds number for two-phase flow.

larger confining pressures), the Reynolds numbers are small and they seem to be independent of the elevated confining pressures (Fig. 4.37). It is evident from the above data that the likelihood of the development of turbulent flow within a rock joint is remote. As all the estimated Reynolds numbers in this study are well below 1000, it can be concluded that the flow within joints remains as laminar flow, even for the rough joints considered here.

4.11 FRACTURE PERMEABILITY OF TWO-PHASE FLOW

In the presence of more than one fluid, one has to consider the relative permeabilities of the two phases apart from the fracture permeabilities. The discussion here is limited to absolute permeability of fractured granite rocks for different boundary conditions including inlet water and air pressures.

For laminar flow conditions ($Re < 1000$), the fracture permeability is expressed by

$$k = \frac{e^2}{12} \tag{4.54}$$

where e = hydraulic aperture or phase height, and k = fracture permeability.

From the above expression, it is clear that the fracture permeability is independent of fluid properties and depends only on the aperture. When a fractured specimen is subjected to various loading conditions, it is relatively easy to estimate the hydraulic aperture from the triaxial test data, using the following expression, assuming that the joint walls are smooth and parallel. In Eqn. 4.55, the effect of gravity on fluid flow is not included because it is negligible over a small length of the specimen when compared to the inlet fluid pressure.

$$e = \left(\frac{12\mu q}{d(\mathrm{d}p/\mathrm{d}x)} \right)^{1/3} \tag{4.55}$$

where q = total flow rate, μ = dynamic viscosity of fluid, d = width of fracture and $\mathrm{d}p$ = pressure difference along the length of $\mathrm{d}x$.

If there are multiple fractures in the tested specimen, the estimated e using Eqn. 4.55 represents the net or equivalent hydraulic aperture. The fracture permeability based on Eqn. 4.54 for two-phase steady-state flow can be examined using the TPHPTA. The individual flow rates of water and air mixture were measured separately, in order to incorporate values for q in Eqn. 4.55. Figure 4.38 shows the permeability of each phase against inlet fluid pressures based on individual flow rates. For a given confining pressure and axial stress, the inlet fluid pressures could be applied in two different ways:

(1) Inlet air pressure = inlet water pressure ($p_a = p_w$) and
(2) Water pressure held constant; inlet air pressure is increased ($p_a \neq p_w$).

(a)   Two-phase fracture permeability when $p_a = p_w$: When the inlet fluid pressure was increased (e.g. $p_a = p_w = 0.2, 0.3$ MPa etc), the permeability of the water phase decreased with the increase in inlet fluid pressures, whereas, a slight increase in permeability of the air phase was observed (Fig. 4.38a and b). In contrast, flow rate of both phases increased with the increase in fluid pressure as shown earlier in Fig. 4.28. This is because, the flow rate is usually linearly proportional to the inlet fluid pressure (in the laminar region). However, the fracture permeability is estimated using $k = e^2/12$, in which the hydraulic aperture e is based on Eqn. 4.55, which does not linearly vary as the flow rate. It is the term e^3 that is linearly proportional to the flow rate.

(b)   Two-phase fracture permeability when $p_a \neq p_w$: The effect of inlet fluid pressure on permeability is also analysed when $p_a \neq p_w$. Figure 4.39 shows the permeabilities in two-phase flow of each phase at different confining pressures when the water pressure is held constant while increasing the inlet air pressure. As an example, in Fig. 4.39, the applied water pressure (p_w) of 0.26 MPa is held constant during the test. Once the air phase is introduced to the specimen, the permeability of water phase decreases, and after a certain magnitude of the inlet air pressure, the air permeability begins to increase. If the magnitude of the inlet air pressure (p_a) is sufficient to replace all the water in the joint, then the water permeability approaches zero. The permeability of both phases becomes equal when the air pressure is approximately 1.1 times the inlet water pressure. In other

(a) Confining pressure 0.5 MPa and $P_a = P_w$

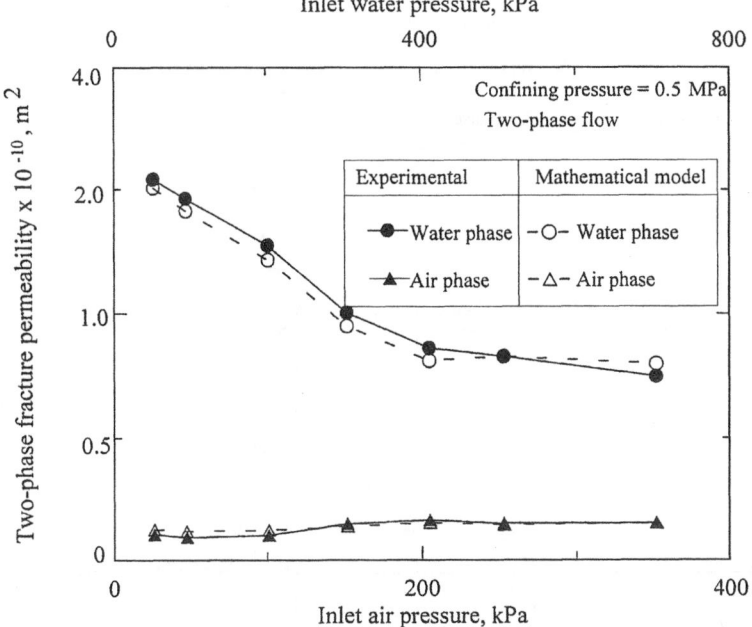

(b) Confining pressure 2.0 MPa and $P_a = P_w$

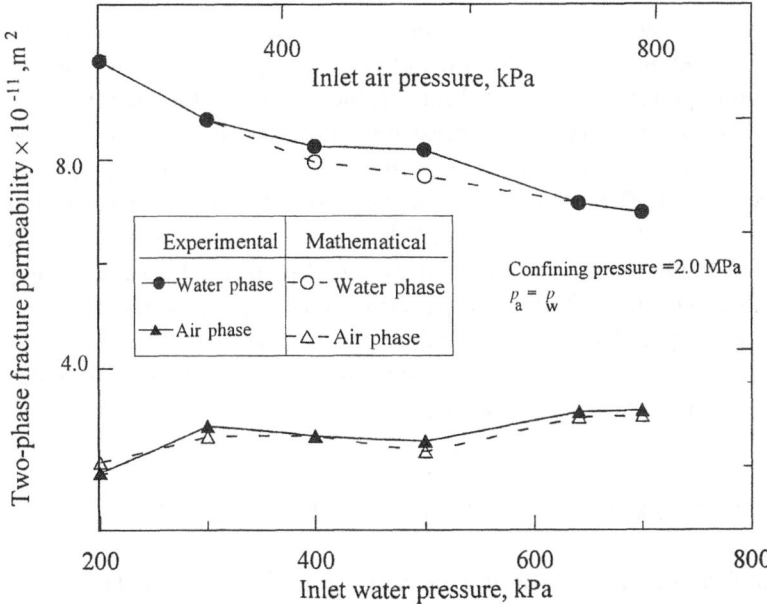

Figure 4.38. Two-phase fracture permeabilities for different inlet fluid pressures when $p_a = p_w$ at different confining pressures.

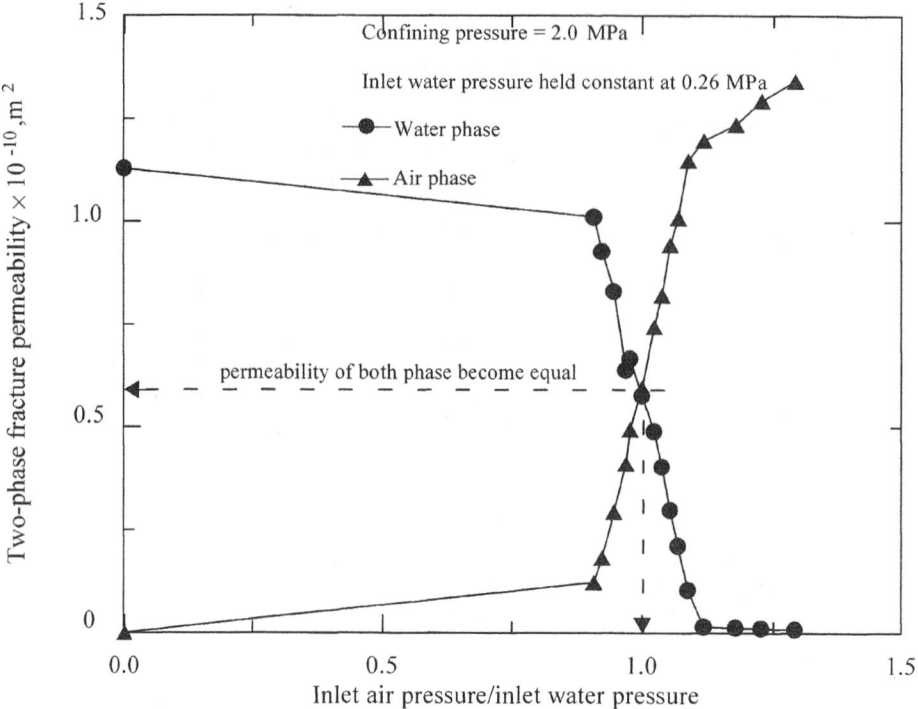

Figure 4.39. Effect on varying inlet air pressure on fracture permeability of two-phase flow through initially water saturated specimen (at 2.0 MPa confining pressure).

words, the equivalent water phase height tends to become equal to the air phase height. When p_a/p_w becomes 1.1, the permeability of both phases is approximately 33% of the single-phase water permeability. Table 4.3 shows the magnitude of phase permeability for given confining pressures and inlet fluid pressures. The values of p_a/p_w are also given, when the permeability of both phases become equal (cross-over point). The ratio of p_a/p_w varies from 0.8 to 1.2 for the tested specimens, at constant inlet water pressure. When p_a/p_w ranges from 0.8 to 1.2, the 'cross-over' two-phase permeability varies 20–70% from the single-phase water permeability. The inlet fluid pressure ratio at which the permeability of both phases become equal depends on the joint aperture and its surface profile. It can be concluded that when p_a/p_w exceeds this limit value, air phase dominates the flow, hence, the effect of water permeability can be neglected.

4.12 VALIDITY OF DARCY'S LAW FOR TWO-PHASE FLOW

In the simplified form of Darcy's law, the hydraulic gradient is assumed to be linear along the fluid path. This assumption is no longer valid when the fluid flow is non-linear. Therefore, in general, the flow can be expressed in terms of hydraulic

Table 4.3. Two-phase fracture permeability for different confining and inlet fluid pressures.

Specimen	Confining pressure, MPa	Inlet water pressure, MPa	Single-phase water permeability × 10^{-11}, m²	Permeability of both Phases becomes equal when p_a/p_w	Two-phase permeability × 10^{-11}, m²	%Reduction of permeability from single-phase flow
Specimen 1	0.55	0.1	30.0	1.1	10.0	33
	1.0	0.25	15.0	1.05	7.0	46
	2.0	0.26	11.7	1.0	6.0	51
	2.0	0.4	11.0	0.82	5.5	50
	3.0	0.26	12.5	1.07	6.0	48
Specimen 2	0.35	0.12	11.0	1.1	3.0	27
	0.5	0.11	1.1	1.1	0.75	68
	0.5	0.19	4.2	0.9	0.9	21
	1.0	0.16	2.5	0.9	0.9	36
	1.0	0.2	2.15	0.9	0.9	42
	1.0	0.3	2.0	1.0	0.8	40
	2.0	0.2	1.7	0.9	0.9	53
	2.6	0.3	1.0	0.85	0.5	50
Specimen 3	0.5	0.1	9.0	1.25	1.9	21
	0.5	0.2	6.5	1.1	1.8	27
	0.5	0.3	6.0	1.06	2.5	42
	1.0	0.3	3.4	1.06	1.9	55
	1.0	0.4	2.0	1.0	1.4	70
	1.0	0.5	1.54	1.0	1.2	78
	2.0	0.4	0.42	1.0	0.24	57
	2.0	0.5	0.38	1.0	0.2	52

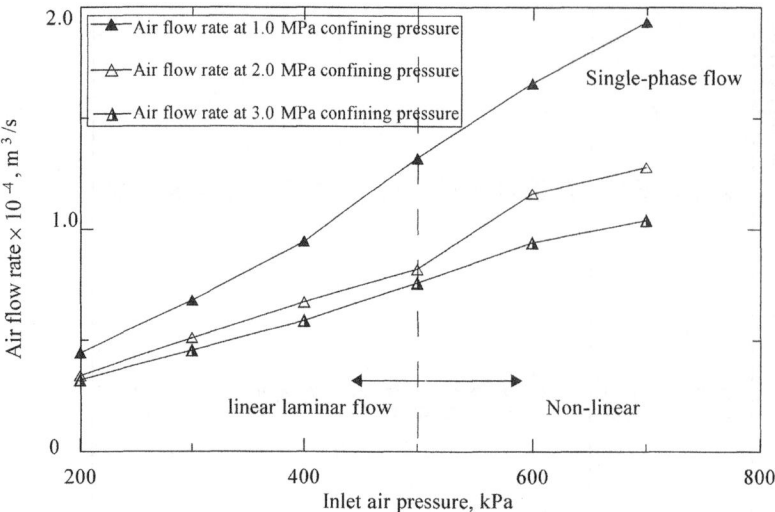

Figure 4.40. Applicability of Darcy's law for single phase flow in natural rock joints.

gradient as follows:

$q = AKi$ for laminar flow   (4.56a)

$q = AKi^{\beta}$ for turbulent flow or non-linear laminar flow   (4.56b)

where q = flow rate, K = conductivity, i = hydraulic gradient, A = cross-section area normal to the flow and β = constant, which varies between 0 and 1. For laminar flow, β becomes unity.

For negligible elevation head and the velocity head, the hydraulic gradient in Eqn. 4.56 is replaced by the pressure gradient (dp/dx) as given below

$$q = AK \frac{dp}{dx}$$ (4.57)

where dp = pressure difference along the length of the fluid flow path, dx.

From the laboratory test results obtained assuming fully saturated flow through natural rock joints, it can be seen that the linear Darcy's law is valid at relatively low confining pressures and at low hydraulic gradients (Fig. 4.40). However, at increased confining stress (e.g., σ_c = 2 MPa) linearity is not followed which can be attributed to the formation of irregular rough joint walls and the changing contact area between joint walls. In the following section, the validity of Darcy's law for two-phase flow through natural rock specimens is examined.

4.12.1   Comparison of single-phase flow rates with two-phase flows

Figure 4.41 shows the measured single-phase and two-phase flow rates against inlet fluid pressures. During the two-phase flow, the applied inlet air and water pressures are equal i.e. $p_a = p_w$. Curves A and B (Fig. 4.41a) represent single-phase flow

(a) Confining pressure 0.5 MPa

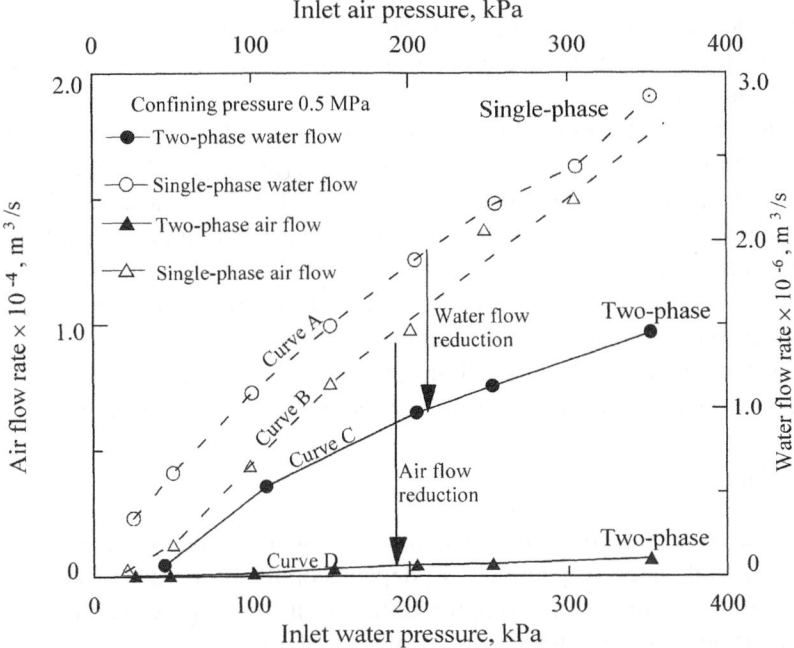

(b) Confining pressure 1.0 Mpa

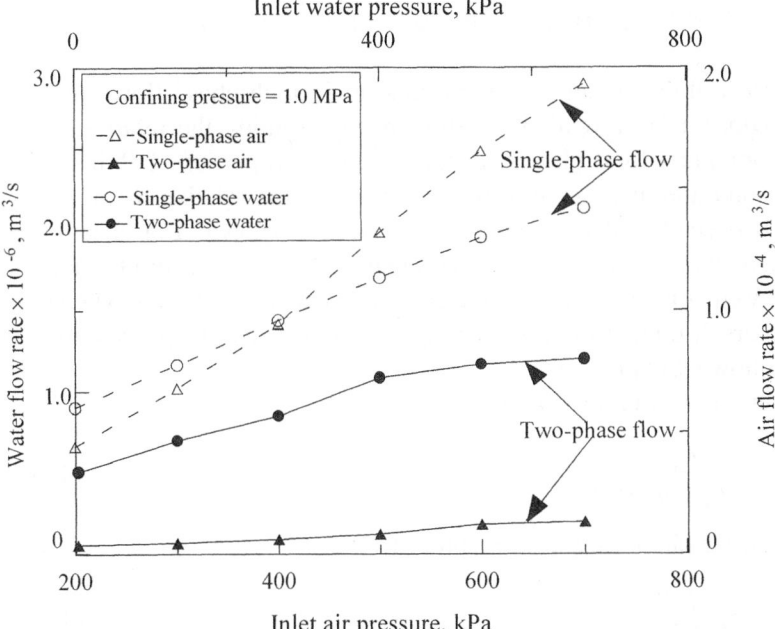

Figure 4.41. Comparison of two-phase with single-phase flow rates for varied inlet fluid pressures.

rates, which vary approximately linearly with the inlet fluid pressure. The individual two-phase water flow (Curve C) and air flow (Curve D) rates also exhibit a linear relationship between flow rates and fluid pressures. A significant reduction of two-phase flow rate (compared to single-phase) is experienced due to the influence of one phase on the other.

For example, at 200 kPa inlet fluid pressure, the individual components of water and airflow rates of two-phase flow have decreased by 50% and 95% respectively, from the single-phase flow rates (Fig. 4.41a). The variation of the observed air and water flows with inlet pressure (two-phase condition) can still be approximated to a linear relationship as in the case of single-phase flow. These findings confirm that two-phase flows also follow Darcy's law for the tested range of confining pressure, and inlet air and water pressures.

Darcy's law can be extended to model unsaturated flow through jointed rocks by introducing the factor, 'relative permeability' (K_r) in Eqn. 4.58 as follows:

$$q = AK_{ri} K \frac{dp}{dx} \tag{4.58}$$

where K_{ri} = relative permeability, the subscript i represents the phase and it takes 'a' and 'w' for air and water, respectively. K_r depends on the properties of permeating fluids such as solubility and compressibility at different stress levels, and the driving (inlet) pressures and temperatures of each phase.

4.13 RELATIVE PERMEABILITY OF TWO-PHASE FLOW

The relative permeability is an important concept in multiphase flow analysis, because it provides a more realistic coefficient representing the relative dominance of the individual fluid phases. Factors such as properties of fluids, fluid pressures and joint apertures govern the magnitude of the relative permeability, which ranges between 0 and 1. In order to examine the behaviour of unsaturated flow through a rock mass, one must incorporate the relative permeability factor in the general flow equations. For negligible elevation and velocity head, when two-phase flow occurs through a joint with an aperture e, the relative permeability is given by the following expressions:

For the water phase (wetting phase),

$$K_{rw} = \frac{q_w \mu_w}{A k_{sw} (dp/dx)_w} \tag{4.59a}$$

Similarly, for the air-phase (non wetting phase)

$$K_{ra} = \frac{q_a \mu_a}{A k_{sa} (dp/dx)_a} \tag{4.59b}$$

where k_s = single-phase permeability = $e^2/12$.

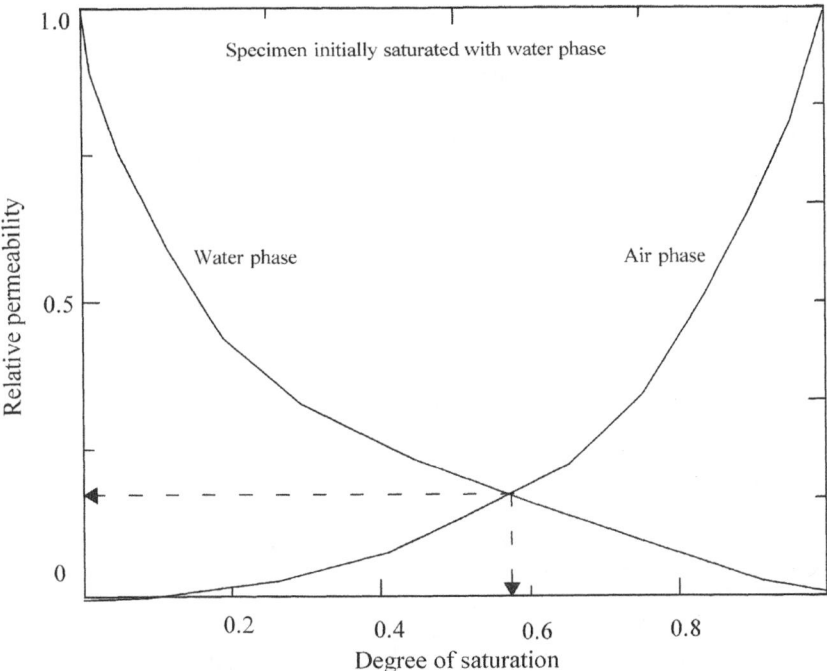

Figure 4.42. The theoretical relationship between relative permeability and the degree of saturation.

K_{ra} and K_{rw} = relative permeability of air and water, respectively. The subscripts 'a' and 'w' represent air and water, respectively.

When the relative permeability with respect to one phase becomes unity, the relative permeability of the other phase becomes zero. For instance, if the medium is considered to be fully saturated with air, then the 'relative permeability of air' is unity, and the 'relative permeability of water' is zero. At intermediate values of saturation, the relative permeability of each fluid phase will be reduced from the corresponding saturated value, due to the reduction in cross-section area of the fluid phase (Fig. 4.42). At a specific degree of saturation, the relative permeability of both phases become equal. Theoretically, the relative permeability of each phase should add up to unity ($K_{ra} + K_{rw} = 1$), but the validity of this result has not been demonstrated experimentally (Fourar et al., 1993; Pyrak-Nolte et al. 1992; Pruess & Tsang, 1990).

(a) *Two-phase flow when $p_a = p_w$:* When the inlet air pressure and water pressure are held approximately equal, the relative permeability coefficients for water and air (Fig. 4.43) are calculated using Eqns. 4.59a and b. When the relative permeability of one phase increases, the relative permeability of the other phase should decrease. For example, at 0.4 MPa inlet fluid pressure (i.e. $p_a = p_w = 0.4$ MPa), the relative permeability of air decreases while increasing the relative permeability of the water phase. This is probably due to fact that the water phase

(a) Confining pressure 0.5 MPa

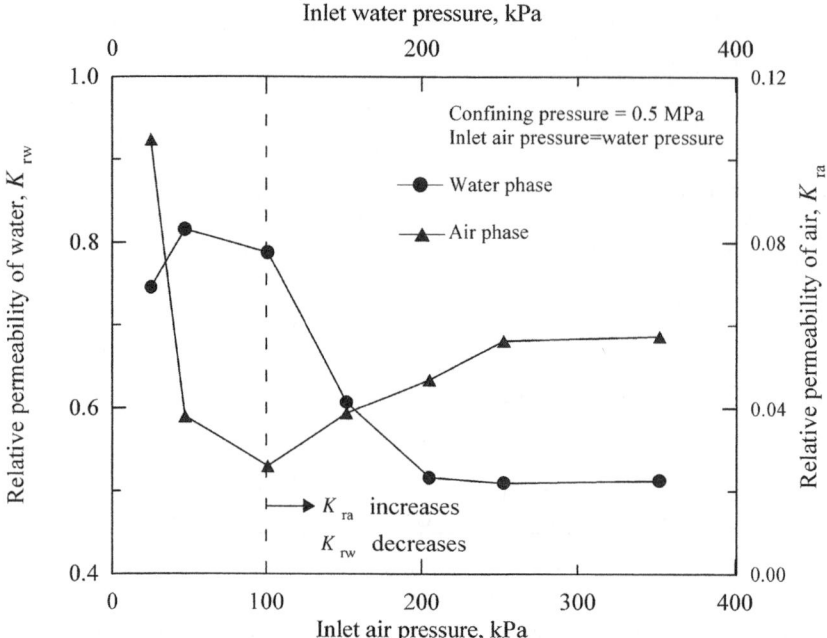

(b) Confining pressure 1.0 MPa

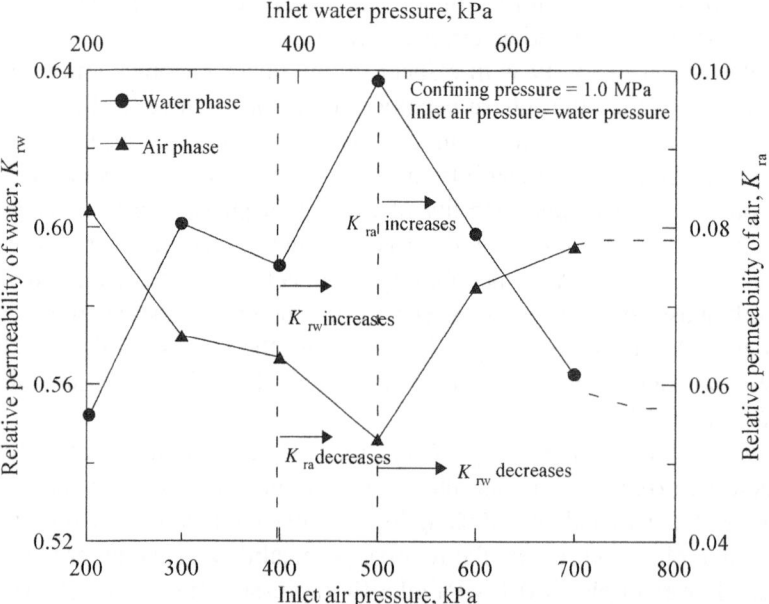

Figure 4.43. Relative permeability of two-phase flow when inlet air pressure equals water pressure.

(a) Confining pressure = 0.5 MPa

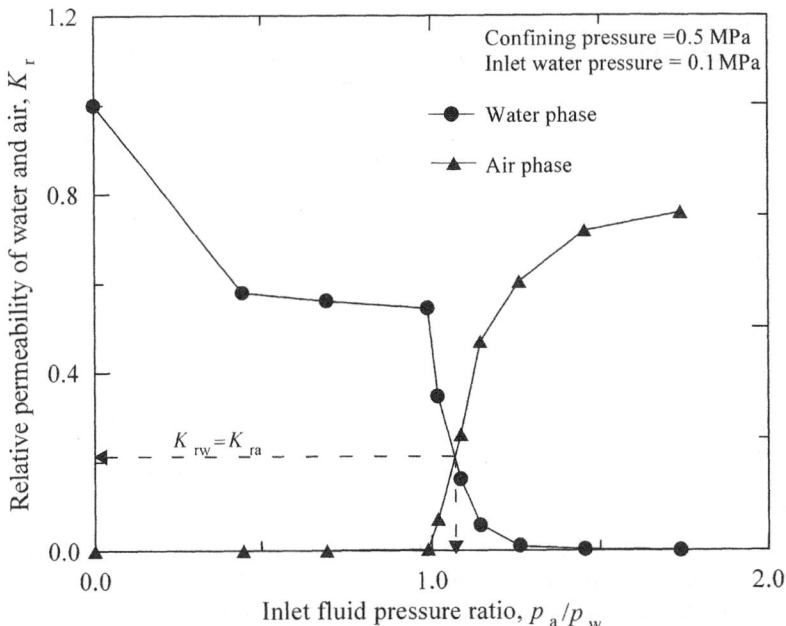

(b) Confining pressure = 2.0 MPa

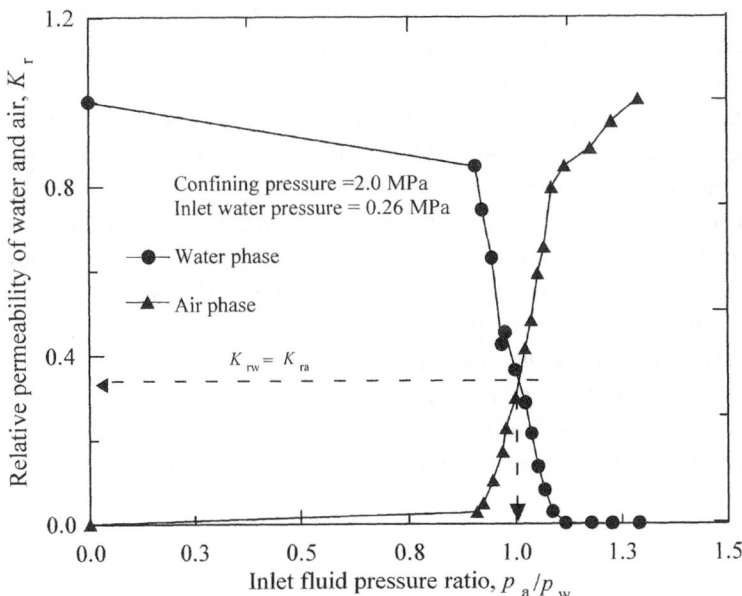

Figure 4.44. Relative permeability for water saturated specimen at various confining pressures.

Table 4.4. Relative permeability factors at different inlet fluid pressure ratios.

Specimen	Confining pressure, MPa	Inlet water pressure, MPa	Relative permeability of both phases becomes equal when p_a/p_w	Relative Permeability
Specimen 1	3.0	0.26	1.1	0.35
	2.0	0.40	0.80	0.37
	2.0	0.26	1.0	0.37
	1.0	0.2	1.05	0.37
	0.5	0.1	1.12	0.2
Specimen 2	2.0	0.2	0.9	0.2
	1.0	0.3	1.0	0.19
	1.0	0.16	0.8	0.28
	0.5	0.19	0.85	0.1
	0.5	0.1	1.2	0.2

dominates when the inlet fluid pressure is below a critical value. A continuous increase in relative permeability of air phase is experienced once the inlet fluid pressure exceeds 0.1 MPa (Fig. 4.43a) and 0.5 MPa (Fig. 4.43b), respectively.

(b)   Two-phase flow when $p_a \neq p_w$: The change in relative permeability of two-phase flow at different inlet fluid pressure ratios is shown in Fig. 4.44. For an initially water saturated specimen, the inlet fluid pressure ratio is defined as the inlet air pressure divided by the inlet water pressure (i.e. p_a/p_w). However, for the initially air saturated specimen, the inlet fluid pressure ratio is defined as inlet water pressure divided by the inlet air pressure (i.e. p_w/p_a). When the ratio of p_a/p_w increases, the relative permeability of air increases while the relative permeability of water decreases (Fig. 4.44). The opposite trend occurs when the p_w/p_a ratio is increased. When the relative permeability of one phase approaches 1, then the joint becomes fully saturated with that fluid phase. For the complete range of confining pressure and axial stress, the relative permeability of both phases tends to become equal when the p_a/p_w ratio is between 0.9 and 1.2. Table 4.4 shows the variation of the relative permeability with the inlet fluid pressure ratios for different fluid pressures and confining pressures.

It is important to note that the range of confining pressures tested by the authors is moderate (1–8 MPa), hence, the findings presented here may not be extrapolated to predict the two-phase flow behaviour at much greater confining pressures.

REFERENCES

Arnold, C.R. and Hewitt, G.F. 1967. Further developments in the photography of two-phase gas–liquid flow. Report AERE-R5318, UKAEA, Harwell.

Brown, S.R. 1987. Fluid flow through rock joints: Effects of surface roughness. *J. Geophys. Res.,* 92(B2): 1337–1347.

Buckingham, E. 1907. Studies of the movement of soil moisture. *U.S.D.A. Bur. Soils* (Bulletin), 38.

Cicchitti, A., Lombardi, C., Silvestri, M., Soldaini, G. and Zavattarelli, R. 1960. Two-phase cooling experiments; pressure drop, heat transfer and burnout measurements. *Energy. Nucl.,* 7: 407–425.

Delhaye, J.M. 1979. Optical methods in two-phase flow. *Dynamic Measurement Unsteady Flow, Proc. Dynamic Flow Conf.,* 1978, pp. 321–343.

Dorsey, N.E. 1940. *Properties of Ordinary Water-Substance.* Hanfer, New York, 673 p.

Dukler, A.E., Wicks, M. and Cleveland, R.G. 1964. Frictional pressure drop in two-phase flow: Comparison of existing correlation for pressure loss and hold-up. *AIChE J.,* 10(1): 38–43.

Fiori and Bergles 1966. A study of boiling water flow regimes at low pressure. M.I.T. Report, 3582-40.

Fourar, M. and Bories, S. 1995. Experimental study of air–water two-phase flow through a fracture (narrow channel). *Int. J. Multiphase Flow,* 21(4): 621–637.

Fourar, M., Bories, S., Lenormand, R. and Persoff, P. 1993. Two-phase flow in smooth and rough fractures: Measurement and correlation by porous medium and pipe flow models. *Water Resour. Res.,* 29(11): 3699–3708.

Fredlund, D.G. and Rahardjo, H. 1993. *Soil Mechanics for Unsaturated Soils.* Wiley, New York, 517 p.

Golan, L.P. and Stenning, A.H. 1969. Two-phase vertical flow maps. *Proc. Inst. Mech. Eng.,* 184(3c): 110–116.

Graham, B.W. 1969. *One-Dimensional Two-phase Flow.* McGraw Hill, New York, 408 p.

Hetsroni, G. 1982. *Handbook of Multiphase Systems.* McGraw Hill, London, 1374 p.

Hewitt, G.F. 1978. Measurement of Two-phase Flow Parameters. *Seminar Int. Center Hat Mass Transfer,* Dubrovnik.

Hewitt, G.F. and Lovegrove, P.C. 1969. A mirror-scanner velocimeter and its application to wave velocity measurement in annular two-phase flow. Report AERE-R3598, UKAEA, Harwell.

Indraratna, B. and Ranjith, P.G. 1999. Deformation and permeability characteristics of rocks with interconnected fractures. *9th Int. Congress on Rock Mechanics,* France, vol. 2, pp. 755–760.

Louis, C. 1976. *Introduction l'Hydraulique des Roches.* PhD thesis, Paris.

Mcadams, W.H., Woods, W.K. and Heroman, L.C. 1942. Vaporization inside horizontal tubes: Benzene-oil mixture, *Trans. ASME,* 64: 193–200.

Mishima, K. and Hibiki, T. 1996. Some characteristic of air–water two-phase flow in small diameter vertical tubes. *Int. J. Multiphase Flow,* 22(4): 703–712.

Pruess, K. and Tsang, Y.W. 1990. On two-phase relative permeability and capillary in rough-walled rock fractures. *Water Resour. Res.,* 26:1915–1926.

Pyrak-Nolte, L.J., Helgeson, D., Haley, G.M. and Morris, J.W. 1992. Immiscible fluid flow in a fracture. *33rd Rock Mechanics Symposium,* pp. 517–578.

Ranjith, P.G. 2000. *Analytical and Experimental Modelling of Coupled Water and Air Flow Through Rock Joints.* PhD thesis, University of Wollongong, Australia.

Rasmussen, T.C. 1991. Steady fluid flow and travel times in partially saturated fractures using a discrete air–water interfaces. *Water Resour. Res.,* 27(1): 67–76.

Richards, L.A. 1931. Capillary conduction of liquids through porous medium. *J. Phys.,* 1: 318–333.

Street, R.L., Gary, Z.W. and John K.V. 1996. *Elementary Fluid Mechanics*. 7th ed., Wiley, New York, 757 p.

Taitel, Y. and Dukler, A.E. 1976. A model for predicting flow regime transitions in horizontal and near horizontal gas–liquid flow. *AICh.E J.*, 21: 47–55.

Tsang, Y.W. and Witherspoon, P.A. 1981. Hydro-mechanical behavior of a deformable rock fracture subjected to normal stress. *J. Geophys. Res.*, 86(B10): 9287–9298.

CHAPTER 5

Fluid flow modelling techniques

5.1 INTRODUCTION

The porous medium is characterized by the manner in which voids are embedded, their interconnectivity, size and shape. Usually, depending on the lattice structure, all natural and artificial materials can be divided into four main categories as follows: (a) porous, (b) non-porous, (c) fractured and (d) combination of fractured and porous media (Fig. 5.1). Various theories have been developed for each application depending on how flow takes place through a porous or fractured media, or a combination of both. For example, in soil science, flow is usually modelled using a porous medium approach. It is usual practice to simulate flow through fractured media, such as in fractured rock media, using modified theories developed for fluid flow through pipes or channels. The selection criterion for type of flow analysis depends on the availability and extent of field data, computer resources and the degree of the accuracy required for the particular application. The available rock mass properties and the geology of the area also influence the initial adoption of the methodology. Techniques of flow analysis technique can be grouped as either one or more of the following:
(1) Continuum approach – for porous media;
(2) Discrete fracture flow theory – fractured media; and
(3) Dual-porosity method – coupled porous and fractured media.
    Flow in porous media can either be single-fluid or multiphase fluid. Modelling of multiphase flow in deformable or undeformable media requires quantification of the interaction between each fluid and the corresponding change of fluid properties. For single-phase flow, the most relevant parameter is the intrinsic permeability, whereas, both relative permeability and intrinsic permeability must be taken into account when modelling multiphase flow.

5.2 CONTINUUM APPROACH

From the point of view of civil engineers, porosity has been recognized as an influencing parameter not only for flow phenomena, but also for deformability

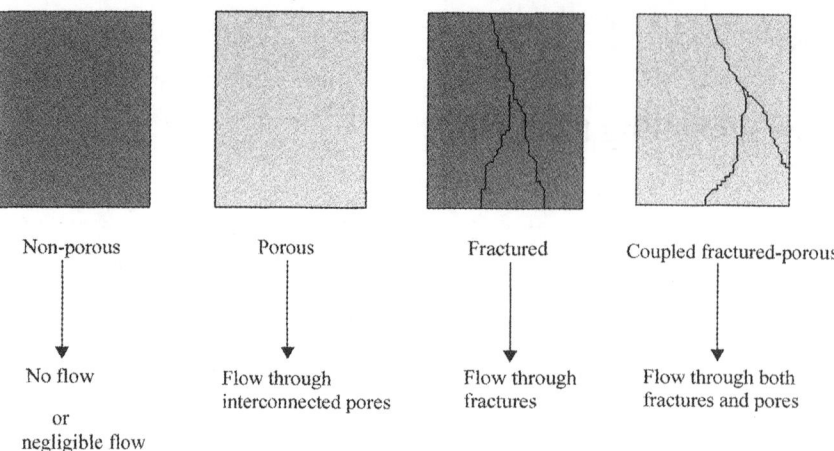

Figure 5.1. Fluid flow mechanism in different materials.

and stability of the material under given loading conditions. In various applications, flow mechanisms are different and complex, therefore, one has to carefully examine the pore structure of the porous medium and how the pore network affects the flow distribution. The size and the shape of solid particles and their arrangement will govern the pore volume. Pore structure and pore volume of the medium may be changed by external loading (e.g. degree of consolidation), chemical reactions, fluid pressures and the solubility of solid in fluid (Fig. 5.2).

A porous medium is broadly classified as either consolidated or unconsolidated. In the analysis of pore structure, there are two types of pores: (a) interconnected and (b) isolated. Although fluid flow takes place through interconnected pores, it is important to consider isolated pores when they are subjected to loads, because with deformation, isolated pores can subsequently become part of the interconnected pore network. Fluid flow takes place only if the following conditions are satisfied:

(1) Existence of an interconnected pore structure and
(2) Fluid particle must pass through the smallest pore in the interconnected pore arrangement.

Pore volume is usually expressed in terms of porosity, which is defined as the ratio of interconnected pore volume divided by the total volume of the sample. It must be noted that the porosity at a single point cannot be measured. What is determined is the average porosity of the sample as a function of ground and fluid stresses. For instance, increase in inlet fluid pressure may result in an increase of the pore volume and interconnectivity of pores, and consequently, the flow rate may be expected to increase. Laboratory experiments of fluid flow through intact sandstone specimen indicate that flow rate is linearly increased up to a certain fluid pressure, and beyond this value the flow becomes non-linear (Fig. 5.3). This

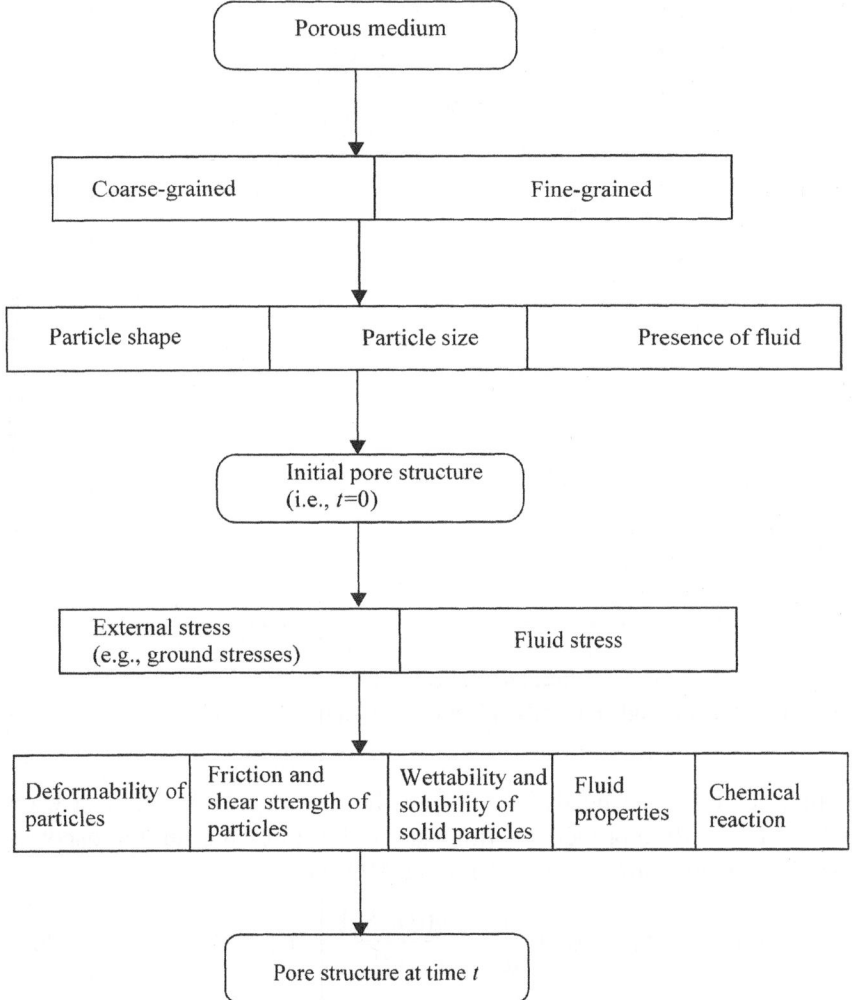

Figure 5.2. Change of pore structure of porous medium.

verifies that further increase in inlet fluid pressure and associated flow does not follow the conventional Darcy's law. Flow through porous rock also depends on the applied normal stress (confining pressure) and these effects were discussed earlier in Ch. 3.

The single-phase fluid movement in porous matrix under steady-state flow can be modelled using the following diffusion equation (Bear, 1979):

$$\nabla \cdot (K \cdot \nabla h) + q(x, y) = 0 \tag{5.1}$$

where K = hydraulic conductivity tensor, ∇ = operator $((\partial/\partial x)i + (\partial/\partial y)j)$, i and j are unit vectors in X and Y directions, q = flow rate of sources, and x, y = cartesian co-ordinates.

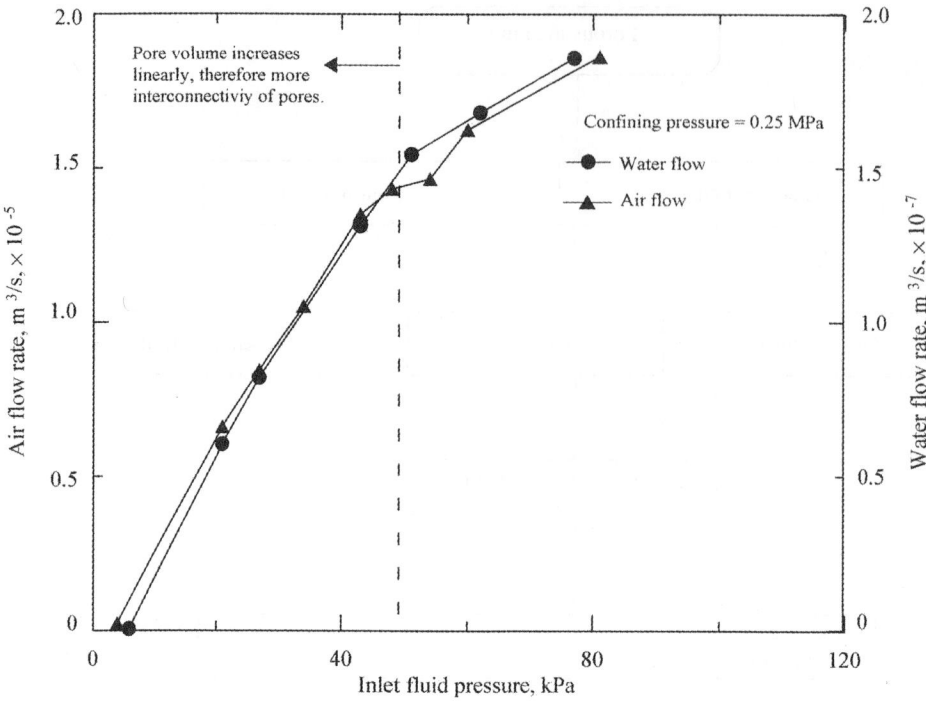

Figure 5.3. Fluid flow through a intact sandstone specimen.

In multiphase flow, one needs to consider the degree of saturation of each phase and change of fluid properties with time in the following fluid transportation equation as given below (Celia & Binning, 1992):

$$\nabla \cdot \left(\rho_\beta k_i k_{r\beta} \cdot (\nabla p_\beta - \rho_\beta g) \right) \frac{1}{\mu_\beta} + \frac{\partial(\phi \rho_\beta S_\beta)}{\partial t} \pm F_\beta = 0 \qquad (5.2)$$

where β = fluid phase, k_i = single phase permeability, p = pressure, $k_{r\beta}$ = relative permeability, μ_β = dynamic viscosity, ϕ = porosity, F = source or sink, S = degree of saturation and g = acceleration due to gravity.

5.3   DISCRETE FRACTURE APPROACH

Fluid flow and solute transport through tight rocks has gained increasing interest in storage of nuclear waste and hazardous liquid toxic waste in deep underground cavities. In the recovery of petroleum products, particularly at secondary recovery process, researchers have been continuously studying the recovery process through tight rocks. Flow through tight tock mainly occurs through interconnected fracture networks. It is not possible to simulate such jointed rock mass using the continuum technique which works well for homogeneous and continuous rock media. Therefore, when the fluid flow is dominated through a defined inter-connected fracture network, the most realistic flow approach is to employ the discrete fracture flow theory. This approach provides several advantages over the continuum approach such as:

(1) Flow properties of the fracture network;
(2) Effect of fracture network on the fluid flow; and
(3) Detailed distribution of stress and deformation patterns around fractures.

The distinct-element method (ITASCA, 1996; Lemos et al., 1985), joint-element method (Goodman, 1976) and block theory (Goodman & Shi, 1985) are based on the discrete fracture theory. The discrete fracture theory has been employed for flow analysis in both research work and in case history applications (Snow, 1969; Long & Witherspoon, 1985; Anderson & Dverstorp, 1987; Indraratna & Ranjith, 1998, 1999; Indraratna et al., 2000). This approach requires detailed information of the distribution of fractures including their geometry, orientation, length, spacing as well as individual relationships for describing flow in fractures when subjected to various stress fields. Although probabilistic methods can be employed to generate fracture geometry based on probabilistic framework, the assumed stochastic model may still not truly represent the actual case. In order to understand the mechanism of fluid flow through rocks fractures, small scale (laboratory) and field experiments have been carried out through single fractures and network of fractures. Earlier studies of fluid flow through single fractures have been reported by Lomize (1951), Snow (1969), Louis (1968) and Wilson (1970). Since small scale experimental studies do not always represent true picture, Ablien et al. (1985) and Neternieks (1985) used field experiments of solute migration in single fractures for obtaining more realistic data.

Discrete fracture flow models can be simulated either in two dimensions (Ranjith, 2000; Indraratna & Ranjith, 1999; ITASCA, 1996; Long & Witherspoon, 1985) or in three dimensions (ITASCA, 1997; Rasmussen, 1988; Dverstorp & Anderson, 1989). Two dimensional models are restricted in several ways because the connectivity of fractures cannot realistically represent the actual spatial joint pattern. In 2-D, fractures, which do not intersect with other fractures may intersect with each other in 3-D. As a result, flow and deformation characteristics will be significantly changed. Unfortunately, three-dimensional joint-flow models are often disadvantaged by the lengthy solution procedures, hence large computing time and memory requirements.

The connectivity is governed by the fracture density, and it is often quantified using the percolation threshold value. The percolation threshold depends on the accuracy of statistical properties of rock, and also on the dimensions (i.e., 2-D or 3-D) of the network. In order to generate a fracture network based on field data, the following parameters are usually used and quantified using statistical methods:

(1) Fracture shape (e.g., rectangular, circular or elliptical);
(2) Fracture size based on trace length or radius;
(3) Fracture orientation;

Table 5.1. Distribution function used to describe joint density.

References	Various Distribution Functions
Joint orientation	
Anderson & Dverstorp (1987)	Fisher distribution, $f(\theta, \varphi) = \dfrac{ke^{k\cos\theta}\sin\theta}{4\pi k(\sin h)}$
Samaniego (1984), Wei et al. (1995)	Normal distribution, $f(\theta, \varphi) = \dfrac{1}{\sigma\sqrt{2\pi}}\,e^{-0.5\left(\frac{\theta-\mu}{\sigma}\right)^2}$
Priest (1993)	Negative exponential distribution, $f(\theta, \varphi) = \dfrac{1}{\mu}\,e^{-\frac{\theta}{\mu}}$
Joint length/size (l)	
Samaniego (1984), & Wei et al. (1995)	Negative exponential distribution
Priest & Hudson (1981)	Exponential
Bridges (1976), McMahon (1971)	Lognormal
Dershowitz (1984)	Gamma
Fracture transmissivity	
Snow (1970)	Lognormal distribution
Spacing (s)	
Priest & Hudson (1981)	Normal
Sen & Eissa (1992)	Exponential and Lognormal
Bridges (1976)	Lognormal, $f(s) = \dfrac{1}{\sqrt{2\pi}\,s\sigma}\exp\left[\left(-0.5\dfrac{1}{s}\ln\dfrac{s}{m}\right)\right]$
Wallis & King (1980)	Exponential

Where θ = dip angle, φ = dip direction, k = dispersion parameter about the mean direction, determined by the maximum likelihood estimator (Mardia, 1972), μ = mean, σ = standard deviation, s = spacing and m = median.

(4) Fracture density; and
(5) Fracture aperture.

It is not feasible to map all the fractures and obtain all the relevant geometrical parameters. In this context, a statistical approach is beneficial, where joints are seen as realizations of stochastic models. As an example, the orientation of joints may be defined using a Fisher distribution or normal distribution depending on the spread of joint inclination. Similarly, for other parameters, different distributions are employed for a given rock mass, and possible distribution functions are used for various applications as listed in Table 5.1. In the discrete approach, the shape of fracture is also considered important, especially when

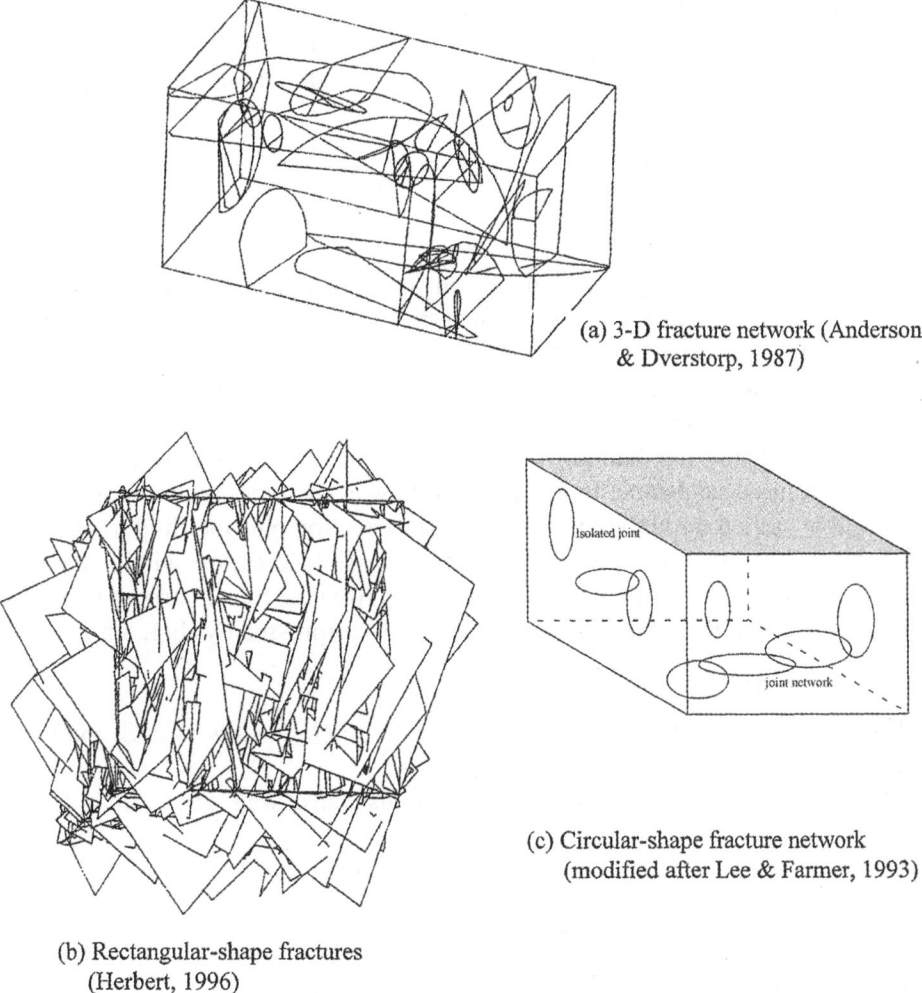

(a) 3-D fracture network (Anderson
& Dverstorp, 1987)

(c) Circular-shape fracture network
(modified after Lee & Farmer, 1993)

(b) Rectangular-shape fractures
(Herbert, 1996)

Figure 5.4. Three-dimensional fracture network models.

3-D analysis is carried out. Fracture shape may be assumed as planar traces, rec-tangular, square, circular, elliptical or polygon (Fig. 5.4). A numbers of studies have treated this problem entirely in two-dimensions by assuming discontinu-ities as linear traces in a planar rock surface (Priest & Hudson, 1981; Pahl, 1981; Indraratna & Ranjith, 1998). Long and Witherspoon (1985), Shapiro and Anderson (1983) and Anderson and Dverstorp (1987) represented fractures as discs of finite radius in their three dimensional models. The data-collection method for joint models (i.e., scanline and window techniques) was discussed earlier in Ch. 3.

In the discrete fracture approach, the fluid flow is assumed to take place via the passage of joint network. While this assumption is generally valid for hard crystalline rocks, continuum approach may still be exercised in sedimentary rocks such as sandstone, shales and other soft (relatively porous) rocks. Fluid-flow concepts and deformation mechanisms of fractures employed in discrete approach were discussed earlier Ch. 3.

5.4   DUAL-POROSITY MODELS

When fluid flow takes place through both rock matrix and discontinuities, the flow behaviour is analyzed using the dual-porosity model. Although this method is not as common as the previous models, several researchers including Fidelibus et al. (1997), Rasmussen (1988) and Shapiro and Anderson (1983) have used dual-porosity models to represent the flow conditions through the rock matrix and joints. For instance, the dual-porosity technique may be applied to simulate fluid flow in fractured sandstone. In this method, flow at each intersection point is due to the flow carried through the joints to the intersection plus the fluid transported through the porous rock matrix. Depending on fluid pressure within the rock matrix and the fractures, flow may be conducted from the rock matrix to the joints or vice versa, and the total flow is mathematically expressed by (Fig. 5.5)

$$Q_i = Q_1 + Q_2 + Q_3 + Q_4 \pm Q_m \tag{5.3}$$

where Q_i = flow at the intersection point i, Q_1 to Q_4 = flow at each joint and Q_m = flow through the rock matrix.

The estimation of flow in fractures and porous matrix were discussed in Ch. 3.

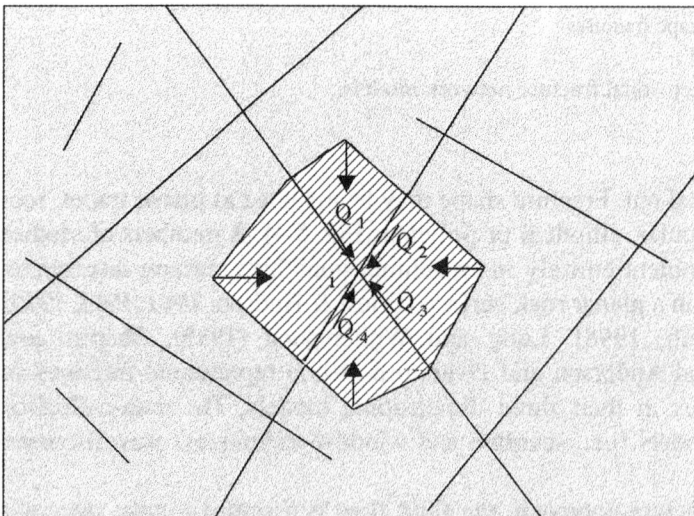

Figure 5.5. Simplified two-dimensional dual porosity approach.

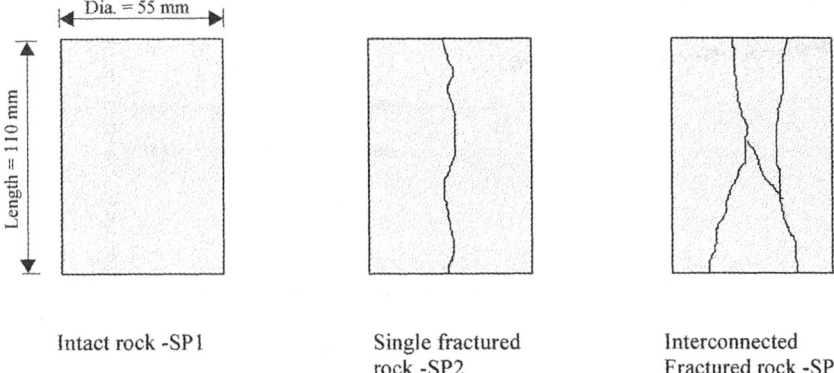

Intact rock -SP1   Single fractured   Interconnected
                                  rock -SP2   Fractured rock -SP3

Figure 5.6. Specimens tested to study the permeability based on continuum and discrete approaches.

5.5   COMPARISON OF PERMEABILITY BASED ON CONTINUUM AND DISCRETE APPROACH

In some field situations, a fractured medium can be modelled using a continuum approach, as long as the number of fractures is large and random. In this simulation, one has to define a suitable representative block with average properties to represent the fractured rock media. While this process is never an easy task, the degree of confidence can be enhanced using experimental results based on two approaches (i.e., continuum and discrete approaches). A series of laboratory tests on different rocks are carried out to study the permeability of rock based on discrete and porous approaches. In this study, granite and sandstones specimens were tested using the triaxial apparatus (Fig. 5.6). The details of the triaxial equipment and the associated experimental procedure were discussed in Ch. 2. For given confining and axial stresses, the steady-state flow rate is observed for different inlet fluid pressures. The fracture permeability (i.e. discrete approach) is estimated using the following Eqn. 5.4, and the continuum permeability is estimated using Eqns. 5.5 and 5.6 for air and water, respectively.

$$k_{\text{d}} = \frac{1}{2.283} \left(\frac{q\mu}{w(\text{d}p/\text{d}x + \rho g h)} \right)^{2/3} \tag{5.4}$$

where q = discharge flow rate, μ = dynamic viscosity, w = fracture width and p = permeating fluid pressure difference along a joint of length x, ρ = density of the fluid, g = acceleration due to gravity, h = hydraulic head.

For air permeability based on porous medium approach

$$k_{\text{ca}} = \frac{8\,q_{\text{e}}p_{\text{e}}\mu L}{(p_{\text{i}}^2 - p_{\text{e}}^2)\pi d^2} \tag{5.5}$$

where k_{ca} = coefficient of air permeability based on continuum approach, q_{e} = exit flow rate of air, p_{e} = exit pressure of air, L = length of the specimen, p_{i} =

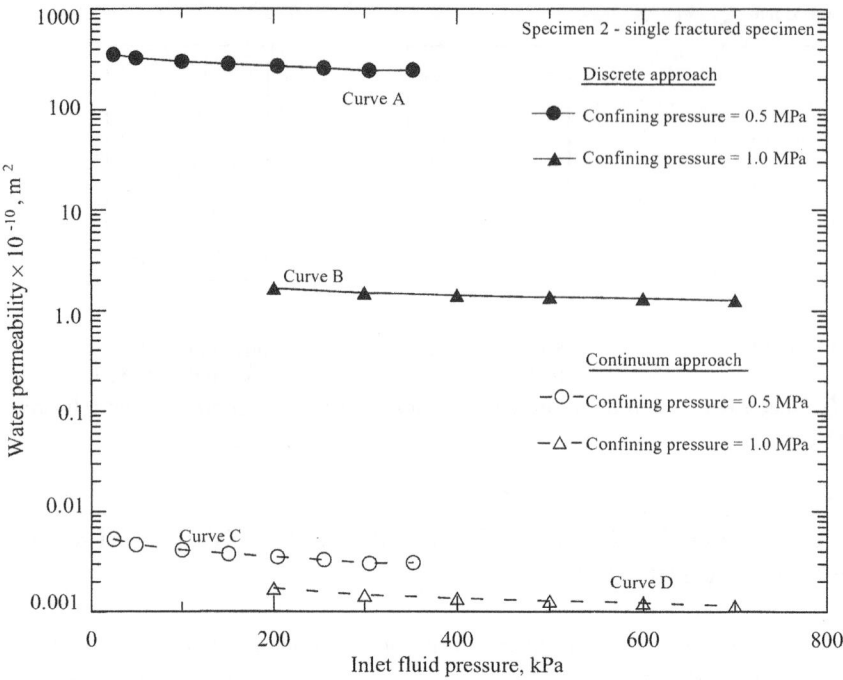

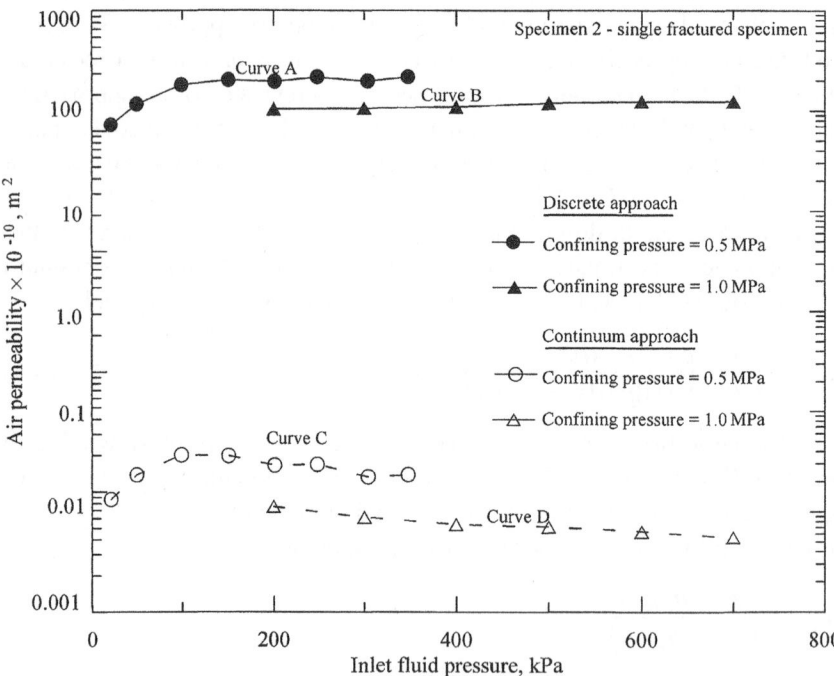

Figure 5.7. (a) Water permeability of rock specimen with a single fracture and (b) air permeability of rock specimen with a single fracture.

entrance pressure of air, μ = viscosity of air at temperature of the test and d = diameter of specimen.

For water permeability based on porous medium approach

$$k = \left(\frac{4q\mu}{\pi d^2 (\mathrm{d}p/\mathrm{d}x + \rho gh)} \right) \tag{5.6}$$

The estimated permeability based on continuum and discrete approaches using Eqns. 5.4–5.6 is shown in Figs. 5.7–5.9. The specimens 2 and 3 presented in Fig. 5.7 are fractured specimens, whereas Fig. 5.8 shows the permeability data for two different intact granite rock specimens. In Fig. 5.7, air and water permeability based on discrete approach are represented by Curves A and B. However, if the fractured rock specimen is simulated as a continuum media, then the estimated permeability data are given in Curves C and D (Fig. 5.7). This shows that the estimated permeability based on the continuum approach is much smaller than that of the discrete approach. Figure 5.8 illustrates the water permeability of rock specimen with multiple fractures, and again, the continuum approach gives smaller permeability values. From Fig. 5.9, the estimated air permeability based on the continuum approach for intact rock is shown in Curve A. If this intact rock specimen is replaced by a single fracture with an equivalent aperture (e) which gives

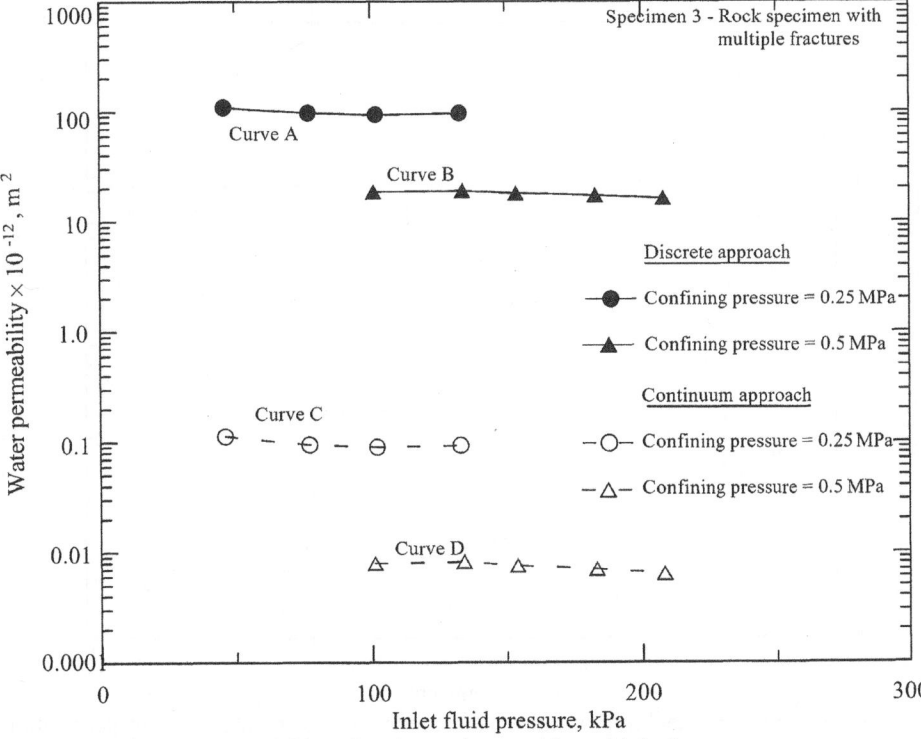

Figure 5.8. Water permeability of rock specimen with multiple fractures.

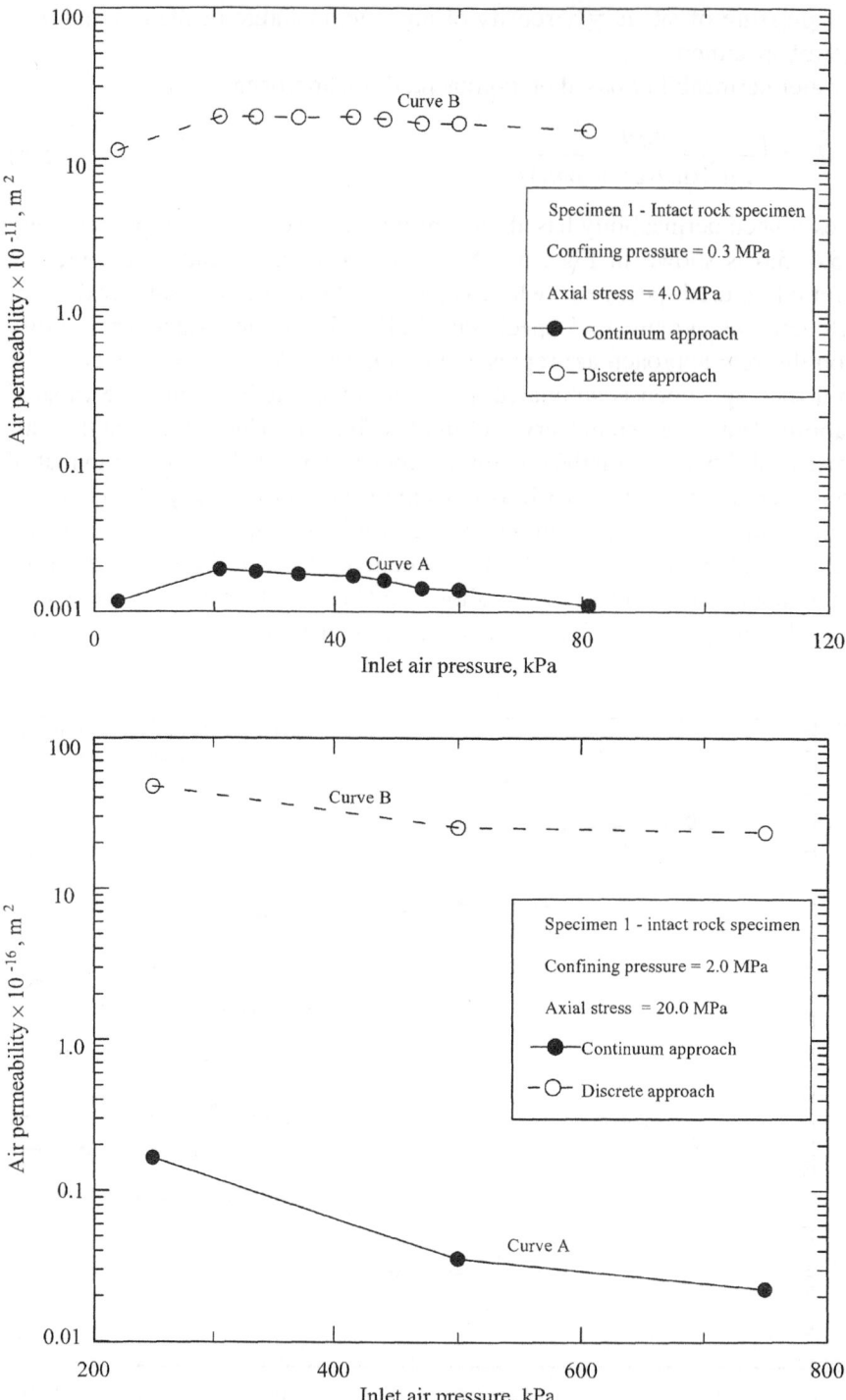

Figure 5.9. (a) Air permeability of intact sandstone rock specimen and (b) air permeability of intact granite rock specimen.

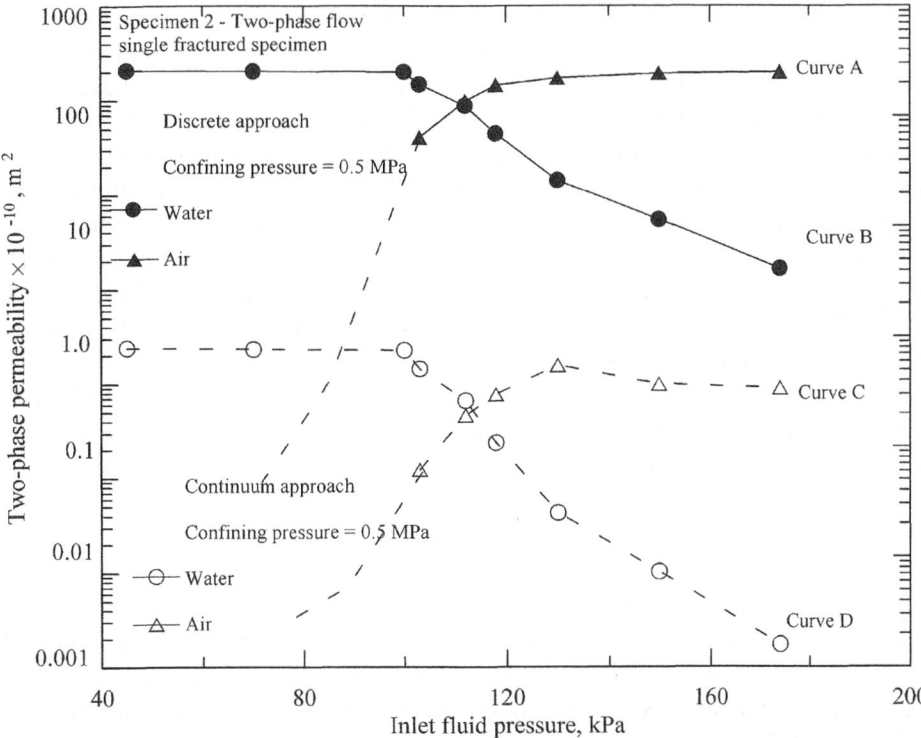

Figure 5.10. Two-phase permeability of rock specimen with a single fracture.

the same quantity of fluid flow under similar conditions as in intact rock specimen, then the value of e can be back calculated using the flow rate as follows:

$$e = \left(\frac{12q\mu}{(dp/dx + \rho gh)w} \right)^{1/3} \tag{5.7}$$

Curve B (Fig. 5.9) shows the fracture permeability based on the equivalent
aperture, derived using Eqn. 5.4. This indicates that the permeability is overesti-
mated if porous medium is simulated as a fractured discrete medium. For given
confining pressure, axial stress and inlet fluid pressures, two-phase permeability
based on continuum and discrete approaches is given in Fig. 5.10. Inlet fluid pressures were held so that they were approximately equal.

Snow (1969), Singh (1973), Long et al. (1982), Oda (1986) and Sietel et al.
(1996) studied the continuum representation of fractured rock media for fluid-
flow calculations. As discussed by Long et al. (1982), fractured media behave
more likely porous media if rock mass has the following characteristics: (a) large
fracture density, (b) approximately constant fracture aperture and (c) variable
orientation of fractures. According to Neuzil and Tracy (1981) and Oda (1986),
the equivalent porous medium concept can be applied to discontinuous rocks, if
the number of discontinuities is sufficiently large to allow the determination of a

statistical average value of flow paths. The main difficulty faced in such a simulation is the estimation of equivalent continuum properties to represent the fractured rock. In other words, the exact mechanical and hydraulic behaviour of fractured rock should be correctly represented by the continuum approach.

The work done in relation to the equivalent permeability tensor can be classified into two groups assuming, (a) infinite length of joints and (b) finite joint length. Most reported work considers infinite length of joints, which is not always realistic in the field. The effects of joint length, orientation, fracture density, joint aperture distribution and their geometrical properties must be included when developing a comprehensive model of the equivalent permeability tensor. Snow (1969) developed an equivalent permeability tensor for a single fracture and then used the method of superimposition to obtain the permeability tensor for a network of joints. The equivalent porosity (ϕ_{eq}) for a single fracture with an aperture e, length l and area A can be expressed by

$$\Phi_{eq} = \frac{el}{A} \tag{5.8}$$

In macroscopic scale, the equivalent porosity of a large rock mass is the summation of individual porosity of each single fracture. Isolated fractures may be ignored as they do not contribute to flow. However, during the deformation stage, these isolated fractures may also influence to change the flow deformation characteristics of the joint network.

The equivalent permeability tensor (k_{ij}) of a joint set with spacing, s and mean orientation, θ, is expressed by the equation given below (Snow, 1969; Sietel et al., 1996):

$$k_{ij} = \left(\frac{\rho g}{12\mu}\right) \sum \left(\frac{e^3}{s}\right)(\delta_{ij} - \cos\theta_i \cos\theta_j) \tag{5.9}$$

where δ = Kronecker delta, θ = orientation, ρ = density of fluid and i, j = unit vectors in perpendicular directions.

From Eqn. 5.9, the equivalent permeability tensor in two-dimensions is given by

$$k_{eq} = \begin{bmatrix} k_{11} & k_{12} \\ k_{21} & k_{22} \end{bmatrix} \tag{5.10}$$

While the joint aperture e plays a major role in flow through fractured rocks, the distribution of joint aperture is not included in Eqn. 5.9. The magnitude of joint deformation depends on the nature of joint asperities, applied stress level and their directions relative to the joint surface.

Apart from equivalent hydraulic properties, it is also important to express the equivalent material properties of jointed rock mass to represent the porous media. Several researchers including Biot (1955), Oda (1986) and Pariseau (1993) discussed equivalent material properties such as shear and normal stiffness, Poisson's ratio and Young's moduli for different joint patterns. Oda (1986) assuming that joints are planar (parallel plate walls and connected by springs), proposed an equation for the equivalent shear stiffness over the domain (Ω) as a function of friction angle, joint compressive strength (*JCS*), applied stress on joint, joint size and the representative elementary volume of rock. The average equivalent shear stiffness was expressed by

$$G_{eq} = \int_{\Omega} \left[\frac{100}{r} \sigma_{ij} n_i n_j \tan(JRC \log_{10}(JCS/\sigma_n) + \phi) \right] E(\dot{n}) d\Omega$$
(5.11)

where *JRC* = joint roughness coefficient, σ_n = normal stress, r = size of the joint, ϕ = friction angle, $E(\dot{n})$ = probability density function, $\dot{n}$ = unit normal vector, and n_i = the component of vector $\dot{n}$ with respect the axis x_i, where $i = 1, 2, 3$.

Similarly, the average equivalent normal stiffness (K_{eq}) over the domain (Ω) was given by Oda (1986) as

$$K_{eq} = \int_{\Omega} \left(\left(h + \frac{r}{e_0} \sigma_{ij} n_i n_j \right) \Big/ r \right) E(\dot{n}) d\Omega$$
(5.12)

where h = constant, depends on number of cycling loadings, e_0 = initial aperture.

A comparison between the principal components of actual and equivalent permeability values for different joint patterns (Fig. 5.11) is shown in Fig. 5.12

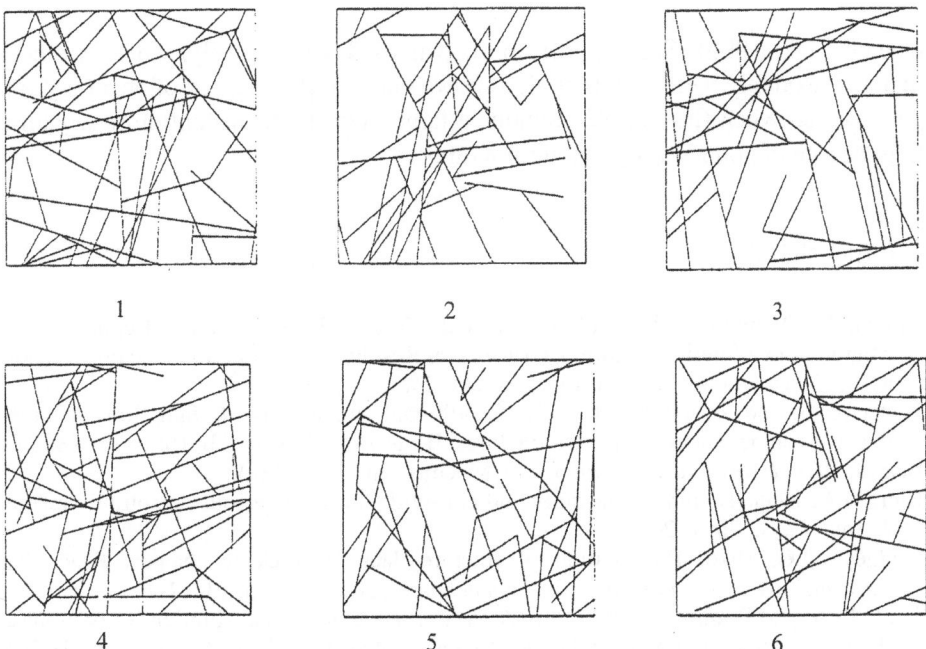

Figure 5.11. Fracture network used by Sietel et al. (1996).

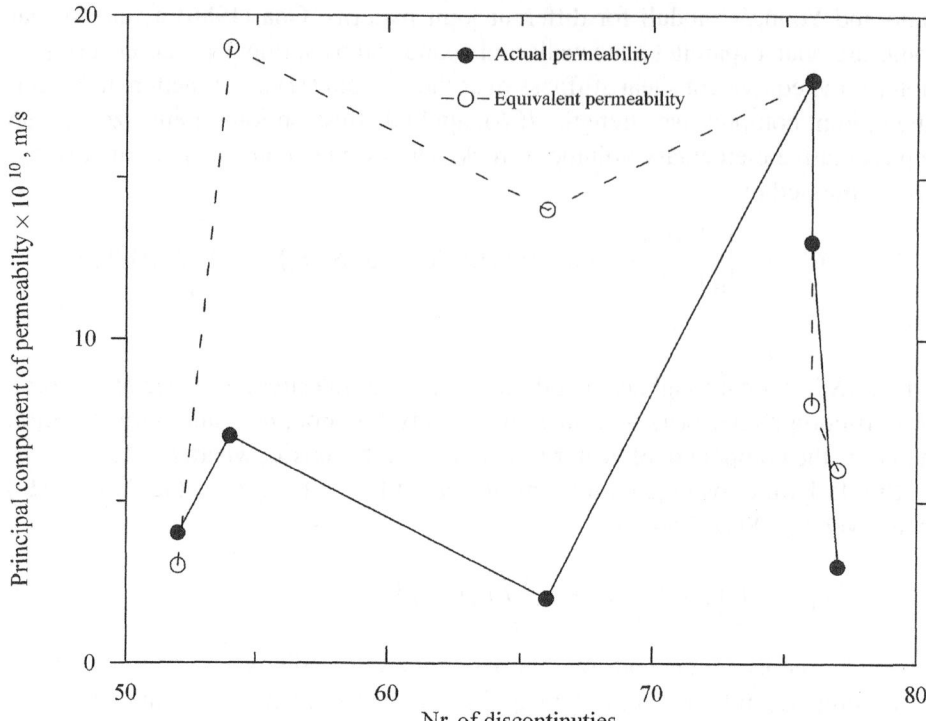

Figure 5.12. Comparison between actual and equivalent permeability (data from Sietel et al., 1996).

(Sietel et al., 1996). Some permeability values deviate largely, whereas others match reasonably well. The plot also shows that a higher degree of accuracy can be expected when the sample contains a large number of discontinuities, with a high degree of connectivity among fractures.

REFERENCES

Ablien, H., Neretnieks, I., Tunbrant, S. and Moreno, L. 1985. Final Report on the Migration in a Single Fracture, Experimental Results and Evaluation. Technical report 85-03, Nuclear, Fuel Safety Project, Stockholm.
Anderson, J. and Dverstorp, B. 1987. Conditional simulations of fluid flow in three-dimensional networks of discrete fractures. *Water Resour. Res.,* 23(10): 1876–1886.
Bear, J. 1979. *Hydraulic of Groundwater.* McGraw-Hill, New York.
Biot, M.A. 1955. Theory of elasticity and consolidation for a porous anisotropic media. *J. Appl. Phys.,* 26: 182–185.
Bridges, M.C. 1976. Presentation of fracture data for rock mechanics. *Proc. 2nd Australia–New Zealand Conf. on Geomechanics,* Brisbane, pp. 144–148.
Celia, M.A. and Binning, P. 1992. A mass conservative numerical solution for two-phase flow in porous media with application to unsaturated flow, *Water Resour. Res.,* 28(10): 2819–2828.

Dershowitz, W.S. 1984. *Rock Joint Systems*. PhD thesis, M.I.T., Cambridge, Massachusetts.

Dverstorp, B. and Anderson, J. 1989. Application of the discrete fracture network concept with field data: Possibilities of model calibration and validation, *Water Resour. Res.*, 25: 540–550.

Fidelibus, C., Barla, G. and Cravero, M. 1997. A mixed solution for two-dimensional unsteady flow in fractured porous media. *Int. J. Numerical Anal. Methods Geomech.*, 21: 619–633.

Fisher, R. 1953. Dispersion of a sphere. *Proc. of the Royal Society of London*, A217: 295–305

Goodman, R.E. 1976. *Methods of Geological Engineering in Discontinuities Rock*. West, St Paul, MN.

Goodman, R.E. and Shi, G.H. 1985. *Block Theory and its Application to Rock Engineering*. Prentice-Hall, Englewood Cliffs, NJ.

Herbert, A.W. 1996. Modeling approaches for discrete fracture network flow analysis. Continuum representation of coupled hydro-mechanic processes of fractured media: Homogeneous and parameter identification. In *Coupled Thermo-Hydro-Mechanical Process of Fractured Rock Media*, Stephenson, O., Jing, L. and Tsang, C.F., eds., Elsevier, New York, 575 p.

Indraratna, B. and Ranjith, P.G. 1998. *Effect of boundary conditions and boundary block sizes on inflow to an underground excavation – sensitivity analysis*. International Mine Water Association, South Africa, pp. 3–11.

Indraratna, B. and Ranjith P.G. 1999. Deformation and permeability characteristics of rocks with interconnected fractures. *9th Int. Congress on Rock Mechanics*, France, Vol. 2, pp. 755–760.

Indraratna, B., Ranjith P.G. and Gale, W. 2000. Single phase water flow through rock fractures. *J. Geotech. Geol. Eng.*, 17: 1–37.

ITASCA 1996. *Universal Distinct Element Code (UDEC), Version 3.0*, USA.

ITASCA 1997. *Three-dimensional Distinct Element Code (3DEC)*, USA.

Lee, C.H. and Farmer, I. 1993. *Fluid Flow in Discontinuous Rocks*. Chapman & Hall, London, 169 p.

Lemos, J.V., Hart, R.D. and Cundall, P.A. 1985. A generalized distinct element program for modeling jointed rock mass – A keynote lecture. *Proc. Int. Symp. on Fundamentals of Rock Joints*, Bjorkliden, pp. 335–343.

Lomize, G.M. 1951. *Filtratsia v Treshchinovatykh, Gosudarstvennoe Energetitcheskoe Izdatelstvo*, Moskva.

Long, J.C.S, Remer, J.S., Wilson, C.R. and Witherspoon, P.A. 1982. Porous media equivalents for networks of discontinuous fractures. *Water Resour. Res.*, 18(3): 645–658.

Long, J.C.S. and Witherspoon, P.A. 1985. The relationship of the degree of interconnectivity to permeability of fracture networks. *J. Geophys. Res.*, 90(B4): 3087–3098.

Louis, C. 1968. Etude des ècoulements d'eau dans les roches fissures et de leurs influences sur la Stabilitè des massifs rocheux. *Bull. De la Direction des tud. Et Rech. EDF*, sèr. A, 3: 5–132.

Louis, C. 1976. *Introduction à l'Hydraulique des Roches*. PhD thesis, Paris.

Mardia, K.V. 1972. *Statistics of Directional Data*, Academic Press, Orlando.

McMahon, B. 1971. A statistical method for the design of rock slopes. *Proc. 1st Australia–New Zealand Conf. on Geomechancis*, pp. 314–321.

Neternieks, I. 1985. Transport in fractured rocks, in proceedings. *Memories of 17th Int. Congress of International Association of Hydrologists*, Vol. XVII, pp. 301–318.

Neuzil and Tracy, J.V. (1981). Flow through fractures. *Water Resour. Res.*, 17: 191–199.

Oda, M. 1986. An equivalent continuum model for coupled stress and fluid flow analysis in jointed rock mass. *Water Resour. Res.*, 22: 1845–1856.

Pahl, P.J. 1981. Estimating the mean length of discontinuity traces. *Int. J. Rock Mech. Min. Geomech. Abstr.,* 18: 221–228.

Pariseau, W.G. 1993. Equivalent properties of a jointed Biot material. *Int. J. Rock Mech. Min. Geomech. Abstr.,* 30(7): 1151–1157.

Priest, S.D. 1993. *Discontinuity Analysis for Rock Engineering.* Chapman & Hall, London, 473 p.

Priest, S.D. and Hudson, J.A. 1981. Estimation of discontinuity spacing and trace length using scanline surveys. *Int. J. Rock Mech. Min. Geomech. Abstr.* 21: 203–212.

Ranjith, P.G. 2000. *Analytical and Experimental Modelling of Coupled Water and Air Flow Through Rock Joints.* PhD thesis, University of Wollongong, Australia.

Ranjith, P.G. and Indraratna, B. in press. Coupled air–water flow through fractured sandstones. *Int. Conf. on Geotechncial & Geological Engineering (Geo Eng 2000),* Melbourne, Australia.

Rasmussen, T.C. 1988. *Fluid Flow and Solute Transport through Three-Dimensional Networks of Variability Saturated Discrete Fractures.* PhD thesis, Department of Hydrology and Water Resources, University of Arizona.

Samaniego, A. 1984. *Simulation of Fluid Flow in Fractured Rock.* PhD thesis, Imperial College, University of London.

Sen, Z. and Eissa, E.A. 1992. Rock quality charts for log-normally distributed block sizes. *Int. J. Rock Mech. Min. Geomech. Abstr.,* 29: 1–12.

Shapiro, A. and Anderson, J. 1983. Steady state fluid response in fractured rock: A boundary element solution for a coupled, discrete fracture continuum model. *Water Resour. Res.,* 19(4): 959–969.

Sietel, A., Millard, A., Treille, E., Vuillod, E., Thoraval, A. and Ababou, R. 1996. Continuum representation of coupled hydro-mechanic processes of fractured media: Homogeneous and parameter identification. In *Coupled Thermo-Hydro-Mechanical Process of Fractured Rock Media,* Stephenson, O., Jing, L. and Tsang, C.F., eds., Elsevier, New York, 575 p.

Singh, B. 1973. Continuum characterization of jointed rock masses. Part 1 constitutive equations. *Int. J. Rock Mech. Min. Geomech. Abstr.,* 10: 311–335.

Snow, D.T. 1969. Anisotropic permeability of fractured media. *Water Resour. Res.,* 5: 1273–1289.

Snow, D.T. 1970. The frequency and apertures of fractures in rock. *Int. J. Rock Mech. Min. Geomech. Abstr.,* 7: 23–40.

Wallis, P.F. and King, M.S. 1980. Discontinuity spacing in a crystalline rock. *Int. J. Rock Mech. Min. Geomech. Abstr.,* 17: 63–66.

Wei, Z.Q., Egger, P. and Descoeudres, F. 1995. Permeability predictions for jointed rock mass. *Int. J. Rock Mech. Min. Geomech. Abstr.,* 32(3): 251–261.

Wilson, C.R. 1970. *An Investigation of Laminar Flow in Fractured Porous Rocks.* PhD thesis, University of California, Berkeley.

CHAPTER 6

Flow analysis through an interconnected fracture network – numerical modelling

6.1 LIMITED USAGE OF ANALYTICAL APPROACHES

Analytical methods may be used to estimate fluid-flow quantities in a given rock mass, provided the rock mass contains a simple fracture network formed with a small number of joints. Typically, there are two approaches based on analytical techniques:
(1) For a given hydraulic conductivity relationship, the flow is estimated;
(2) The hydraulic head at each intersection point of the given fracture network is first estimated, and then the flow is quantified.

The first approach is viable when the hydraulic conductivity for different boundary conditions including stresses or depths is known. The second approach is tedious and often involves lengthy calculation procedures, but it yields fairly accurate results for a given fracture network consisting of a few joint sets. Goodman et al. (1965) employed an analytical approach (Eqn. 6.1) for steady-state flow into a horizontal drift assuming that the drawdown curve of the water table is negligible. The application of this method is not realistic in most cases, except for undersea tunnels. The steady-state flow rate (Q) is given by

$$Q = \frac{2K_s\pi H_0}{\ln(4l/d)} \tag{6.1}$$

where K_s = constant, H_0 = the depth of water table to the centre of the tunnel, d = diameter of the tunnel and l = depth of the ground surface to the centre of the tunnel.

Zhang and Franklin (1993) used an analytical approach to estimate inflow to a tunnel assuming that the hydraulic conductivity decreases exponentially with depth. There are several empirical and theoretical relationships for estimating hydraulic conductivity as a function of depth or stress. Zhang and Franklin (1993) adopted the simple approach of Louis (1974) to give

$$K = K_s \exp(-Ah) \tag{6.2a}$$

where K = hydraulic conductivity at depth h, K_s = constant and A = hydraulic conductivity gradient.

The flow rate (Q) represented by Eqn. 6.2b derived by Zhang and Franklin (1993) does not account for some of the most important aspects related to excavation, such as the change of joint spacing, joint apertures and development new fractures.

$$Q = \frac{2K_s\pi(\gamma_m - \gamma_w)\exp\left[-\left(\frac{\gamma_m A}{\gamma_m - \gamma_w} - \frac{\gamma_w a_3}{\gamma_m - \gamma_w}\right)L\right]\left[\exp\left(\frac{\gamma_w(A - a_3)H_0}{\gamma_m - \gamma_w}\right) - 1\right]}{\gamma_w(A - a_3)(K_0 Ad/4 - K_0 AL)}$$

(6.2b)

where γ_m = unit weight of rock, γ_w = unit weight of water, H_0 = the depth of water table to the centre of the tunnel, K_0 = the hydraulic conductivity at the centre of the tunnel, d = diameter of the tunnel, L = depth of the ground surface to the centre of the tunnel and a_3 = the stress-independent factor, which may be taken as approximately $0.25A$, in which A is the hydraulic conductivity gradient.

In most analytical approaches employed by various researchers, it is assumed that the joints are of infinite length in a given volume, and that they are orientated along perfectly parallel planes (Sharp, 1970; Maini, 1971). Consider the following simple fracture network intersecting a circular tunnel periphery (Fig. 6.1). It is assumed that the two joint sets completely extend over the given area in order to apply simplified 'pipe network' theory. The network has basically three types of nodal points depending on the interconnectivity and the boundary conditions, as listed below:
(1) Internal nodal points in which two joint sets are intersected;
(2) Tunnel periphery nodal points, intersected at the tunnel periphery; and
(3) Nodal points on the top of the bed rock, where hydraulic pressure is known.

The following section elucidates an extension of the procedure for flow computation, as proposed by the authors, based on an analytical approach. Assuming that the rock material is impermeable, the fluid remains incompressible and continuous along the joint, and the net inflow to an internal node equals the net outflow from the same node, as written below.

At node i, for laminar steady-state flow

$$q_{i-1,i} + q_{i+2,i} = q_{i,i+1} + q_{i,i+3}$$

(6.3)

where $q_{i-1,i}$ = flow rate from node $(i - 1)$ to node i.

Assuming parallel plate flow theory, the flow rate can be expressed in terms of the hydraulic gradient and the joint aperture to yield

$$q_{i-1,i} = \frac{e_{i-1,i}^3}{12\mu}\left[\frac{h_{i-1} - h_i}{l_{i-1,i}}\right]$$

(6.4)

where $e_{i-1,i}$ = joint aperture of the joint length $l_{i-1,i}$, $h_{i-1} - h_i$ = hydraulic head difference along the length of $l_{i-1,i}$ and μ = dynamic viscosity of the fluid.

Similarly, flow rate of $q_{i+1,i}$, $q_{i,i+2}$ and $q_{i,i+3}$ can be expressed in terms of the hydraulic head and the aperture of each segment. Using Eqns. 6.3 and 6.4, the

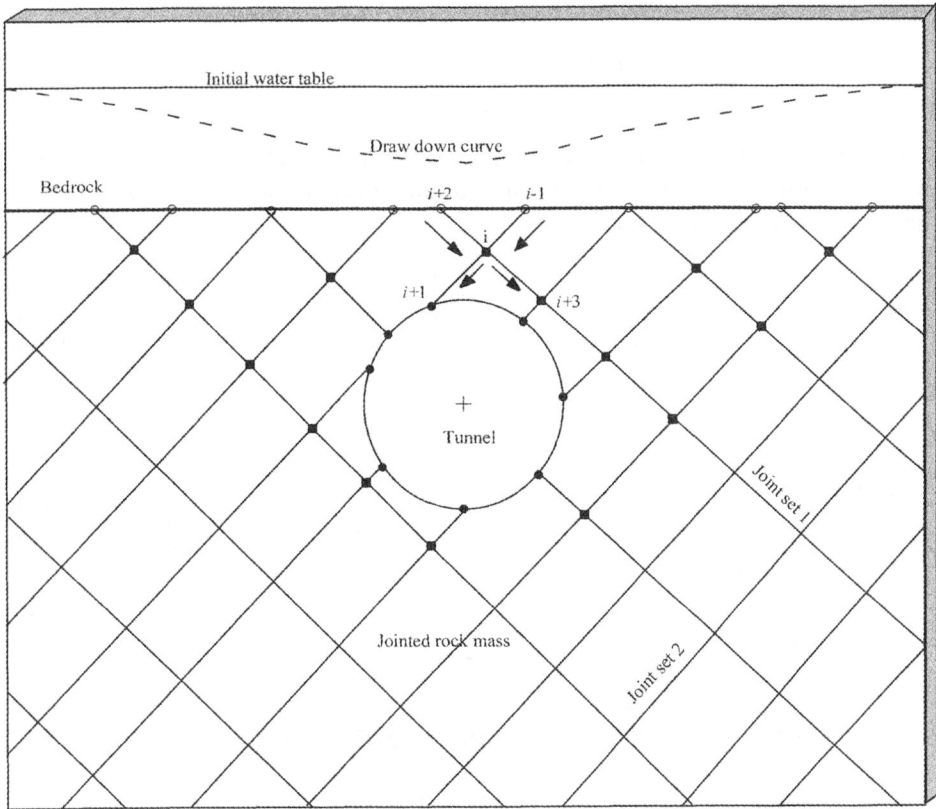

Figure 6.1. A circular tunnel in a rock mass with two sets of regular joints.

following equation is derived for the flow rate at node i.

$$\frac{e_{i-1,i}^3}{12\mu}\left[\frac{h_{i-1}-h_i}{l_{i-1,i}}\right] + \frac{e_{i-2,i}^3}{12\mu}\left[\frac{h_{i+2}-h_i}{l_{i+2,i}}\right]$$

$$= \frac{e_{i-1,i}^3}{12\mu}\left[\frac{h_i-h_{i+1}}{l_{i,i+1}}\right] + \frac{e_{i-1,i}^3}{12\mu}\left[\frac{h_i-h_{i+3}}{l_{i,i+3}}\right] \qquad (6.5)$$

If

$$c_{i-1,i} = \frac{e_{i-1,i}^3}{12\mu}\left[\frac{1}{l_{i-1,i}}\right],$$

then Eqn. 6.5 may be rewritten as follows:

$$c_{i-1,i}(h_{i-1}-h_i) + c_{i+2,i}(h_{i+2}-h_i)$$

$$= c_{i,i+1}(h_i-h_{i+1}) + c_{i+3,i}(h_i-h_{i+3}) \qquad (6.6)$$

Making $c_{i,i} = -(c_{i-1,i} + c_{i+2,i} + c_{i,i+1} + c_{i+3,i})$, Eqn. 6.6 may be rearranged in a simplified form:

$$c_{i,i-1}h_{i-1} + c_{i,i}h_i + c_{i,i+1}h_{i+1} + c_{i,i+2}h_{i+2} + c_{i,i+3}h_{i+3} = 0 \qquad (6.7)$$

At a given node i, therefore, the hydraulic head can be written as

$$h_i = -\left(\frac{c_{i,i+1}h_{i+1} + c_{i,i+2}h_{i+2} + c_{i,i+3}h_{i+3} + c_{i,i-1}h_{i-1}}{c_{i,i}}\right) \qquad (6.8)$$

If there are n numbers of nodes in the model, it is feasible to arrange Eqn. 6.7 in matrix form, hence, the solution for hydraulic head at each node is obtained by the following matrix iteration.

$$\begin{bmatrix}
c_{2,1} & c_{2,2} & c_{2,3} & c_{2,4} & c_{2,5} & 0 & 0 & 0 & 0 & 0 & \cdot & \cdot \\
0 & c_{3,2} & c_{3,3} & c_{3,4} & c_{3,5} & c_{3,6} & 0 & 0 & 0 & 0 & \cdot & \cdot \\
0 & 0 & c_{4,3} & c_{4,4} & c_{4,5} & c_{4,6} & c_{4,7} & 0 & 0 & 0 & \cdot & \cdot \\
\cdot & \cdot & \cdot & \cdot & \cdot & \cdot & \cdot & \cdot & \cdot & \cdot & \cdot & \cdot \\
\cdot & \cdot & \cdot & \cdot & \cdot & \cdot & \cdot & \cdot & \cdot & \cdot & \cdot & \cdot \\
\cdot & \cdot & \cdot & \cdot & \cdot & \cdot & \cdot & \cdot & \cdot & \cdot & \cdot & \cdot \\
0 & 0 & 0 & 0 & 0 & & & & & 0 & & 0 \\
0 & 0 & 0 & 0 & 0 & 0 & & & & & & 0 \\
0 & 0 & 0 & 0 & 0 & 0 & 0 & c_{n,n-1} & c_{n,n} & c_{n,n+1} & c_{n,n+2} & c_{n,n+3}
\end{bmatrix}
\begin{bmatrix}
h_1 \\ h_2 \\ h_3 \\ h_4 \\ h_5 \\ h_6 \\ \cdot \\ \cdot \\ h_n
\end{bmatrix}
=
\begin{bmatrix}
0 \\ 0 \\ 0 \\ 0 \\ 0 \\ 0 \\ 0 \\ 0 \\ 0
\end{bmatrix}$$

$$(6.9)$$

6.2 NUMERICAL MODELLING IN ROCK MECHANICS

The usage of numerical modelling in soil and rock engineering has been expanded in recent years, in order to handle complex problems efficiently. There have been numerous computer programs (codes) developed for research work, as well as for practicing engineers. On the basis of the type of flow problem dealt with, these computer codes can be classified into three categories:
(1) Domain method;
(2) Boundary formulations; and
(3) Lattice structure method.

The main difference between the domain and the boundary methods is that in the former, the interior media is discretised, while in the latter, external surface is discretised. Some of the widely used domain and boundary techniques in rock mechanics are listed below (ISRM, 1988):
(1) Finite Element Method (FEM) – domain formulation;
(2) Finite Difference Method (FDM) – domain formulation;
(3) Boundary Element Method (BEM) – boundary formulation;
(4) Discrete Element Method (DEM); and
(5) Combination of above (e.g. coupled FEM and BEM)

The third category (i.e., lattice structure method) is not as popular as others. For example, the Lattice-gas and Lattice-Boltzmann methods can be used to model fluid flow through porous and fractured media. The conventional methods will dominate until the lattice structure technique gains further development for handling complex problems.

Techniques such as FEM or FDM, can be used to model non-linear behavior and non-homogeneous materials (e.g. non-linear flow behavior, stress-deformation and unsaturated flows). The FEM has gained increased popularity in solving rock mechanics problems because of the capability of having finer mesh arrangements at the edges and corners. Also, in most cases, the system matrices are symmetric, thereby, yielding a efficient solution approach (Neuman, 1973; Wangen, 1997). Basically in FEM, the region is subdivided into small elements and the equilibrium of each element is described in an implicit manner. FEM is efficient when the ratio of volume to surface area is small and when the boundary stresses are not of primary importance. There are a number of commercially available computer codes based on FEM and FDM. Fast Lagrangian Analysis of Continua (FLAC) developed by ITASCA (1993) is an explicit finite difference continuum code, which is commonly used for the analysis of soil and rock problems. The NAPSAC fracture network code developed by AEA Technology, based on FEM, is used to simulate flow through interconnected fracture networks (Wilock, 1996).

Boundary Element Method (BEM) is a relatively recent technique compared to FEM. It is also used for analyzing rock mass by discretising the surface into boundary elements (Beer & Poulsen, 1994; Crouch & Starfield, 1983; Crotty & Wardle, 1985). As discussed by Elsworth (1987), BEM is suited for analyzing situations where the ratio of volume to surface area ratio is high, and for ensuring high accuracy of boundary stresses. One of the main demerits of BEM software is that the rock is often assumed to behave as a homogeneous and elastic medium, which is not realistic in practice, especially where fractured rock is encountered.

Discrete element method (DEM) is best suited for discontinuous media such as fractured rock mass, which is in direct contrast to continuum techniques such as FEM and FDM. In DEM, there are two main advantageous over the continuum approaches, as described below:

(1) Large deformations due to joint slip and block rotations are allowed;
(2) Both material and discontinuous (i.e. joints) properties are used to simulate the actual rock mass.

The shapes of rock blocks depend on the orientation of joints, discontinuity length and their spacing. Distinct element method was initially developed for mechanical analysis of solid blocks by Cundall (1971), and then further extended by co-workers (Cundall & Strack, 1979). The commercially available Universal Distinct Element Code (UDEC) is a two-dimensional program based on the distinct element approach, in which the rock blocks are assumed as either deformable or rigid (ITASCA, 1996).

Coupled boundary element and finite element methods (BEM–FEM) have been successfully used in rock mechanics to optimize the solution efficiency for complex problems (Elsworth, 1985, 1986, 1987; Zienkiewicz et al., 1977). Complexity of problems occurs, because, the rock is often inhomogeneous, non-linear, anisotropic and discontinuous (faults and fractures). Moreover, an infinite boundary is often assumed for modelling rock engineering problems. Under these circumstances, one may couple FEM and BEM to obtain high precision in modelling. Hybrid distinct element-boundary method (DEM–BEM) was used by Lorig et al. (1986) to study the stresses and displacements in highly jointed rock mass surrounding an underground cavity. The distinct element method was applied to model the jointed rock close to the cavity, while far field rock was modelled using the boundary element method. One advantage of this coupled technique is that the equilibrium conditions at the interface between the two domains are obtained explicitly.

6.3   FLUID FLOW THROUGH FRACTURED NETWORK IN A ROCK MASS

The experimental study of fluid flow through fractured rock network was described in Ch. 3, and the following part of this Chapter is aimed to describe flow through an interconnected fracture network, using numerical modelling techniques. Analytical methods cannot be employed for such analysis of groundwater flow problems in fractured media, because of the large number of geo-hydraulic variables involved.

In a comprehensive study of flow analysis, one has to consider an array of geo-hydrological factors as previously illustrated in Fig. 3.1, Ch. 3. Depending on the availability of geological data and the required accuracy of flow estimation corresponding to the availability of time, numerical techniques and computer resources, the most appropriate flow approach can be selected for the particular study (i.e. discrete, continuum or combination of both). The discrete method is preferred for fractured rock, in which fluid-flow mainly takes place through a network of fractures. The details of fluid-flow models were presented earlier in Ch. 5. Once the flow approach is chosen, the next step is to identify whether the flow is in saturated or unsaturated state, based on field data. Generally, flow through most discontinuous media will be in an unsaturated form (e.g. water + air, water + solid and water + air + solid). However, most numerical techniques currently available are based on saturated flow, in the absence of comprehensive unsaturated flow models for jointed rock. The scope of this Chapter is limited to saturated water flow analysis based on UDEC.

When fluid flow is governed by fractures, the interconnectivity and the density of fractures play a very important role, since they provide the multiple flow paths that conduct water through the rock mass (Lee & Farmer, 1993; Brady & Brown 1994; Tsang & Stephansson, 1996; Herbert, 1996). Apart from this, the stability of rock mass decreases with the increase in degree of interconnectivity. Particularly in underground constructions, catastrophic mine roof and longwall failures may occur, if the fractures transport abundant water to generate excess internal water pressures that substantially reduce the effective stresses at the boundaries of the mine opening. In order to describe a pattern of fracture interconnectivity, one needs to know the fracture lengths, their orientation and location. The trace length, orientation, joint apertures and joint roughness were discussed earlier in Chs. 1 and 3. A network of fractures is formed by connecting several fractures. Some fractures may be isolated because their length, orientation and location are not suitable to connect with the existing network. The volume of water conducted in the jointed rock mass is a function of the degree of joint interconnectivity, the joint apertures and the magnitude of driving (fluid) pressures. Long and Witherspoon (1985) investigated how the permeability varies in a fracture network with the degree of interconnectivity. Recent studies have shown that permeability increases with the increase of fracture connectivity (Zhang et al., 1996).

Excavation induces stress relief and stress re-distribution, forming new fractures or opening up of existing fractures along the cavity surface. The extent of these fractures depends on the magnitude of the stresses, fluid pressures within the rock mass and the rock properties. Excavation technique itself controls fracture patterns and the magnitude of joint apertures. As an example, rock blasting using explosives generates high magnitude stress waves and significant gas pressures, which change the initial fracture pattern significantly. However, the effect on fracture pattern due to mechanical excavation (e.g., tunnel boring machines) is much less. Therefore, as shown in Fig. 6.2, one also needs to consider the prior and post-conditions of the rock mass associated with the type of excavation, before any numerical modelling technique can be implemented, in order to simulate the fracture patterns correctly.

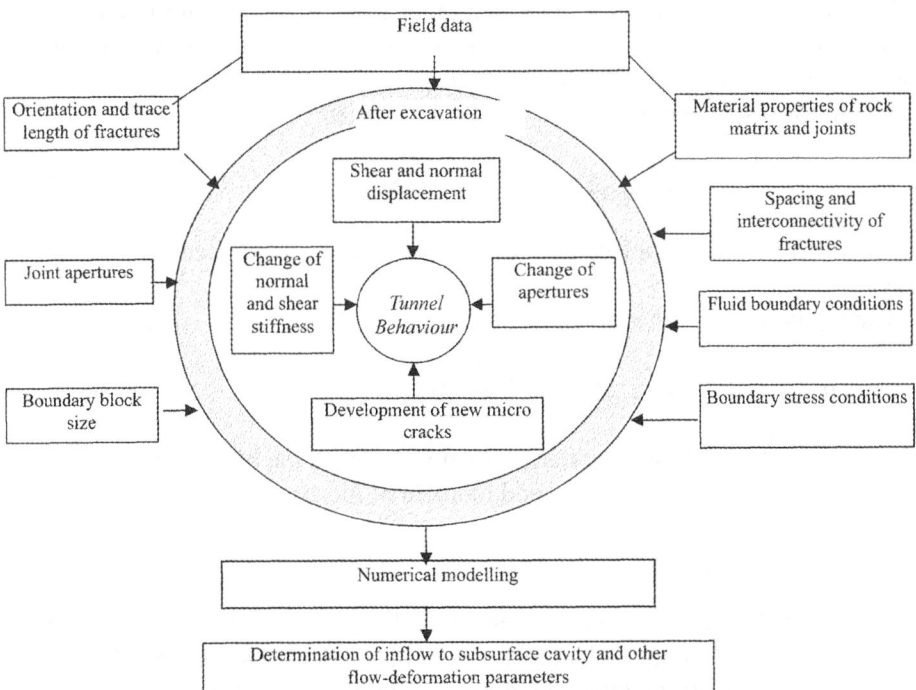

Figure 6.2. Factors affecting the water ingress to underground cavity.

6.4 INTRODUCTION TO UDEC

The following is an attempt to investigate how the total inflow towards a mine cavity in a jointed rock media changes with the boundary conditions such as block dimensions, joint properties, effect of excavation and ground stress ratio. In this study, UDEC (ITASCA, 1996) was employed to simulate water flow through joints adopting a fully coupled hydro-mechanical analysis. UDEC is a two-dimensional numerical program based on the distinct element method for discontinuum modelling. For coupled hydro-mechanical flow analysis, UDEC code is suitable when the flow is mainly governed by a well defined network of fractures.

6.4.1 Block motion theory

The motion of loaded blocks is due to the magnitude and direction of the resultant out-of-balance moments and forces. The type of motion includes both linear and angular movement. Considering the Newton's second law, the motion of a body under a given force F is given by

$$\frac{\mathrm{d}u}{\mathrm{d}t} = \frac{u^{t+\Delta t/2} - u^{t-\Delta t/2}}{\Delta t} = \frac{F}{m} \tag{6.10}$$

where u = velocity, t = time, Δt = small time increment and m = mass of the block.

Considering the effect of gravity and other external forces on the blocks, the angular and linear velocity can be used to estimate the linear displacement and the rotation of each block as described below:

The linear displacement at time $t + \Delta t$, is given by

$$x_{t+\Delta t} = x_t + u_{t-\Delta t/2}\Delta t + \left(\frac{\Sigma F_t}{m} + g\right)\Delta t^2 \tag{6.11}$$

where x_t = the initial displacement of the block, and g is the acceleration due to gravity.

The rotation at $t + \Delta t$ is given by, $\theta_{t+\Delta t}$

$$\theta_{t+\Delta t} = \theta_t + \theta_{t-\Delta t/2}\Delta t + \left(\frac{\Sigma M_t}{I}\right)\Delta t^2 \tag{6.12}$$

where θ = rotation of the block, M = total moment of the block due to the external forces and gravity and I = second moment of inertia.

6.4.2 Joint models

UDEC assigns two types of models separately for the behaviour of rocks: (a) joint model and (b) block model. The block model may be rigid or deformable depending on the situation. In practice, the degree of deformation of intact rock depends

Table 6.1. Different constitutive models used in UDEC (ITASCA, 1996).

Block models	Applications
Null model	To represent material either removed or excavated
Elastic model	For homogeneous and isotropic material (e.g., steel, soil layers)
Drucker–Prager plasticity model	Soft clay with low friction
Mohr–Coulomb plasticity model	Used for rock/soil mechanics applications (e.g., coarse grain sand stone, soil, rock)
Strain-hardening model	Progressive failure of structures (e.g., concrete beam)
Double-yield model	Hydraulically placed backfill

Table 6.2. Joint models incorporated in UDEC (ITASCA, 1996).

Joint models	Applications
Point contact – Coulomb slip	For highly fractured and unstable rock
Joint area contact – Coulomb slip	General rock mechanics
Joint area contact – Coulomb slip with residual strength	General rock mechanics
Continuously yielding	For dynamic loadings
Barton & Bandis model	To estimate variable hydraulic apertures

on the magnitude and direction of stress, the porosity of rock and the existing fluid-flow condition. There are five block models which are included in UDEC, and the practical applications of these constitutive models are listed in Table 6.1. Out of these models, the Mohr–Coulomb model is more popular mainly because of the ease of obtaining the relevant strength parameters by simple laboratory tests, such as the direct shear box.

In order to represent the joint characteristics, the user has the choice of five joint models, as given in Table 6.2. The application of point contact method is remote in reality, as joints have several discrete points along the joint path. The area contact method is preferable because any two blocks defining a particular joint are usually in contact due to the external loads. The main advantage of using the residual strength is that this approach is capable of simulating displacement-weakening of joints due to the loss of friction or cohesion. The analysis of a single joint including variable apertures associated with different stress conditions can be investigated thoroughly using the Barton–Bandis model.

Several researchers including Herbert (1996), Zhang et al. (1996) and Liao and Hencher (1997) have used the UDEC code to investigate fluid flow through jointed rock media. In this study, the blocks surrounded by the discontinuities were modelled as deformable material. Fluid-flow analysis was performed in which the joint conductivity was directly related to the mechanical deformation associated with the joint (domain) water pressures. Each domain (filled with water) was separated by contact points at which mechanical interaction between the blocks was established.

Depending on the type of contacts, UDEC employs two types of flow equations.

$$q = k_c \Delta p \tag{6.13}$$

where k_c = point contact permeability factor, and

$$\Delta p = p_2 - p_1 + \rho_w g(y_2 - y_1) \tag{6.14}$$

where p_1 = pressure in joint domain 1, p_2 = pressure in joint domain 2, ρ_w = density of water, y_2 and y_1 = y (vertical) co-ordinates of domain centres and g is the acceleration due to gravity.

The domain pressures are updated by taking into account of the net flow into each domain, as well as the changes in domain volume due to the incremental motion of the surrounding blocks. As a result, the new domain pressure becomes

$$p = p_0 + K_w Q \frac{\Delta t}{V} - K_w \frac{\Delta V}{V_m} \Delta t \tag{6.15}$$

where p_0 is the domain pressure in the proceeding time step, Q is the sum of the flow rate into the domain from all surrounding contacts and K_w is the bulk modulus of the fluid. $\Delta V = V - V_0$ and $V_m = (V + V_0)/2$, where V and V_0 are the new and old domain areas, respectively.

In the case of edge to edge contact, the cubic law for flow in a planar fracture was used for estimating flow rate, as given by the following expression:

$$q = -k_j a^3 \frac{\Delta p}{l} \tag{6.16}$$

where k_j = joint permeability factor, a = hydraulic conductivity aperture, l = length of the joint and q = flow rate per meter width.

The simple relationship between the mechanical and hydraulic apertures of the joints used in this analysis is as follows:

$$a = a_0 + u_n \tag{6.17}$$

where a_0 = joint aperture at zero normal stress, and u_n = joint normal displacement.

At each time step in the mechanical calculation, UDEC employs updated geometry of the system, thus prescribing new values of apertures for all domain contacts and volumes. It is postulated that the discontinuities which do not form connectivity with the main fracture network may not contribute to any flow.

Consequently, UDEC ignores isolated fractures for fluid-flow calculations, but these fractures still contribute to a reduced overall modulus.

6.5 ROCK MASS PROPERTIES AND BOUNDARY CONDITIONS

A square boundary block size ($a \times a$) was selected to analyze the deformation and permeability characteristics of the jointed rock mass, associated with the induced stresses. A boundary block contained regular joints, irregular joint networks and isolated joints depending on their location, orientation and the lengths. The analysis presented here is based on the regular joints (e.g. horizontal bedding planes with cross joints, continuous joints) with various boundary conditions (Figs. 6.3–6.5). The joint pattern shown in Fig. 6.5 is particularly applicable to a variety of sedimentary rock types with systematic bedding planes intersecting with cross-joints.

In order to simulate the jointed rock mass, joint geometrical parameters such as, the orientation, spacing, joint aperture, gap length and spacing were assigned for all joint sets and for isolated joints, separately. Joint aperture is one of the most difficult parameters to measure. The geometrical properties of discontinuities can be incorporated using two different approaches: (a) direct technique and (b) stochastic method. The direct approach is suitable for small fracture network, such as fractures in laboratory (small) scale specimens. However, this approach is not feasible when there is a large number of fracture sets, making it difficult to predict the correct fracture flow path within the rock mass. Alternatively, the stochastic modelling method, based on a statistical description of fracture network may be used to generate the fracture network. This method does not describe the location, spacing and orientation of joints deterministically. Also, the stochastic technique does not represent the actual fracture network as described by the direct or conventional approach. In reality, fractures may have complex geometry and variable apertures, but for the current analysis, the geometry of fractures was simplified. The joint geometrical properties of Cases 1, 2 and 3 (Figs. 6.3–6.5) are given in Table 6.3.

Once the joint geometrical properties are assigned, the next task is to incorporate the material properties of intact rock, discontinuities and fluid, as presented in Table 6.4. For simplicity, the variation of the material properties within the boundary block has not been taken into account. For each joint set, mean value of material properties and their standard deviations (e.g. friction angle, normal and shear stiffness) are incorporated in the model separately. Properties such as density, cohesion and bulk modulus of the intact rock were used to characterize the rock matrix. In order to quantify the effect of fluid flow on the deformation characteristics, the density, dynamic viscosity and bulk modulus of the fluid were also incorporated in the model. Naturally, the bulk modulus of the fluid plays a much bigger role for highly compressible fluids, such as CO_2, CH_4 and air, in underground coal mining.

(1) Case 1a

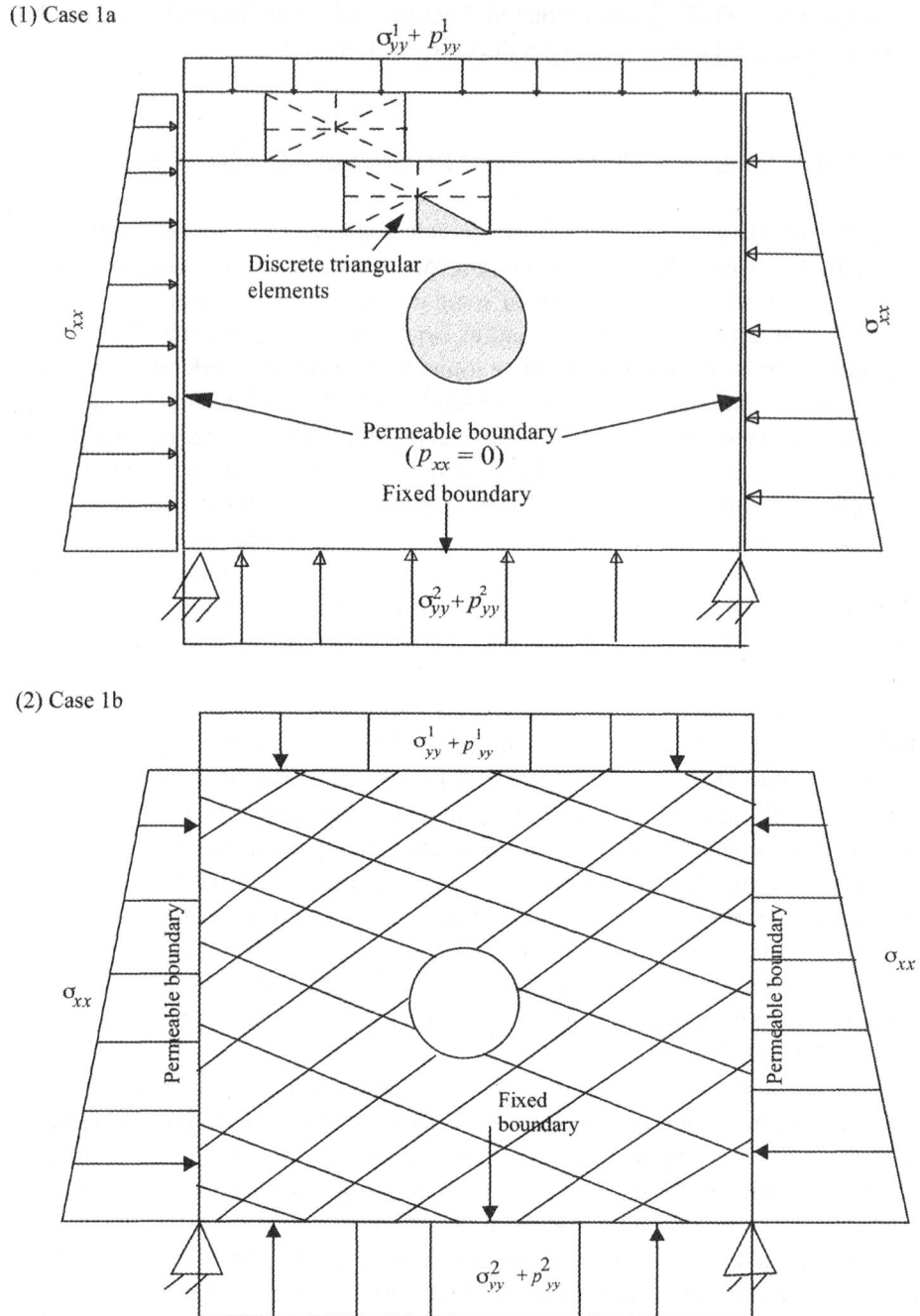

Figure 6.3. Boundary conditions applied in the model (Case 1).

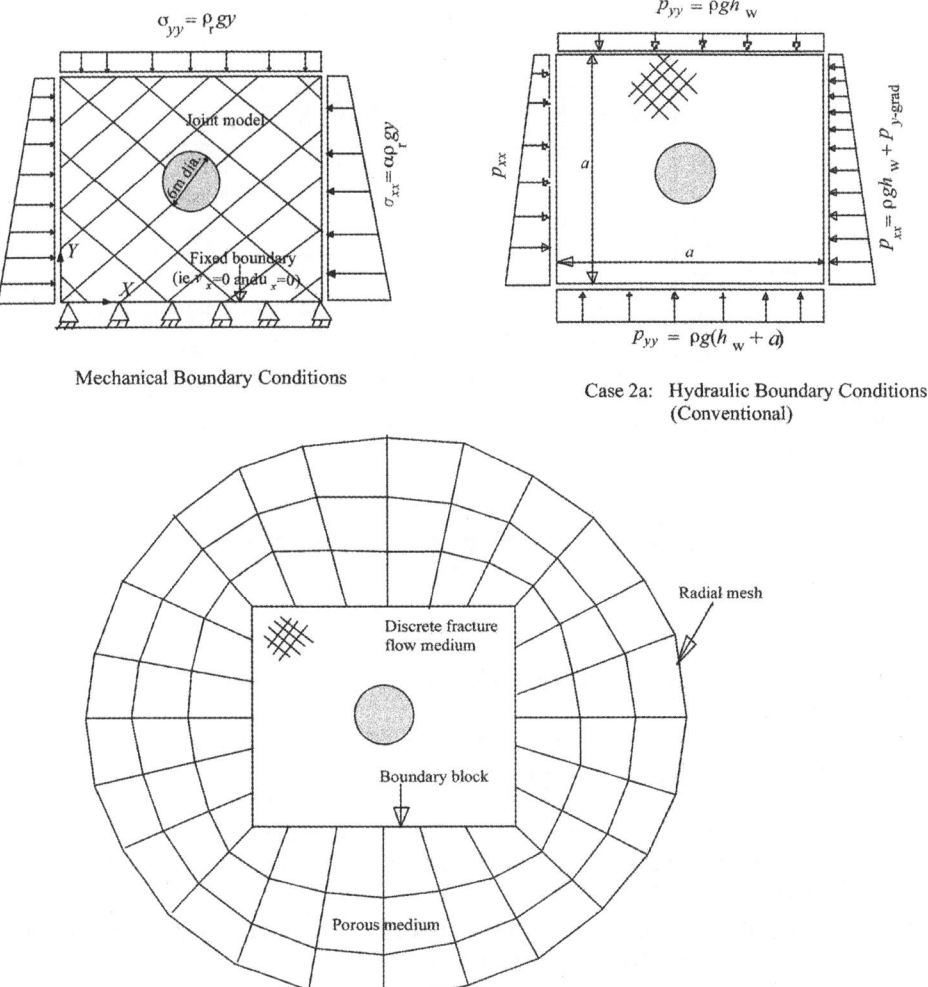

Mechanical Boundary Conditions

Case 2a: Hydraulic Boundary Conditions (Conventional)

Case 2b. Porous medium grid is wrapped around the the external boundary block to represent hydraulic boundary conditions (i.e. large scale mesh discretisation).

Figure 6.4. Boundary conditions applied in the analysis (Case 2).

The assumed hydraulic boundary conditions (Figs. 6.3–6.5) for Cases 1–3 are listed below:
(1) Constant water pressure;
(2) Linearly varying water pressure;
(3) Permeable/impermeable boundaries; and
(4) Wrapping a porous medium around the boundary block.

Constant water pressures act along the top and bottom boundary surfaces, while linearly varying fluid pressure acts along the left and right vertical surfaces

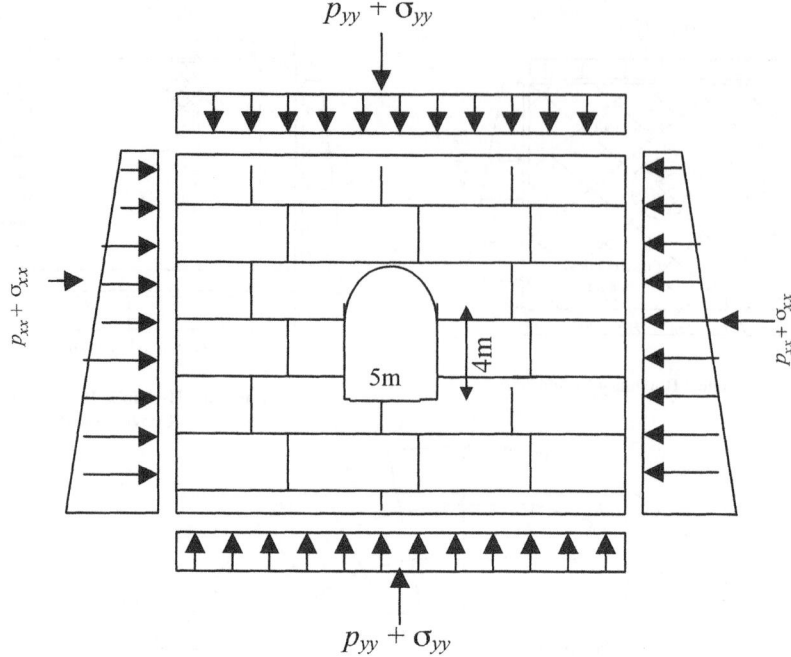

Case 3a: Static water pressures are applied on all four sides as shown in Figure 7.

Case 3b: Same as case 3a, except the ground water table is made to coincide with top of the boundary block.

Case 3c: Same as case 3a, except bottom boundary is made fully permeable, such that excess pore pressures are dissipated to zero. This practically simulates a sand aquifer or sand lense underlying the fractured rock.

Figure 6.5. Horizontal and vertical stresses associated with gravity and pore pressure due to static water head applied into the model (Case 3).

(Cases 1, 2a and 3). Permeable boundary conditions may arise naturally when some boundary surfaces coincide with two faults of highly permeable nature (Fig. 6.3, Case 1a). Impermeable boundary surfaces can include fluid filled joints but with no flow. The constant and linearly varying pressures are given by the Eqns. 6.18 and 6.19, respectively:

$$p_{yy} = \rho_w g h_w \qquad (6.18)$$

$$p_{xx} = \rho_w g h_w + p_{y\text{-grad}} y \qquad (6.19)$$

where

p_{xx} = fluid pressure along the vertical boundaries,
p_{yy} = fluid pressure along the top and bottom boundaries,
ρ_w = density of water,

Table 6.3. Joint parameters used in Cases 1, 2 and 3 in Figs. 6.3–6.5.

Parameters	Units	Joint set 1	Joint set 2	Joint set 3
Case 1A				
Orientation	deg.	0	90	90
Spacing	m	3.0	3.5	3.5
Gap length	m	0	3.0	3.0
Trace length	m	5.0	3.0	3.0
Case 1B				
Orientation	deg.	30	150	
Spacing	m	4.0	3.5	
Gap length	m	0	0	
Trace length	m	5.0	3.0	
Case 2A & 2B*				
Orientation	deg.	45	135	
Spacing	m	3.5	4.0	
Gap length	m	0	0	
Trace length	m	4.5	3.5	
Case 3				
Orientation	deg.	0	90	0
Spacing	m	3.0	2.5	4.5
Gap length	m	0	3.0	2.8
Trace length	m	5.0	3.0	4.5

*Permeability tensor [m^3/(Pa·s)] for Case 2b: $K_{11} = 2.33 \times 10^{-10}$; $K_{12} = 1.33 \times 10^{-10}$; $K_{22} = 2.33 \times 10^{-10}$.

h_w = depth of the water table,
$p_{\text{y-grad}}$ = hydraulic gradient in Y-direction and
y = vertical depth from the ground surface.

A porous medium is wrapped around the boundary block in order to simulate flow on a large scale. Having created the radial mesh around the block, then the fluid pressure is imposed (Fig. 6.4, Case 2b) on the created mesh.

Darcy's law applied to fluid flow in an anisotropic medium is represented by the following equation:

$$v_i = K_{ij}\left(\frac{\partial p}{\partial x_j}\right) \tag{6.20}$$

where v_i is the velocity vector, p is the pressure and K_{ij} is the permeability tensor given by $\begin{bmatrix} K_{11} & K_{12} \\ K_{21} & K_{22} \end{bmatrix}$.

The permeability tensor for a continuous joint set was calculated using the following equations:

$$K_{11} = K_j \cos^2 \alpha \tag{6.21a}$$

$$K_{22} = K_j \sin^2 \alpha \tag{6.21b}$$

Table 6.4. Material properties of jointed rock mass and fluid.

Material	Parameter	Units	Rock matrix	Discontinuities (Joint models 1 and 2)			Fluid (water)
				Joint set 1	Joint set 2	Joint set 3	
Rock matrix	Block modulus	N/m^2	2.7×10^{10}				
	Block shear modulus	N/m^2	0.7×10^{10}				
	Density	kg/m^3	2500				
	Cohesion	N/m^2	6.72×10^6				
	Friction angle	deg.	27.0				
Rock fractures	Normal stiffness	N/m^2		3.0×10^{10}	1.2×10^{10}	1.2×10^{10}	
	Shear stiffness	N/m^2		2.0×10^{10}	1.0×10^{10}	1.0×10^{10}	
	Joint permeability factor	Pa^{-1} s^{-1}		100	100	100	
	Friction angle	deg.		32	31	31	
	Initial aperture	m		1.0×10^{-3}	2.0×10^{-3}	2.5×10^{-3}	
	Residual aperture	m		3.0×10^{-4}	4.0×10^{-4}	3.0×10^{-4}	
Fluid	Density	kg/m^3					1000
	Dynamic viscosity	Pa·s					8×10^{-4}
	Bulk modulus	N/m^2					2×10^9

$$K_{12} = K_j \cos \alpha \sin \alpha \qquad (6.21c)$$

$$K_j = \frac{a^3}{12\mu s} \qquad (6.21d)$$

where α = orientation of joint set, a = aperture of joint set, s = spacing of joint and μ = dynamic viscosity of water.

In order to ascertain the permeability tensor, the contribution of permeability of each joint set is summed together. Initially, there is no flow into the region, and the hydrostatic pressure prevails all around the excavation. Once the tunnel is excavated, a constant atmospheric (zero) pressure is applied around the tunnel surface. Therefore, the resulting inward hydraulic gradient causes fluid to flow from the boundary towards the underground cavity.

The initial compressive stress is defined by the isotropic stress associated with gravity, and is represented by the following equations:

$$\sigma_{yy} = \rho_r g y \qquad (6.22a)$$

$$\sigma_{xx} = \alpha \rho_r g y + \sigma_{y\text{-grad}} y \qquad (6.22b)$$

where ρ_r = density of rock, y = vertical distance (depth), measured downward from the ground surface, σ_{yy} and σ_{xx} = insitu stress components in Y and X directions, $\sigma_{y\text{-grad}}$ = insitu stress gradient and α = insitu stress ratio factor.

The vertical component of compressive stress (σ_{yy}) is applied along the top and bottom boundary surfaces, while the horizontal component (σ_{xx}) is applied to the vertical surfaces. In order to prevent the boundary being displaced, the bottom boundary was fixed in the X and Y directions (i.e. $v_x = v_y = 0$). The hydraulic and insitu stress boundary conditions are listed in Table 6.5. The input variables used in UDEC for the three cases of analysis are summarized in Table 6.6.

Table 6.5. Stresses and hydraulic boundary conditions applied in the model.

Cases	Vertical stress	Horizontal stress	Fixed boundaries	Hydraulic stress	Permeable boundaries
Case 1	$\sigma_{yy} = \rho_r g h$	$\sigma_{xx} = \alpha \rho_r g h$	Bottom boundary $v_x = v_y = 0$	$p_{yy} = p_w g h_w$	Left and right boundaries
Case 2a	$\sigma_{yy} = \rho_r g h$	$\sigma_{xx} = \alpha \rho_r g h$	Bottom boundary, $v_x = v_y = 0$	$p_{yy} = p_w g h_w$ $p_{xx} = p_w g h_w$ with $p_{y\text{-grad}}$	No
Case 2b	$\sigma_{yy} = \rho_r g h$	$\sigma_{xx} = \alpha \rho_r g h$	$\sigma_{yy} = \rho_r g h$	Porous medium is wrapped around the block	No
Case 3	$\sigma_{yy} = \rho_r g h$	$\sigma_{xx} = \alpha \rho_r g h$	$\sigma_{yy} = \rho_r g h$	$p_{yy} = p_w g h_w$ $p_{xx} = p_w g h_w$ with $p_{y\text{-grad}}$	No

Table 6.6. Input variables used to study the influence on fluid flow.

Case 1 – Fig. 6.3

Variables	Case 1A	Range	Case 1B	Range
Boundary block size	Variable	20–60 m	Variable	20–80 m
Excavation dimension	Constant	6 m dia.	Constant	6 m dia.
Depth of the excavation	Constant	100 m below the ground surface		
Orientation of joint sets	Constant	$\theta_1 = 0$ and $\theta_2 = 90$	Variable only for joint set 2	$\theta_1 = 0$ $\theta_2 = 0 < \theta < 108°$
In-situ stress ratio,α (α = horizontal stress/vertical stress	Variable	0.5–2.25	Variable	0.5–2.25
Fluid pressures	Constant	$P_{yy} = \rho g h_w$ left and right boundaries = permeable	Constant	$P_{yy} = \rho g h_w$ left and right boundaries = permeable

Case 2 – Fig. 6.4

Variables	Case 2A	Range	Case 2B	Range
Boundary block size	Variable	25–400 m	Variable	25–400 m
Excavation dimension	Constant	6 m dia.	Constant	6 m dia.
Depth of the excavation	Constant	200 m	Constant	200 m
Orientation of joint sets	Constant	Joint set 1 = 45° Joint set 2 = 45°	Constant	Joint set 1 = 45° Joint set 2 = 45°
In-situ stress ratio,α	Variable-	0.4–2.0	Variable	0.4–2.0
Fluid pressures	Variable	$P_{xx} = \rho g h_w$ with $p_{y\text{-grad}}$ $P_{yy} = \rho g h_w$	Constant	A porous medium is wrapped around the block

Table 6.6. Continued

	Case 3A	Case 3 – Fig. 6.5 Range
Boundary block size	Variable	25–100 m
Excavation dimension	Constant	Horseshoe shape tunnel
Depth of the excavation	Constant	100 m below the ground surface
Orientation of joint sets	Constant	$\theta_1 = 0$ and $\theta_2 = 90$
In-situ stress ratio,α	Variable	0.5–2.0
Fluid pressures		
Case 3A	Variable	$P_{xx} = \rho g h_w$ with $p_{y\text{-grad}}$ and $P_{yy} = \rho g h_w$
Case 3B	Variable	$P_{xx} = \rho g h_w$ with $p_{y\text{-grad}}$, $P_{yy} = \rho g h_w$ and P_{yy} at top is zero
Case 3C	Variable	$P_{xx} = \rho g h_w$ with $p_{y\text{-grad}}$, $P_{yy} = \rho g h_w$ and bottom boundary is permeable

*α = horizontal stress/vertical stress.

6.6 MODEL DESCRIPTION AND BEHAVIOUR

For a given boundary block size (e.g. 100 × 100 m), let us assume that the centre of the boundary block is 50 m below the ground surface and the water table coincides with the top of the boundary block. Having defined the co-ordinates and the joint geometrical parameters of this boundary block, a 6 m diameter tunnel is simulated in the jointed rock. It is important to note that the excavation of the tunnel is simulated at a later stage, i.e. once the model has reached the initial equilibrium condition. Assuming the rock is deformable, the rock blocks formed by the joints are discretised into smaller triangular blocks for a finer analysis (Fig. 6.6). Having assigned the material properties of intact rock, joints and fluid (Table 6.4), the fluid and ground stresses were then imposed, and the model behaviour under these loading conditions were observed.

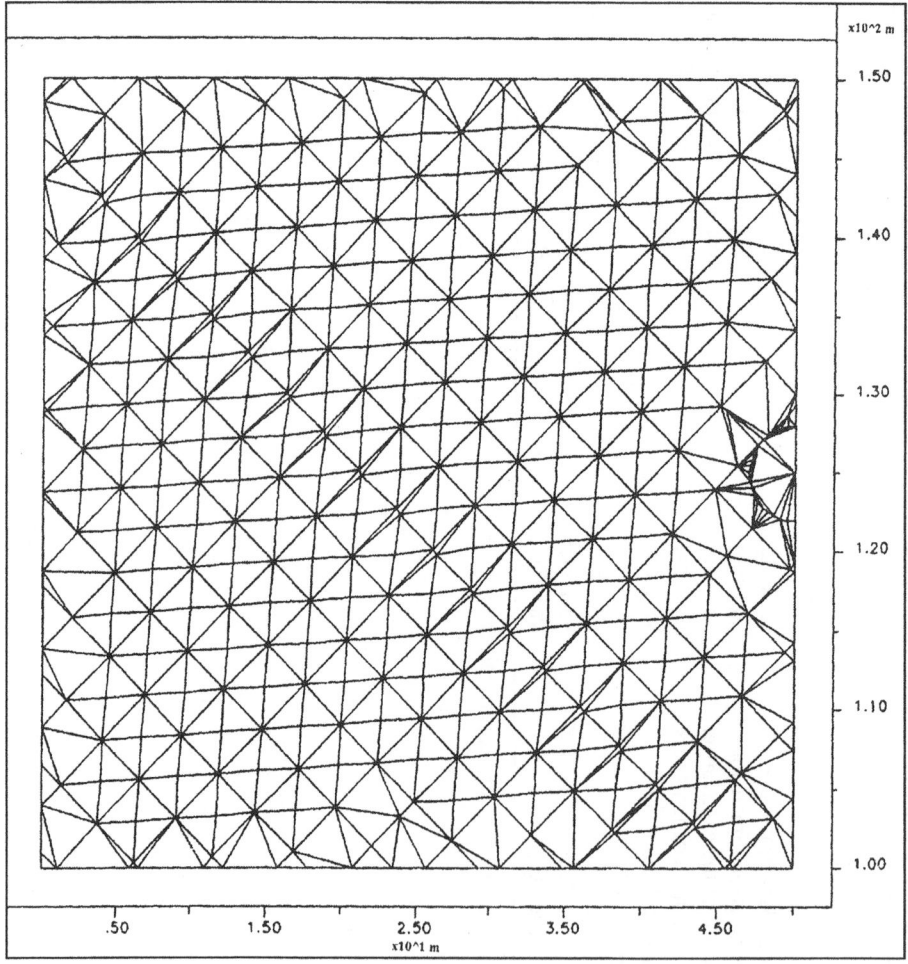

Figure 6.6. Discretised finer mesh arrangement.

Figure 6.7a shows the ground stress (principal stress levels) and fluid pressure variation within the model. Fluid pressures within joints greater than 0.1 MPa are shown in Fig. 6.7b. Not surprisingly, the magnitudes of these stresses increase with depth. The analysis may be carried out for different stages as follows:

(1) Undrained conditions, i.e. mechanical deformation of rock mass only;
(2) Drained conditions with no mechanical deformation of rock mass; and
(3) Coupled flow and mechanical deformation of rock mass.

The total unbalanced force, the total net fluid flow and the deformation at several locations are essential parameters to be considered for equilibrium.

Forces are accumulated at each grid point of the deformable blocks, and the algebraic summation of these forces should approach zero at static equilibrium. However, in practice, the total unbalanced force may never reach zero, hence an engineering judgement is required for the acceptable magnitude of unbalanced

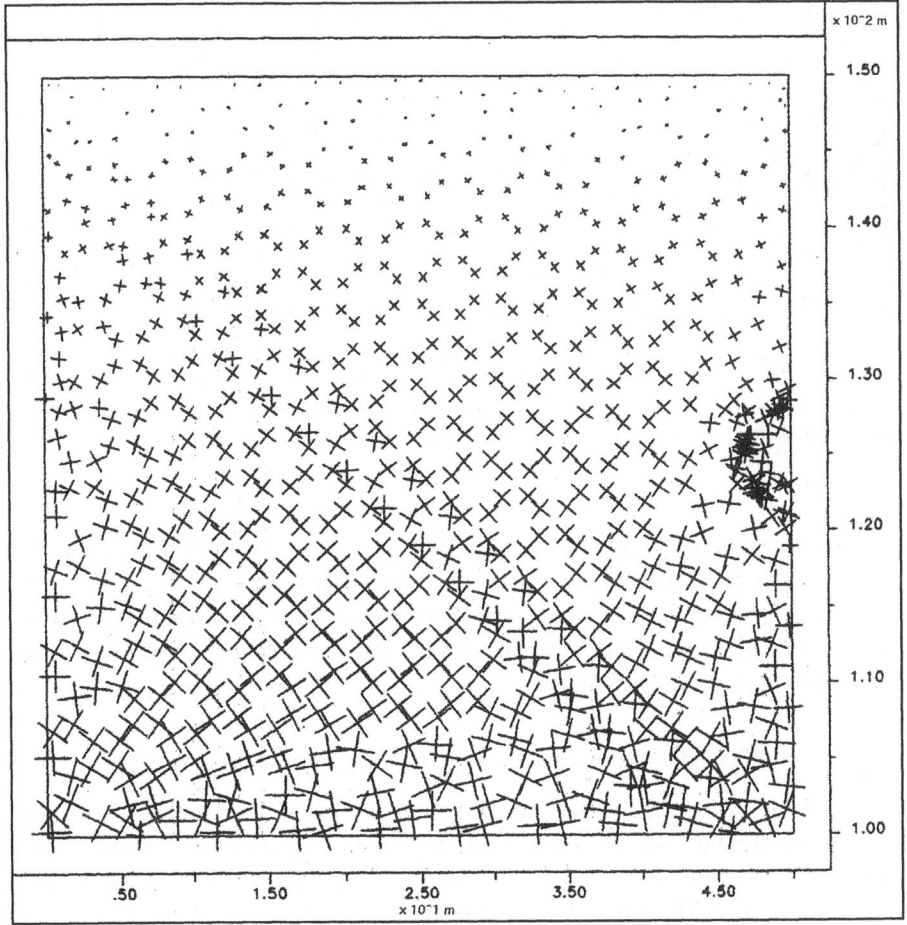

Figure 6.7a. Initial principal stress distribution.

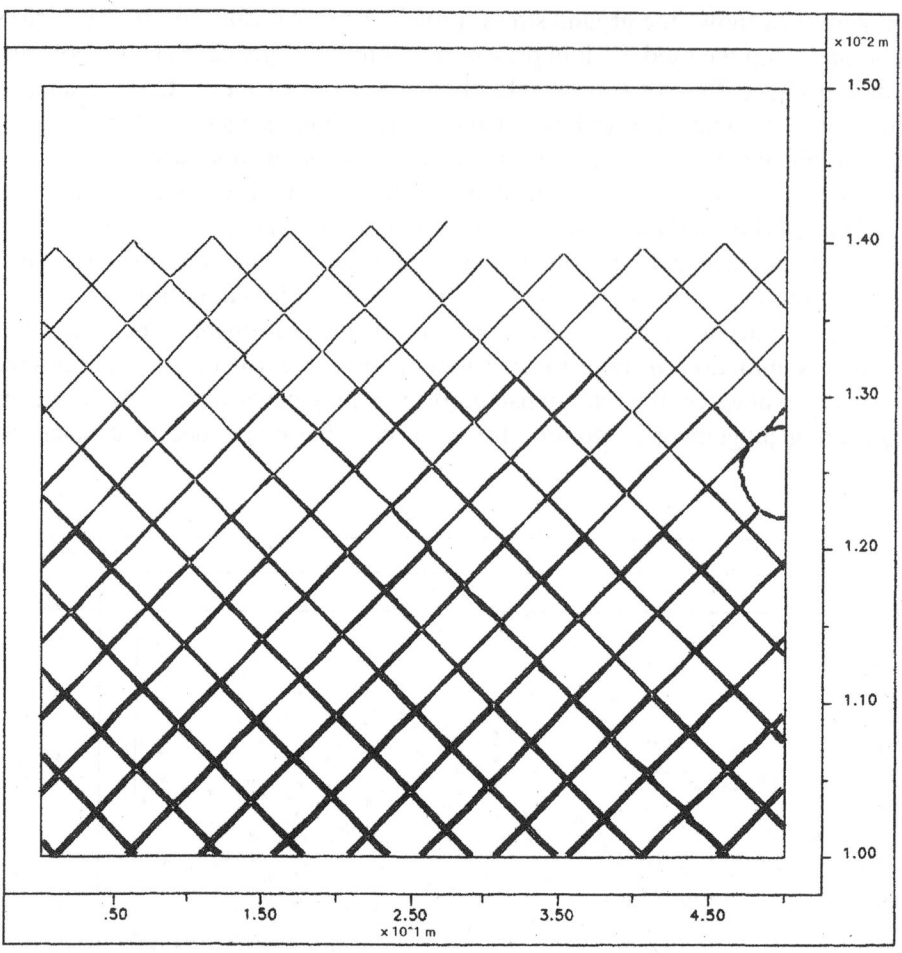

Figure 6.7b. Initial water pressure distribution along the joints.

forces at given time steps. Under coupled flow-mechanical deformation stage, the excavation induces deformation of the rock mass and affects the magnitude of joint apertures, which in turn affects the flow rate and the redistribution of stresses. In the next iterative time step, the current stress levels are employed in the coupled flow-deformation analysis.

6.6.1   Effect of excavation on the flow and deformation characteristics

The concept of a disturbed zone around the excavation is important in the design of piers, evaluating the tunnel stability and for predicting mine inundation and gas outbursts. As described earlier, the degree of deformation depends on the excavation technique, material properties of the rock, the presence of geological features, ground stresses and fluid-flow conditions.

In UDEC, once the model was brought to equilibrium under the initial field conditions, the excavation of the tunnel was made instantaneously, and again the equilibrium of the model was attained. Any disequilibrium of the system was reflected by large unbalanced nodal forces and a significant discrepancy between the total inflow and outflow of the model. In this regard, one needs to ensure that the total net flow to several randomly selected internal nodal points is compatible with the principle of conservation of energy. In order to study the effects of fluid pressure on the deformation of joints after the excavation of the cavity, two types of analyses were carried out:

(1) Undrained flow; and

(2) Drained flow (i.e. steady-state flow).

6.6.2 Undrained flow analysis

During the undrained flow analysis method, only the mechanical deformation of the model was executed with the flow mode switched off. The deformation of the joints is due to the tunnel excavation, and a high hydraulic gradient is generated towards the tunnel boundary. Undrained displacement vectors show that a different state of deformation close to the tunnel periphery occurrs, as presented in Fig. 6.8. The arrowheads represents the direction, and the length of arrow indicates the magnitude of the displacement vectors of the rock blocks. The rock mass above and below the tunnel appears to have undergone significant deformation. It is evident that close to the tunnel periphery, deformation is initiated and subsequently propagates toward the top and bottom of the boundary block. As expected, the tunnel periphery undergoes large deformations. The largest deformations occur along the joint planes (e.g. BD, BE and AC) as indicated in Fig. 6.9. By considering the general pattern of the displacement vectors, and identifying the probable deformed zone, the potential unstable zone can be estimated (Fig. 6.8). The unstable rock zone is more vulnerable when the rock joints carry water.

The fluid pressure distribution along the joints is shown in Fig. 6.9, in which the line thickness shows the magnitude of fluid pressures within joints. The fluid pressure at point A is approximately 5 times the static fluid pressure (i.e. ρgh, where h depth of water table to the point A). The effective stress within the joints is the difference between normal component of initial total stress in the joint and the initial hydrostatic pressure. The elevated fluid pressure exceeds the normal stress in the joint, thereby resulting with a negative effective pressure. The negative effective pressure at point A and B (Fig. 6.9) results in dilation of the joints. These effective stresses are used to calculate the flow deformation parameters in the model.

6.6.3 Drained flow analysis (steady-state flow)

During drainage, a coupled hydro-mechanical analysis was carried out under the steady-state flow to observe deformation due to drainage of groundwater. The

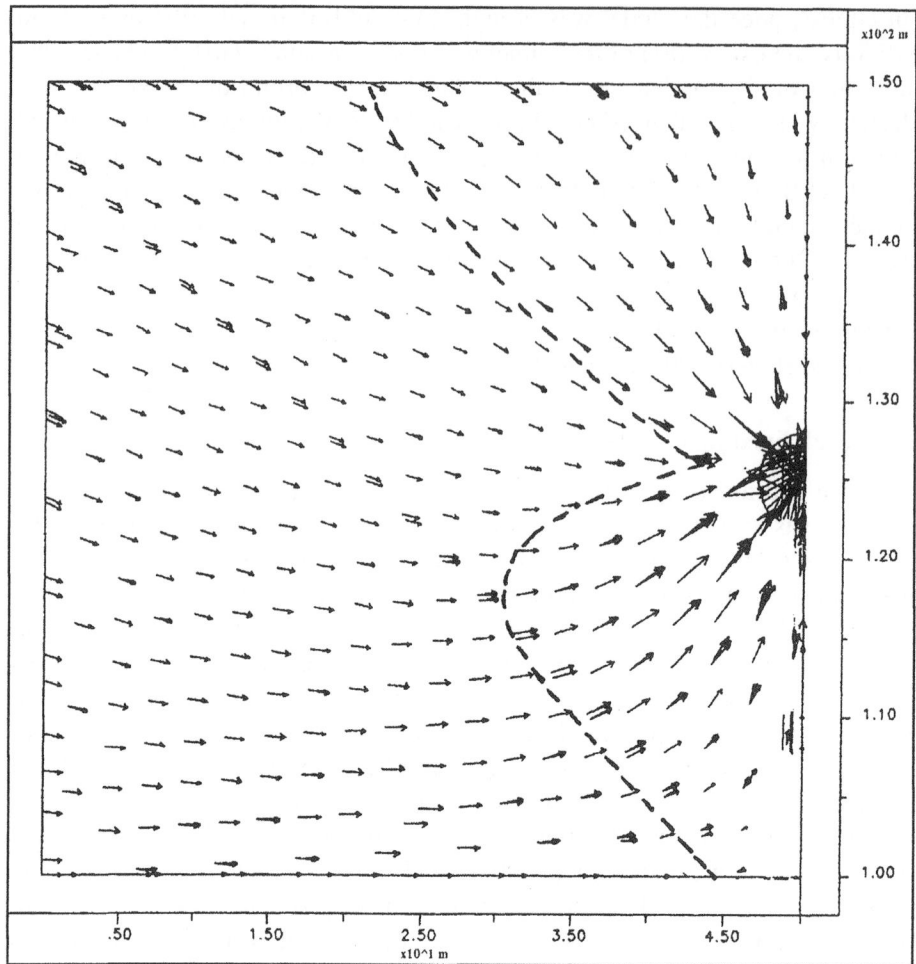

Figure 6.8. Undrained deformation vectors after excavation of the tunnel.

drainage of water flow through joint network causes a large deformation close to the tunnel periphery. As seen in Fig. 6.10, a significant deformation has occurred above the tunnel. This is because the reduced shear strength due to the flow of water has increased the movement of rock blocks towards the tunnel periphery. As shown in Fig. 6.11, a large shear stress concentration has developed close to the tunnel. The magnitude of shear stress less than 0.8 MPa is not plotted in here. Due to excavation, the increased hydraulic gradient towards the tunnel causes fluid flow into the cavity, and as a result, pore pressures within the joints close to the tunnel are dissipated. During steady-state flow, the pore pressure distribution within the joints is shown in Fig. 6.12, in which the line thickness indicates the magnitude of the fluid pressure. The minimum fluid pressure shown in the plot is

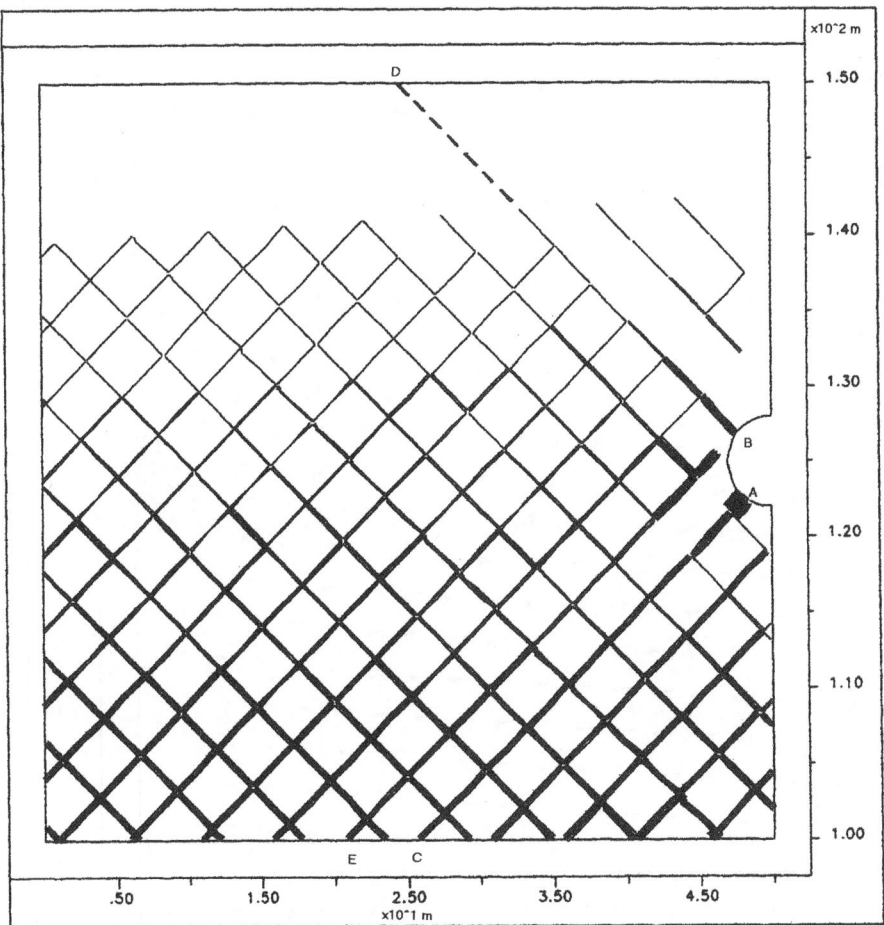

Figure 6.9. Pore pressure distribution after undrained deformation.

0.1 MPa. After excavation, fluid pressure variation within joints BD and CA (see Fig. 6.9) is shown in Fig. 6.13a and b, respectively. Fluid pressure increases away from the excavation for drained analysis, whereas as expected, a large fluid pressure is observed close to the tunnel, during undrained analysis.

It is important to note how the effective stress varies along the joints which intersect the tunnel boundary, as the effective stress controls the quantity of fluid carried along these joints towards the tunnel. Before and after excavation, the effective stress along the two joints (i.e. AC and BD) intersecting the tunnel boundary is shown in Fig. 6.14. As expected, the effective stress before the excavation at points C and B is greater than that at points at A and D, respectively. However, after the excavation, the effective stress at A and B has significantly increased due to dissipation of pore pressure at the tunnel boundary. As an example, the effective stress at point A after the excavation of the tunnel is

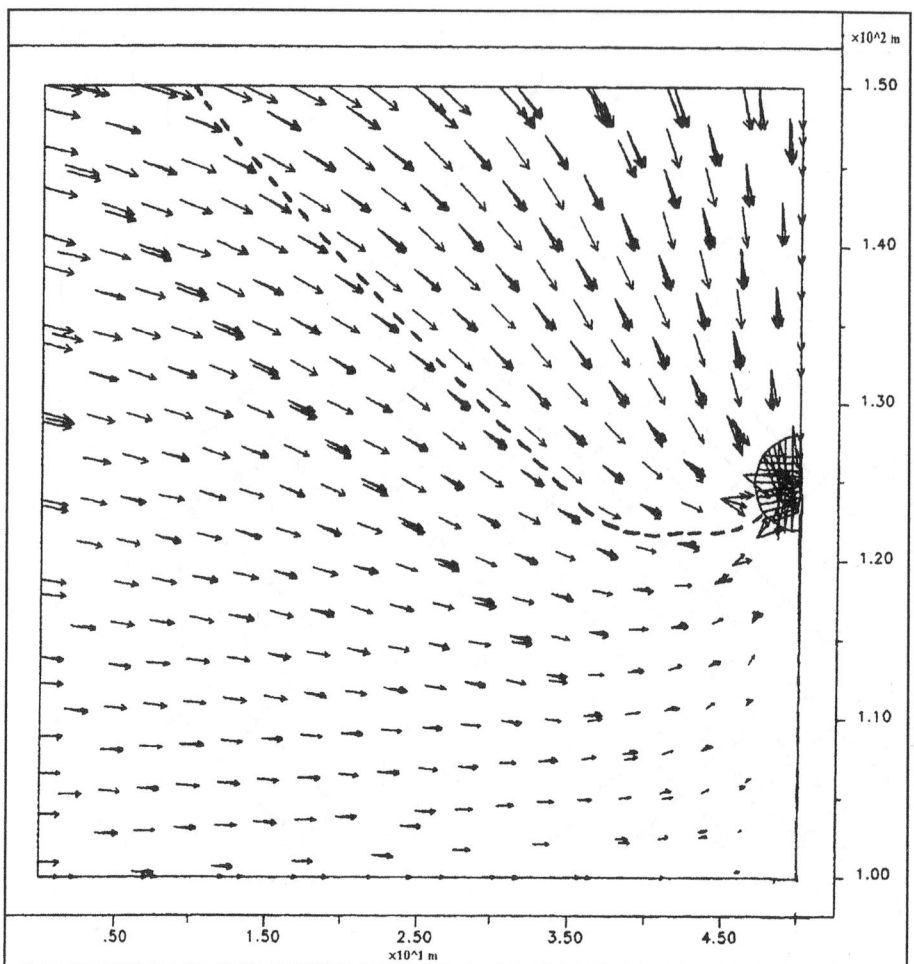

Figure 6.10. Displacement vectors during steady-state flow.

approximately 7 times that of the initial effective stress at A before the excavation. The increased effective stress causes change of joint apertures, and in turn affects the flow rate. If one compares the drained and undrained analyses, it is seen that the presence of fluid greatly influences the deformation of joints. During the undrained analysis, the maximum fluid pressure and the least value of the effective stress (can be negative) exists close to the tunnel, while the opposite trend is observed in the drained analysis.

The induced stresses associated with stress relief significantly influence the pore pressures and flow rates within the joints. In the following section, the parameters which influence the flow rates to the tunnel are discussed. They are: (a) different sizes of representative blocks, (b) orientation of joint sets and (c) vertical and horizontal stresses.

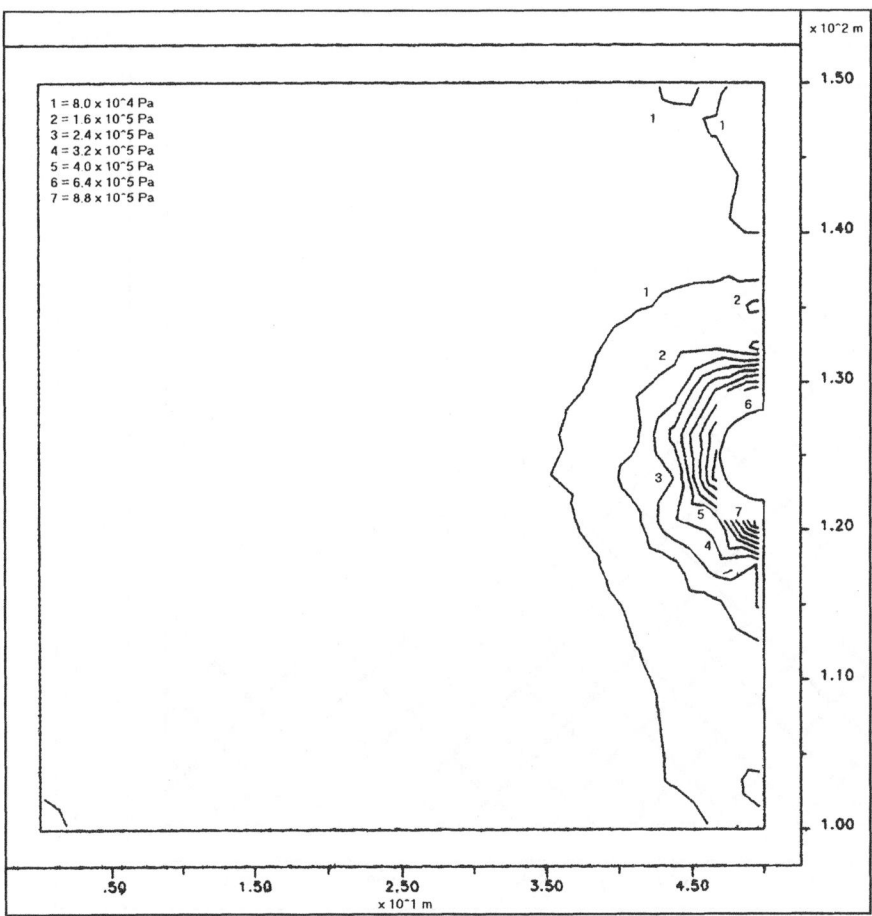

Figure 6.11. Shear stress contours after excavation.

6.7 DETERMINATION OF FLOW RATES BASED ON EFFECTIVE STRESS ANALYSIS

The effective stresses on the joints influence the change of joint aperture and the permeability. For given boundary conditions, the flow direction and quantity in each joint close to the tunnel are shown in Fig. 6.15, in which the line thickness indicates the magnitude of flow rate. The joints with flow rates below 2×10^{-4} m^3/s are not plotted here. For a given joint pattern, boundary block size and boundary conditions, the total water ingress is estimated by the summation of the flow quantity of each joint intersecting the tunnel periphery. The numerical results are based on steady-state flow computations only. As an example, the total inflow to the tunnel is the summation of the flow rates at points A and B, which are the intersections of the joints AC, BD and BE with the tunnel boundary (Fig. 6.15). It is evident that the flow rate at the excavation boundary

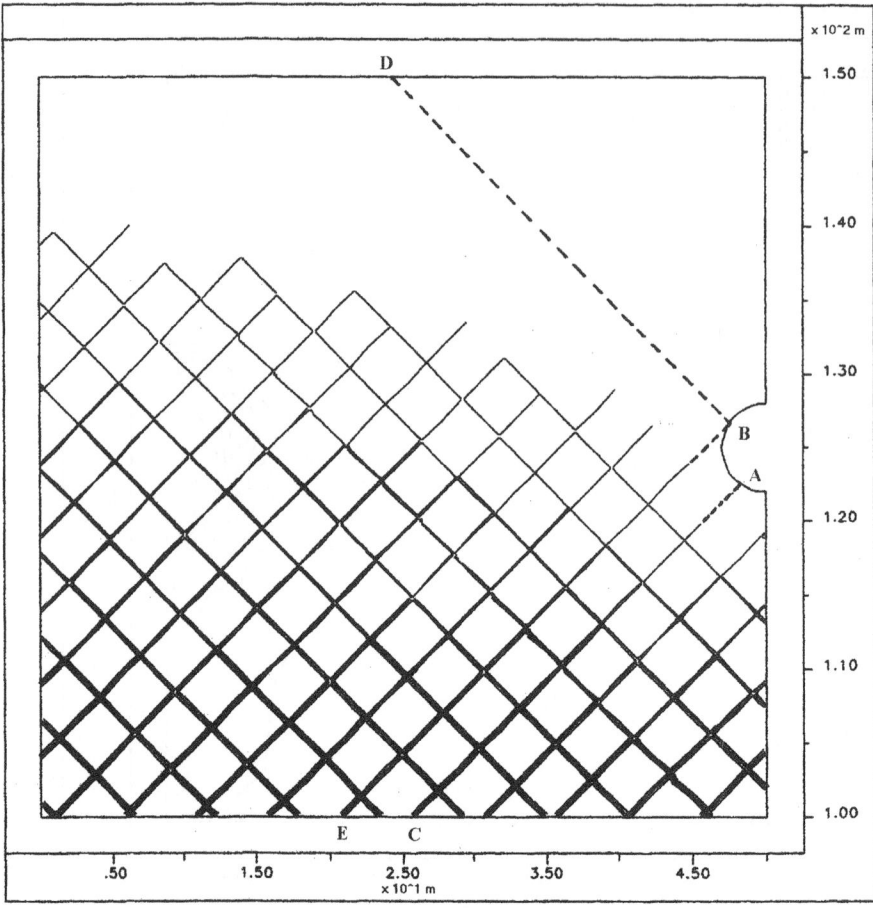

Figure 6.12. Pore pressure distribution during the steady-state flow.

has increased in all joints because of the high hydraulic gradient developed towards the tunnel.

6.7.1   Effect of representative block sizes

For the joint pattern and boundary conditions illustrated in Fig. 6.3, the relationship between the total water ingress and various block sizes is shown in Fig. 6.16, which indicates that increasing the block sizes will result in a decreasing flow rate. On one hand, it may be argued that the flow rate should increase with the increasing block size, because, the discontinuities which intersect the tunnel periphery have a greater degree of intersection with other discontinuities away from the cavity. On the other hand, if the block size is increased, the effects of water pressure on discontinuities that are intersected by the tunnel boundary become less, consequently, a lower value of water flow can then be expected due to a decreased hydraulic head.

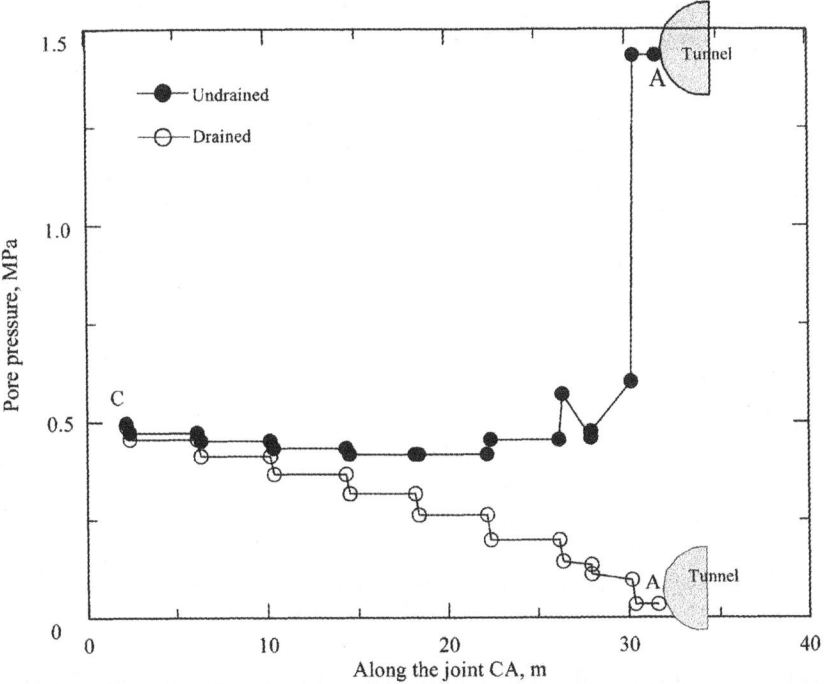

Figure 6.13a. Pore pressure variation along the joint (CA) during drained and undrained conditions.

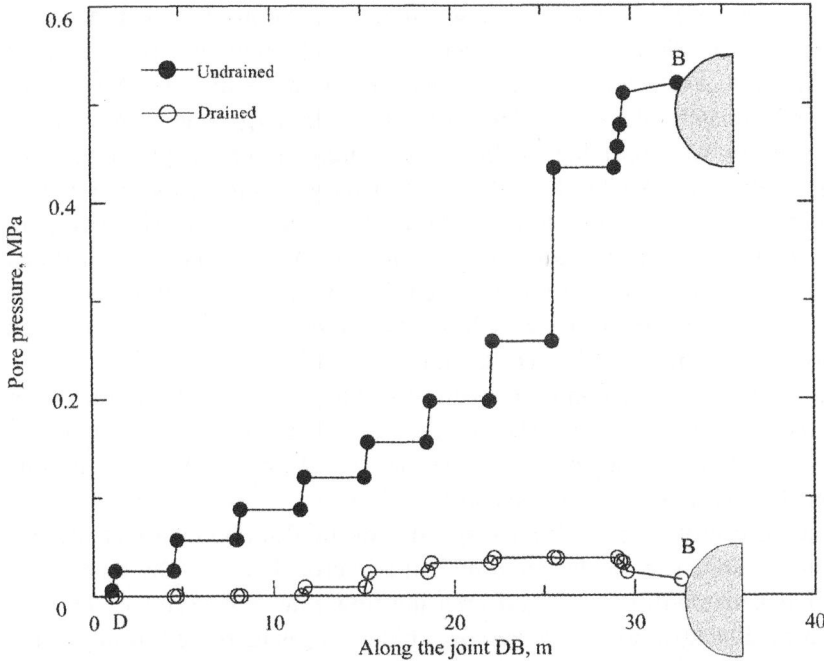

Figure 6.13b. Pore pressure variation along the joint (DB) during drained and undrained conditions.

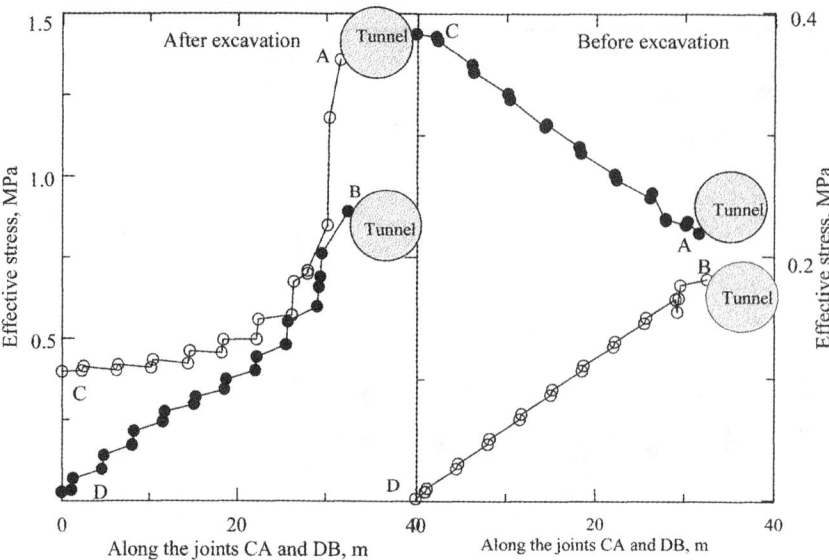

Figure 6.14. Effective stresses along typical joints which intersect the tunnel periphery.

Figure 6.17 illustrates the effect of joint surface area on the flow rate for the joint model shown earlier in Fig. 6.3. The joint surface area increases when the boundary block size increases, if the joint spacing is kept the same. For a larger block size, as the length of joints increases, the hydraulic gradient will decrease for the same external fluid pressure applied to the block boundary. Therefore, the flow rate to the excavation is expected to decrease with the increasing joint surface area or increasing block size. This is because, the larger the representative block size (i.e. fracture area), the smaller the hydraulic gradients applied towards the opening, hence, the smaller the inflow to the cavity. It is more useful to develop a relationship between the total water ingress and a dimensionless ratio defined by total joint surface area/excavation area (A_c/A_e). It can be demonstrated that the total water ingress is very high when the ratio A_c/A_e is within the range 10–40. It can be also shown that very low water ingress can be expected when A_c/A_e ratio varies from 80 to 200, as shown in Fig. 6.17.

The hydraulic boundary conditions can also play a major role on the water ingress to the subsurface cavity as shown in Figs. 6.18 and 6.19. A very high flow rate can be expected when the block size lies between 15 m and 50 m. If one side of the boundary is treated as permeable (i.e., Fig. 6.5, Case 3c), the flow rate increases significantly (Fig. 6.19b). Irrespective of the fluid boundary conditions or insitu stress ratios (i.e. insitu horizontal stress/vertical stress), the increase in boundary block size will result in a decreasing flow rate. For example, the flow rate is decreased by 80% when the boundary block size is increased from 15 m to 70 m (Fig. 6.18). The change in flow becomes marginal when the boundary block size exceeds 70 m in all three cases. A sensitivity analysis such as this provides a

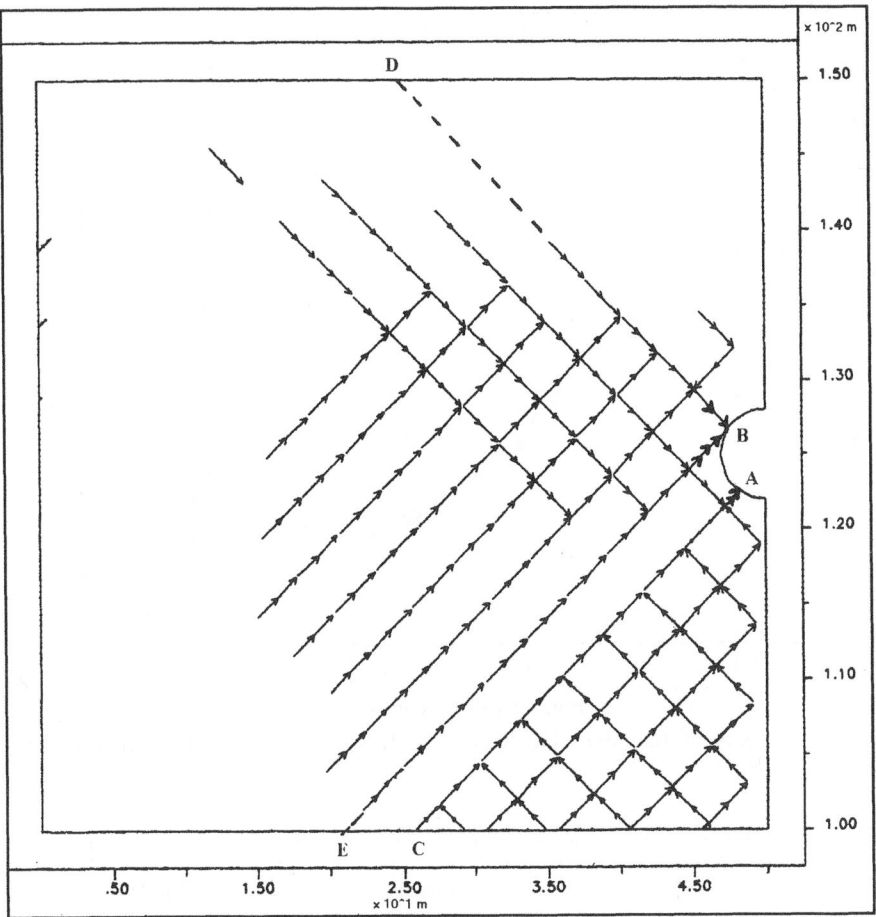

Figure 6.15. Water flow towards the tunnel.

most appropriate block size to be selected for numerical analysis. Based on the current distinct element method, the optimum boundary block size is about 10–12 times the maximum width of the excavation.

6.7.2 Effects of orientation of discontinuities

Figure 6.20 shows the relationship between the total flow rate to the tunnel and the orientation ratio of joint set 1 to joint set 2. In this analysis, orientation of joint set 1 was kept constant ($\theta_1 = 30°$ relative to the X-axis), while the orientation of joint set 2 (θ_2) was varied for an in-situ stress ratio (σ_h/σ_v) of 2.0, which is typical of the Wollongong region, NSW, Australia. Figure 6.20 clearly demonostrates that the maximum water flow to the tunnel occurs when θ_2/θ_1 equals 3.0. Not surprisingly, this indicates that the vertical discontinuities ($\theta_2 = 90°$), which intersect the tunnel periphery carry more water than any other discontinuity in the model. Naturally,

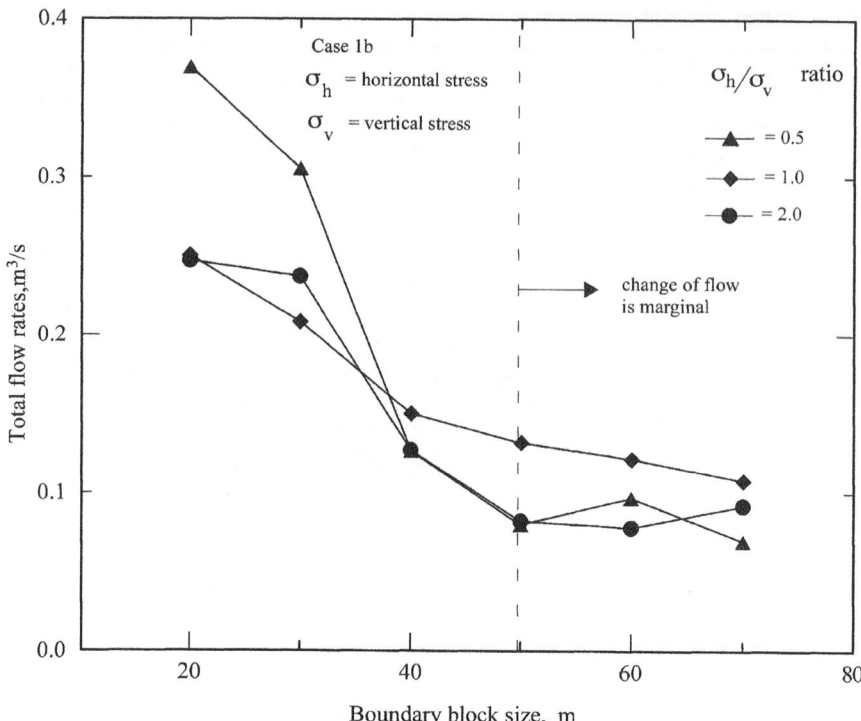

Figure 6.16. Effects of boundary block size on flow rate for different in-situ stress (Fig. 6.3, Case 1b).

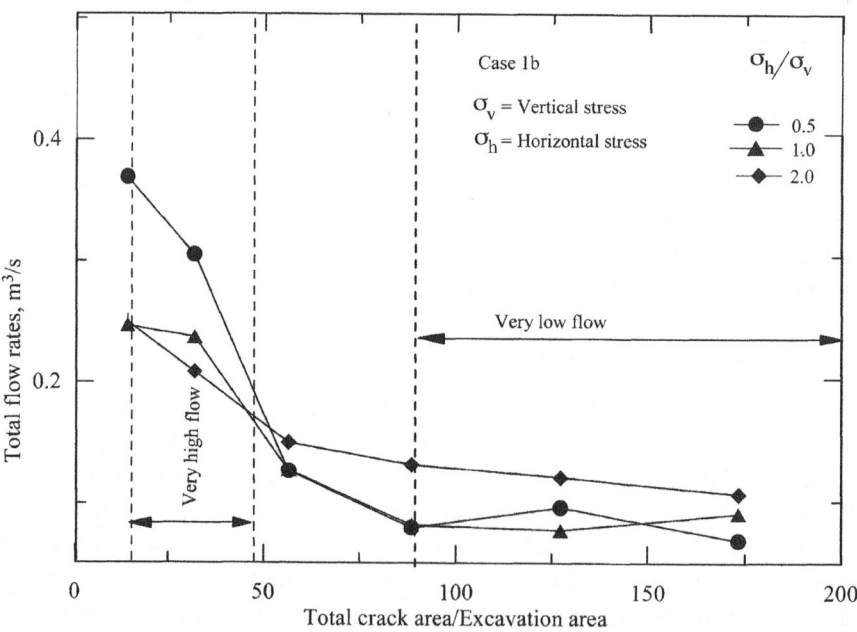

Figure 6.17. Results of total flow against the total crack area/excavation area ratio.

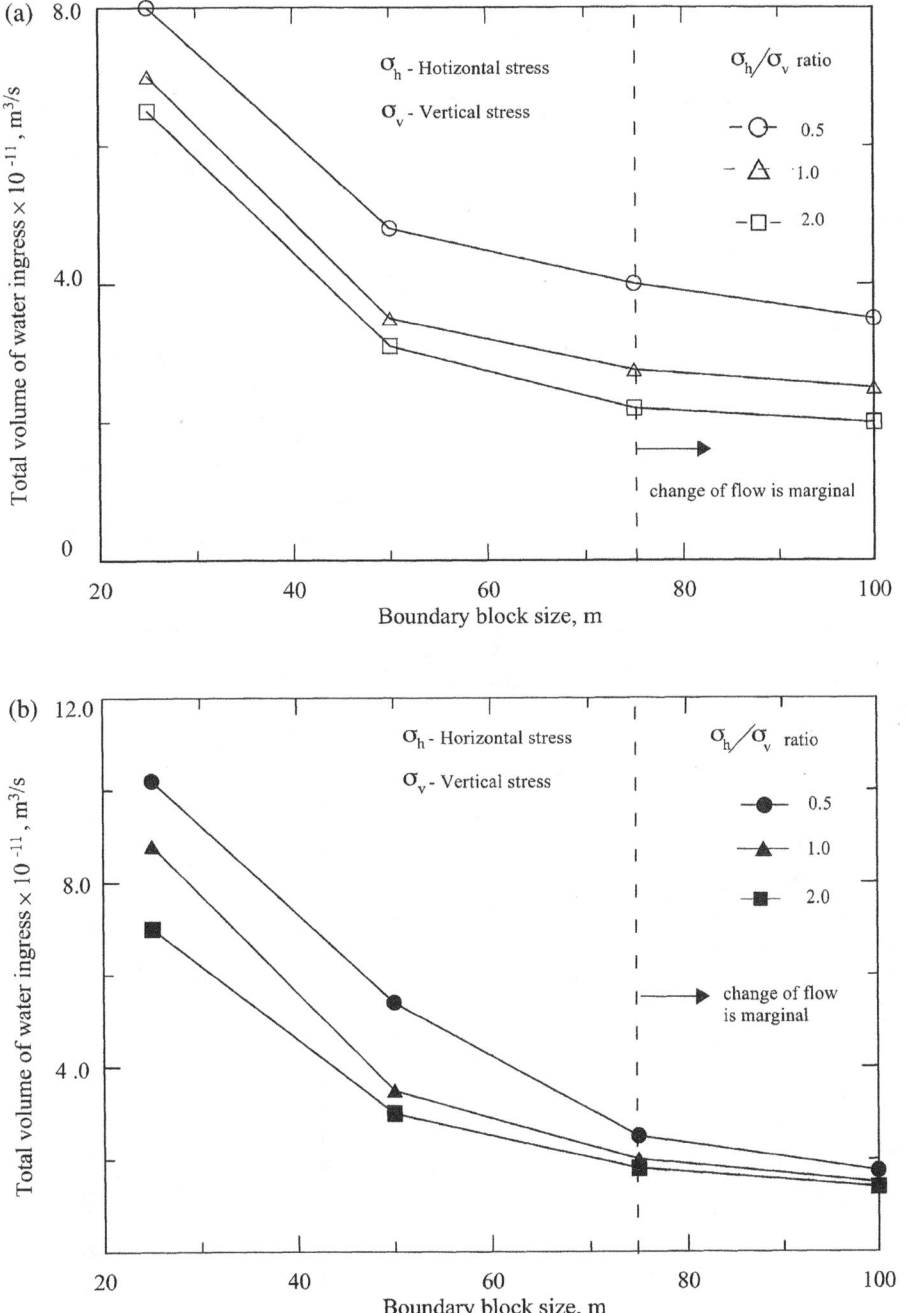

Figure 6.18. Effects of boundary block size on flow rate for different in-situ stress. (a) Case 2a. (b) Case 2b.

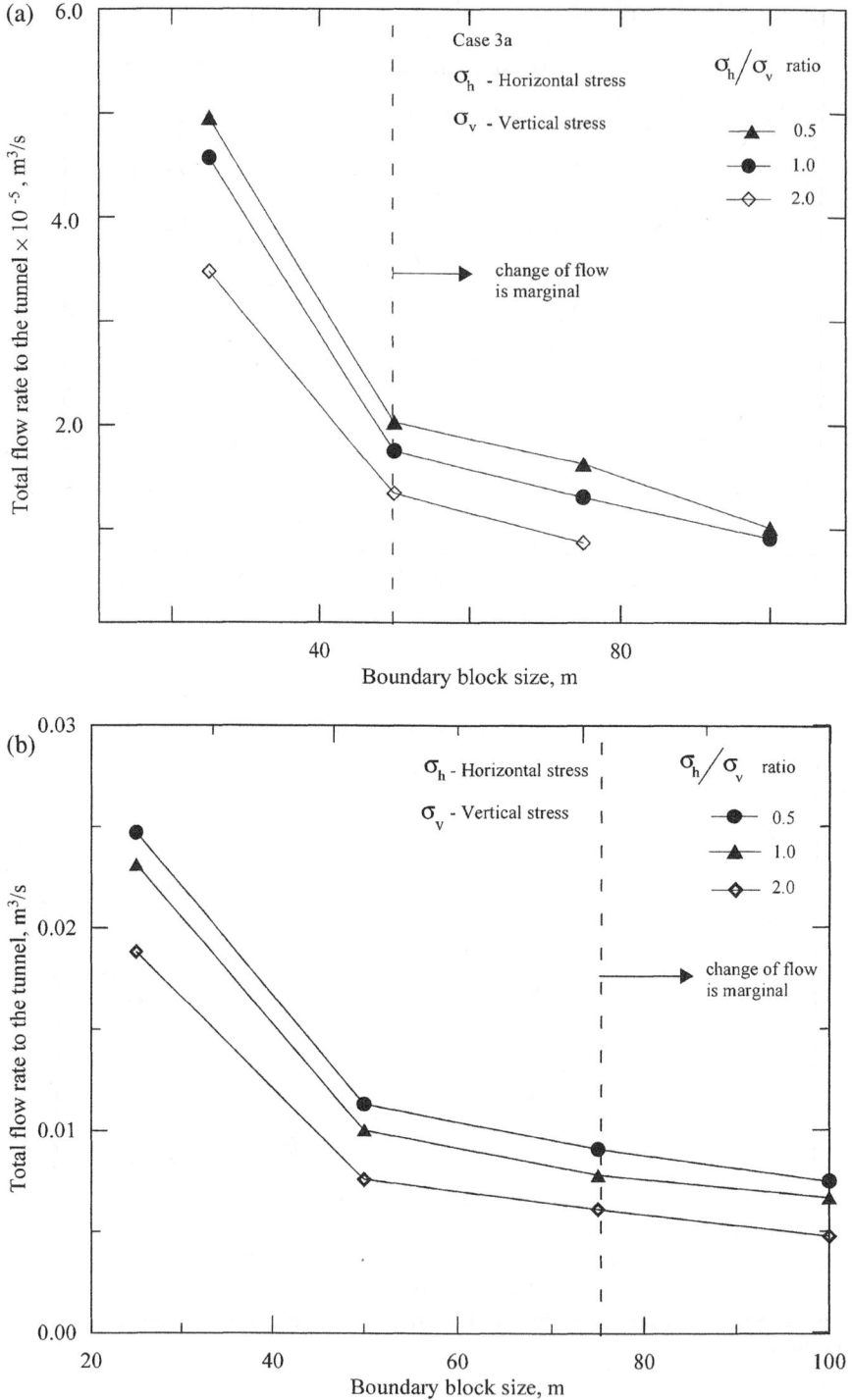

Figure 6.19. Effects of boundary block size on flow rate for different in-situ stress. (a) Case 3a. (b) Case 3c.

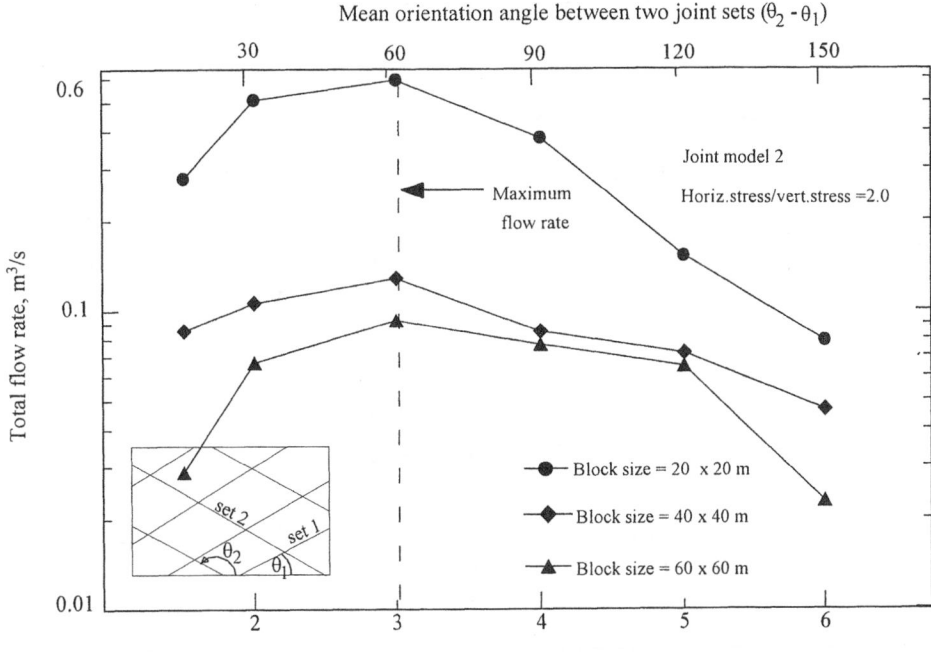

Figure 6.20. Effects of orientation of joint sets on total inflow towards the tunnel (Case 1b).

the effects of gravity flow are optimized in this situation. The maximum flow rate occurs when the angle between the two joint sets is around 60° ($\theta_2/\theta_1 = 3$), whereas Zhang et al. (1996) show that the permeability is maximum if the angle between the joint sets is around 30°. This is to be expected because, the joint pattern considered by Zhang et al. (1996) is more interconnected. Therefore, depending on fracture orientation and density, and the extent of connectivity of fluid-flow paths in a given joint pattern, the resulting flow rate can be significantly different even for the same boundary stress levels.

6.7.3 Effects of horizontal stress and vertical stress

Figure 6.21a and b demonstrate the distinctly different flow rates for various joint models, as a function of the σ_h/σ_v ratio. As expected, the flow rate decreases with the increasing horizontal stress. However, for large boundary block sizes, the decrease in flow rate is insignificant with the increase in insitu ratio. As the ratio of σ_h/σ_v increases from 0.5 to 2.25, the reduction in the percentage of water ingress to the mine cavity lies in the range 40–60% (Fig. 6.21a). When the in-stress ratio exceeds 1.2, the change of joint flow rate is marginal, because, most conducting fractures have by then reached their residual apertures (Fig. 6.21b).

These results are generally in agreement with Indraratna and Wong (1995) who observed that a flow reduction of up to 70% was possible for a systematic joint

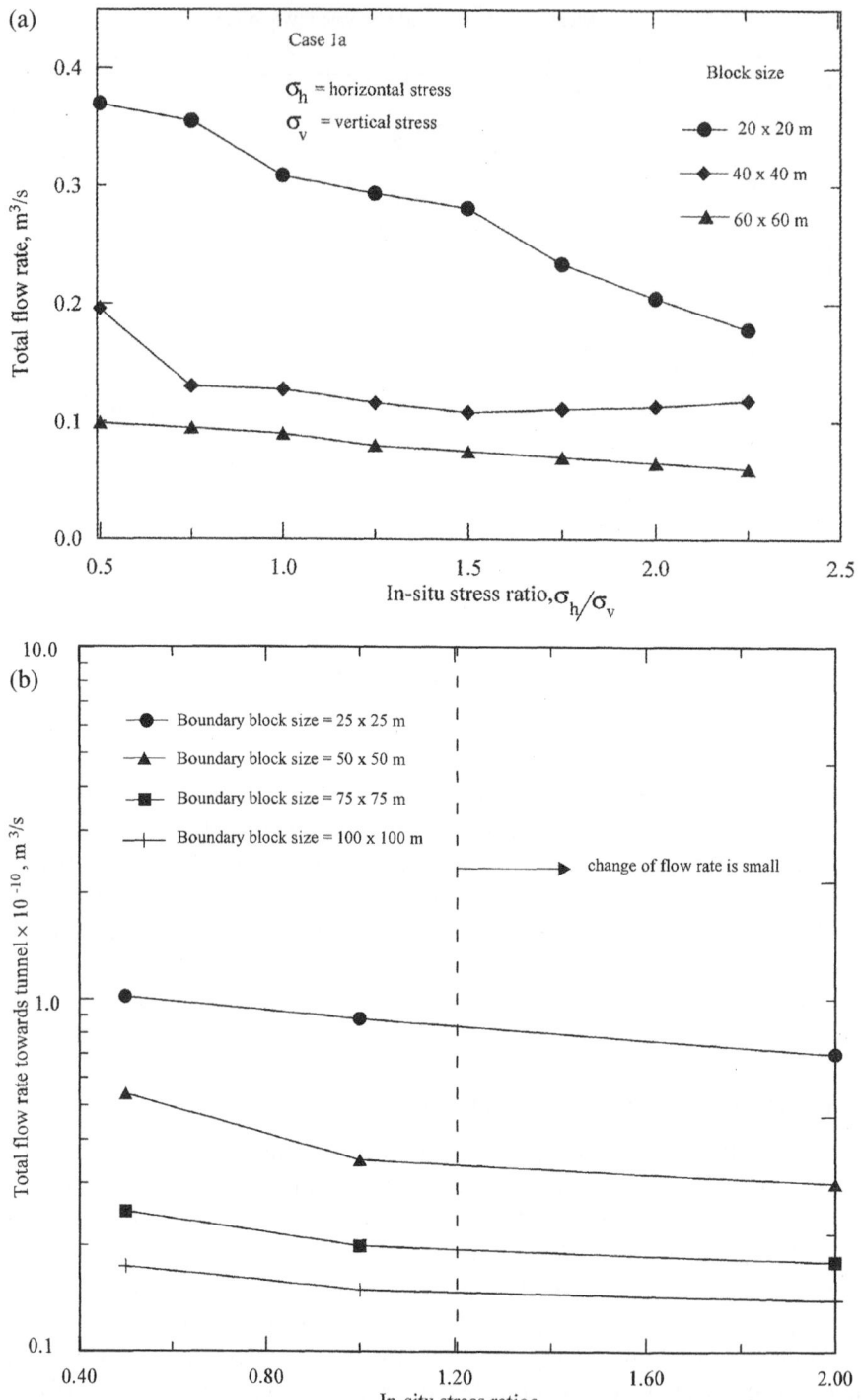

Figure 6.21. Effects of in-situ stress ratio on (a) flow rate (Case 1a) and (b) water inflow (Case 2a).

pattern, such as in Case 1a. Using UDEC, Liao and Hencher (1997) investigated the effect of horizontal stress, boundary conditions and block sizes on the permeability characteristics of jointed rock mass. They found that the overall permeability decreased with the increase in horizontal to vertical stress ratio. These findings are also in accordance with the analysis presented here, the difference being that this analysis has concentrated more on the variation of flow rates rather than the permeability variations. According to the numerical flow analysis carried out by Zhang et al. (1996), permeability decreases significantly with the increase in vertical stress for a given joint pattern and horizontal stress. In the current study, the predicted inflow decreases with the increase in horizontal stress for a given joint pattern. This is because, the near vertical joints that conduct significant amount of water ($\theta = 90°$) become compressed (reduced aperture) when the horizontal stress is increased. This shows that the influence of horizontal stress on the closure of near vertical joints is greater than that of vertical stress, resulting in reduced flow through a fracture network.

REFERENCES

Beer, G. and Poulsen, B.A. 1994. Efficient numerical modeling of faulted rock using the boundary element method. *Int. J. Rock Mech. Min. Sci. Geomech. Abstr.*, 31(5): 485–506.

Brady, B.H.G. and Brown, E.T. 1994. *Rock Mechanics for Underground Mining*. 2nd ed., Chapman & Hall, London, 570 p.

Crotty, J.M. and Wardle, L.J. 1985. Boundary integral analysis of piecewise homogeneous media with structural discontinuities. *Int. J. Rock Mech. Min. Sci. Geomech. Abstr.*, 22: 419–427.

Crouch, S.L. and Starfield, A.M. 1983. *Boundary Element Methods in Solid Mechanics*. Allen & Uwin, London.

Cundall, P.A. 1971. A computer model for simulating progressive, large scale movements in block rock systems, paper II-8. *Proc. Int. Symp. on Rock Fracture*, ISRM, Nancy, France.

Cundall, P.A. and Strack, O.D.L. 1979. A discrete numerical model for granular assemblies. *Geotechnique*, 29: 47–65.

Goodman, R., Moye, D., Schalkwyk, A. and Javandel, I. 1965. Ground water inflow during tunnel driving. *Eng. Geol.*, 2: 39–56.

Elsworth, D. 1985. Coupled finite element/boundary element analysis for non-linear flow in rock fractures and fracture networks. *Proc. 26th U.S. Symp. on Rock Mechanics*, Balkema, Netherlands, pp. 633–641.

Elsworth, D. 1986. A hybrid boundary element-finite element analysis procedure for fluid flow simulation in fractured rock masses. *Int. J. Numer. Anal. Methods Geomech.*, 10(6): 569–584.

Elsworth, D. 1987. A boundary element-finite element procedure for porous and fractured media flow. *Water Resour. Res.*, 23(4): 551–560.

Herbert, A.W. 1996. Modeling approaches for discrete fracture network flow analysis. In *Coupled Thermo-Hydro-Mechanical Process of Fractured Rock Media*. Stephenson, O., Jing, L. and Tsang, C.F., eds., Elsevier, New York, 575 p.

Indraratna, B. and Wong, J.C. 1995. Effects of stress change on water inflows to underground excavation. *Aust. Geomech.*, 29: 99–114.

ISRM, 1988. List of computer programs in rock mechanics. *Int. J. Rock Mech. Min. Sci. Geomech. Abstr.*, 25(4): 183–252.

ITASCA Consulting Group (1993). *FLAC-Fast Lagrangian Analysis of Continua*, User's Manual.

ITASCA Consulting Group (1996). *UDEC-Universal Distinct Element Code*, Version 3.0, Vols. 1–3, User's Manual.

Lee, C.H. and Farmer I. 1993. *Fluid Flow in Discontinuities Rocks.* Chapman & Hall, London, 169 p.

Liao, Q.H. and Hencher, S.R. 1997. Numerical modeling of the hydro-mechanical behavior of fractured rock masses *Int. J. Rock Mech. Min. Sci. Geomech. Abstr.*, 34(3–4). Paper no. 117 (CD ROM).

Long, J.C.S. and Witherspoon P.A. 1985. The relationship of the degree of interconnectivity to permeability of fracture networks. *J. Geophysical Res.*, 90(B4): 3087–3098.

Lorig, L.J., Brady, B.H.G. and Cundall, P.A. 1986. Hybrid distinct element-boundary element analysis of jointed rock. *Int. J. Rock Mech. Min. Sci. Geomech. Abstr.*, 23(4): 303–312.

Louis, C.A. 1974. In *Rock Hydraulics in Rock Mechanics.* Muller, L., ed., Springer Verlag, Vienna, pp. 299–382.

Louis, C. 1976. *Introduction l'Hydraulique des Roches*, PhD thesis, Paris.

Maini, Y.N.T. 1971. *Insitu Hydraulic Parameters in Jointed Rock – Fluid Measurement and Interpretation.* PhD thesis, Imperial College, University of London.

Neuman, S.P. 1973. Saturate-unsaturated seepage by finite elements. *J. Hydraulic Eng.*, 99: 2233–2251.

Sharp, J.C. 1970. *Flow Thorough Fissured Media.* PhD thesis, Imperial College, University of London.

Tsang, C.F. and Stephenson, O. 1996. A conceptual Introduction to coupled thermo-hydro-mechanical processes in fractured rocks. In *Coupled Thermo-Hydro-Mechanical Process of Fractured Media*, Elsevier, Amsterdam, 576 p.

Wangen, M. 1997. Two-phase oil migration in compacting sedimentary basins modeled by the finite element method. *Int. J. Numer. Anal. Methods Geomech.*, 21: 91–120.

Wilock, P. 1996. The NAPSAC fracture network code. In *Coupled Thermo-Hydro-Mechanical Process of Fractured Media*, Elsevier, New York. 576 p.

Witherspoon P.A, Wang, J.S.Y., Iwai, K. and Gale, J.E. 1980. Validity of cubic law for fluid flow in a deformable rock fracture. *Water Resour. Res.*, 16(6): 1016–1024.

Zhang, L. and Franklin, J.A. 1993. Prediction of water flow into rock tunnels: An analytical solution assuming a hydraulic gradient, *Int. J. Rock Mech. Min. Sci. Geomech. Abstr.*, 30(1): 37–46.

Zhang, X., Sanderson, D.J., Harkness, R.M. and Last, N.C. 1996. Evaluation of the 2-D permeability tensor for fractured rock mass, *Int. J. Rock Mech. Min. Sci. Geomech. Abstr.*, 33(1): 17–37.

Zienkiewicz, O.C., Kelly, D.W. and Bettes, P. 1977. The coupling of the finite element method and boundary element procedures. *Int. J. Numer. Method Eng.*, 11: 355–375.

CHAPTER 7

Highlights of hydro-mechanical aspects of fractured rocks and recommendations for further development

7.1 SUMMARY

Two-phase flow behaviour through fractured rock has been studied analytically and experimentally by a number of investigators (Fourar & Bories, 1995; Pruess & Tsang, 1990; Indraratna et al., 1999). The authors' research studies have specifically included (a) design of a novel, two-phase triaxial equipment, (b) study of fully saturated flow through fractured and intact rock and (c) two-phase flow measurement and analysis through natural and artificially created rock joints. Naturally fractured rock specimens were obtained from the underground coal mines in the Wollongong region. The analytical phase included the development of a mathematical model to estimate two-phase flow through a single joint, and the validation of the model was carried out experimentally.

Chapter 1 was devoted for discussing the mechanisms of formation of discontinuities and their classification systems. The measurement and effects of physical properties of intact rocks and discontinuities (length, orientation, shape and infill materials) on permeability were highlighted. Field investigation of discontinuities in various sites in Australia was also included.

Chapter 2 included the illustration of various types of triaxial apparatus and their application in soil and rock property measurement. The design concept of a unique two-phase triaxial equipment (TPHPTA) was presented in detail. A critical review of literature on fluid flow through rock masses was presented in Ch. 3, introducing the numerical and analytical approaches employed for fluid-flow estimations. The applicability of fluid-flow models for various field conditions was also highlighted. In order to determine water ingress to an underground cavity, a fully coupled, hydro-mechanical analysis was also carried out using UDEC, with various geo-hydraulic and joint parameters (Ch. 6).

An analytical procedure based on mass, momentum and energy balance principles for the estimation of individual flow rates for different boundary conditions was presented in Ch. 4. The model predictions for various boundary conditions were compared with the experimental data in Ch. 4. The laboratory findings verified the acceptable accuracy of the mathematical model developed by the authors.

The following is a summary of the important findings in relation to fluid flow through rock joints.

7.1.1 Type of triaxial equipment

Significant efforts to study the stress–strain and permeability characteristics of soil and rocks under laboratory conditions are clearly evident from the numerous items of triaxial apparatus (Ch. 2). However, only a few of the existing triaxial facilities are capable of modelling realistic fluid flow through jointed rock mass. The newly designed triaxial apparatus at University of Wollongong is capable of simulating the actual fluid-flow field in jointed rock mass. In this apparatus, two-phase (water-air) fluid flow through soft and hard rocks can be simulated (i.e., saturated or unsaturated flow conditions) under triaxial stress state. For 3D stress applications, another triaxial facility available at University of Strathclyde (Smart, 1995) can be used for single phase flow in rock joints.

7.1.2 Failure implications

Based on authors' research studies, failure stress of water- and air-saturated intact granite attained 150 MPa at 0.035 strain, and 190 MPa at 0.02 strain, respectively. For vertically fractured conditions, failure stress of water- and air-saturated specimens were 140 MPa at 0.04 strain, and 155 MPa at 0.03 strain, respectively. It is evident from this experimental data, that a higher failure load at a lower strain is expected, if the permeating fluid is air. In contrast, for water-saturated samples, a lower ultimate strength is attained. This concludes that, (a) ductility of rock mass increases and (b) deformation modulus decreases, when the water content is increased in relation to the air content.

From a practical point of view (e.g., the stability of tunnel roofs and mine long-walls under different fluid pressures), when gas flow dominates, sudden instability of mine roof can be expected at a critical gas pressure. In contrast, if the jointed rock is saturated with water, the failure process is more gradual and also predictable.

7.1.3 Permeability aspects and effects of loading

Flow through a single joint is a function of the magnitude of the joint aperture, external stress field and its loading and unloading history, applied fluid pressures, the joint surface roughness and the relative orientation of the joint in relation to the principal stresses. Effect of loading and unloading on fractured rock permeability is significant. There is a marked reduction of flow rate during the first loading cycle, but the contribution from the second and third loading and unloading cycles is very small. Based on the authors' experience, once fractures attain their residual apertures at relatively high confining pressure, subsequent dilation (due to unloading) and compression (due to reloading) seem to be insignificant. According to recent findings, above 8 MPa effective confining pressure, the average permeability decreases by almost 90% from the coefficient of permeability at zero confining pressure. This reduced permeability associated with the residual aperture will always prevail once the joints are loaded above the threshold value (Brace et al., 1968; Ranjith, 2000)

Intact (matrix) permeability of granite specimens is in the order of 10^{-19} m^2, whereas fracture permeability can vary from 10^{-12} to 10^{-15} m^2 depending on the magnitude of joint aperture and the interconnectivity of fractures. For numerical analysis, matrix permeability of granite can be neglected, in relation to the fracture permeability.

7.1.4 Joint roughness

From the experimental work, it is verified that the joint roughness coefficient (*JRC*) varies considerably from one location to another along the same specimen (*JRC* = 3–12). This indicates that field fractures can be highly irregular, and it is not always appropriate to model them as parallel plates. For practical purposes, Barton's standard profile method (Barton & Choubey, 1977) and alternative techniques based on the maximum amplitude of the joint yield roughness coefficients reasonably accurately.

When the asperity heights (k) between the maximum and minimum of joint surface approach the magnitude of joint aperture (e), the pressure drop coefficient is reduced by about eight times from the pressure drop coefficient of the smooth joint. Increase in *JRC* will result in an exponential decrease in flow rate. Even at very high *JRC* values (e.g., 15) and high normal stresses, there is always a minimum flow corresponding to the residual aperture. Although roughness is important in fluid-flow estimates, in reality, it is not feasible to incorporate roughness of each joint separately. Having identified these limitations, the cubic law can still be used in caution in numerical modelling applied to practical situations (Witherspoon et al., 1980).

7.1.5 Mathematical modelling of unsaturated flow

Assuming that the fluid-flow pattern in a rock joint is stratified, a simplified mathematical model can be formulated incorporating joint deformation, effect of solubility of air in water, compressibility of water and air associated with joint deformation, and change of fluid properties such as density. For the above conditions, the proposed model can predict equivalent phase heights of water $h_w(t)$ (e.g., Eqn. 4.23) and air $h_a(t)$ (e.g., Eqn. 4.24). Subsequently, these phase heights can be used to estimate flow rates, permeability and relative permeability of each phase. The current model shows that almost 95% of the magnitude of $h_a(t)$ and $h_w(t)$ is due to the normal joint deformation (δ_n), the rest being the combined effect of ξ_{ac} (air compressibility) and ξ_{ad} (solubility of air in water). The term, ξ_{ac} is more significant than the component ξ_{ad}, and ξ_{ac} amounts to about 4–5% of the value of δ_n. Once the joint aperture reaches its residual value (confining pressure exceeding 6 MPa), the role of ξ_{ac} and ξ_{ad} becomes increasingly pronounced.

7.1.6   Experimental simulations

The authors suggest that at low inlet fluid pressures, the two-phase flow within rock joints can be best described by bubble flow. Further increase in inlet fluid pressure may result in annular flow. At very high inlet fluid pressures, two-phase flow in highly irregular joints should be modelled using complex flow patterns. Recent studies have verified that for flow of low viscosity fluid through a high viscosity media (e.g., air injected to water-saturated joints), steady-state conditions result in a shorter period of time ($<$ 4 hrs). A significantly larger period of time is taken to observe steady-state flow, when water is injected to air-saturated joints ($>$10 hrs). Reynolds numbers measured for two-phase flow are usually well below 1000. Therefore, laminar flow is considered appropriate for both single and two-phase flows observed under the boundary conditions tested in most laboratory environments (Ranjith, 2000).

At relatively small inlet fluid pressure ($<$ 0.5 MPa), both experimental and predicted results show that two-phase flow rates vary almost linearly with inlet fluid pressures, when the inlet pressures of both phases are approximately equal ($p_a = p_w$). However, non-linear changes take place when the inlet fluid pressure is increased beyond say 0.5 MPa or when $p_a \neq p_w$. Increase in inlet pressure of one phase usually results in the increase of flow rate of the same phase. Nevertheless, the flow rate of the air phase is always higher than that of the water phase.

Increase in confining pressure leads to a decrease in the two-phase flow rates. At elevated confining pressures exceeding 6 MPa, change in flow rate becomes marginal, because the joints have then attained their residual aperture.

Effect of inlet air to inlet water pressure ratio, p_a/p_w
For granite specimens, the authors have verified that the fracture permeability of both air and water phases becomes equal when the ratio of p_a/p_w is between 0.9 and 1.2. For initially water-saturated specimens, the air permeability plays the most dominant role at high p_a/p_w ratios.

Darcy's law can be extended to represent two-phase flow, based on the relative permeability concept. The relative permeability of the air phase increases almost exponentially with the increase in p_a/p_w ratio, while decreasing the relative permeability of the water phase, at the same time. The opposite trend occurs when the p_w/p_a ratio is increased.

7.1.7   Numerical analysis

Distinct element modeling (DEM) is popular for predicting flow through jointed rock, and in this regard Universal Distinct Element Code (UDEC) has been widely employed (Ranjith, 2000; Indraratna & Ranjith, 1998; Zhang et al., 1996).

Research conducted by the authors indicate that in the UDEC analysis of ground-water ingress to an underground cavity, the flow rate becomes marginal when the boundary block size exceeds 50 m for selected joint patterns. Based on DEM, the optimum block size for flow analysis should be 10–12 times the maximum width of excavation to obtain numerical convergence.

Numerical results generally confirm that the flow rate decreases with the increasing horizontal stress. The reduction in the percentage of water ingress to a mine cavity may lie in the range of 40–60%, when the ratio of horizontal stress to vertical stress (σ_h/σ_v) increases from say 0.5 to 2.25. When σ_h/σ_v ratio exceeds a critical value, then the rate of change of flow becomes insignificant.

7.2 RECOMMENDATIONS FOR FURTHER RESEARCH

7.2.1 Theoretical aspects

Mathematical models can be applied to characterize two-phase flow in an inclined joint using a numerical procedure. It is suggested that coupled finite element and boundary element methods can be applied to model the interface between fluids, when the joint is subjected to deformations. In such numerical analyses, the rough-ness profiles of both top and bottom joint surfaces must be considered to evaluate correctly the effects of surface irregularity on flow. It is anticipated that a much larger computer memory will be required in this case.

At given boundary conditions, it is of interest to study the relationships between joint permeability and the fracture (void) volume. This information can then be used to extend the current mathematical model and the scope of numeri-cal analysis. Current theories may be extended to model the flow through fracture networks by considering energy losses and change of fluid properties at each joint intersection. Special attention should also be given to the change of the interface conditions at the intersection, which will probably lead to complex flows that have not been modelled earlier.

At very high inlet fluid pressures, fluid flow within fractures may become tur-bulent. Under such conditions, possible flow may take a complex pattern. General flow equations should be developed for each phase to model such complex flows, and subsequently, jump flow equations can be written at the boundary. By design-ing synthetic joints with transparent material (e.g., perspex) and coloured fluids, the occurrence of complex flow patterns may be studied more comprehensively. These complex flow patterns can then be contrasted with simplified mathematical models.

Further studies should be undertaken to include gases other than air. Gases such as CH_4, CO_2 and mixtures of CO_2/CH_4 are of particular significance to coupled flow in coal measures rocks. The properties of gases can be significantly different to those of air, and therefore, the coupled water/gas flow analysis should provide new avenues for investigating problems of gas explosion in underground mines.

7.2.2 Modifications to laboratory procedures

Pore pressure distribution within joints cannot be measured precisely under laboratory conditions, unless very small pressure transducers can be installed within a joint. Although micro-transducers for hydraulic applications may be useful, budget limitations may prevent the use of such expensive instrumentation in most studies.

The current two-phase high pressure triaxial facilities can be modified in the following ways:

(1) Inclusion of high inlet fluid pressures (exceeding 2 MPa) and high velocity flows to investigate the effect of elevated fluid pressures on the application of Darcy's law for natural rock fractures.
(2) Less frictional volume change devices may be built employing 'teflon'-based materials withstanding high pressure. However, the cost of such devices will be at least a factor of 2–3 greater than the commonly used stainless steel volume change devices.
(3) Extension of current triaxail facilities to include three-phase flows, with special reference to petroleum engineering (i.e., oil–gas–water) will be most beneficial.

REFERENCES

Barton, N. and Choubey, V. 1977. The shear strength of rock joints in theory and practice. *Rock Mech.,* 10: 1–54.

Brace, W.F., Walsh, J.B. and Frangos, W.T. 1968. Permeability of granite under high pressure. *J. Geophys. Res.,* 73(6): 2225–2236.

Fourar, M. and Bories, S. 1995. Experimental study of air–water two-phase flow through a fracture (narrow channel). *Int. J. Multiphase Flow,* 21(4): 621–637.

Indraratna, B. and Ranjith, P.G. 1998. *Effect of Boundary Conditions and Boundary Block Sizes on Inflow to an Underground Excavation – Sensitivity Analysis.* International Mine Water Association, South Africa, pp. 3–11.

Indraratna, B., Ranjith P.G. and Singh, R.N. 1999. Laboratory investigation of two-phase flow in jointed rock media. *Proc. 37th US Rock Mechanics Symposium – Rock Mechanics for Industry,* Colorado, USA, pp.827–833.

Pruess, K. and Tsang, Y.W. 1990. On two-phase relative permeability and capillary in rough-walled rock fractures. *Water Resour. Res.,* 26: 1915–1926.

Ranjith, P.G. 2000. *Analytical and Experimental Modelling of Coupled Water and Air Flow Through Rock Joints.* PhD thesis, University of Wollongong, Australia.

Smart, B.G.D. 1995. A true triaxial cell for testing cylindrical rock specimen. *Int. J. Rock Mech. Min. Sci. Geomech. Abstr.,* 32(3): 269–275.

Witherspoon, P.A., Wang, J.S.Y., Iwai, K. and Gale, J.E. 1980. Validity of cubic law for fluid flow in a deformable rock fracture. *Water Resour. Res.,* 16(6): 1016–1024.

Zhang, X., Sanderson, D.J., Harkness, R.M. and Last, N.C. 1996. Evaluation of the 2-D permeability tensor for fractured rock mass. *Int. J. Rock Mech. Min. Sci. Geomech. Abstr.,* 33(1): 17–37.

Subject index

T - #0641 - 101024 - C0 - 254/178/16 - PB - 9789058093103 - Gloss Lamination